D1559691

Delta Sugar

Creating the North American Landscape

GREGORY CONNIFF
EDWARD K. MULLER
DAVID SCHUYLER
Consulting Editors

GEORGE THOMPSON
Series Founder and Director

Published in cooperation with the Center for American Places,
Santa Fe, New Mexico, and Harrisonburg, Virginia

DELTA SUGAR

THE JOHNS HOPKINS UNIVERSITY PRESS
Baltimore & London

Louisiana's Vanishing Plantation Landscape

JOHN B. REHDER

Printed in the United States of America on acid-free paper
9 8 7 6 5 4 3 2 1

The Johns Hopkins University Press
2715 North Charles Street
Baltimore, Maryland 21218-4363
www.press.jhu.edu

Library of Congress Cataloging-in-Publication Data will be found
at the end of this book.
A catalog record for this book is available from the British Library.

ISBN 0-8018-6131-4

Frontispiece: Pellico Plantation, Ascension Parish, Louisiana. Photograph, 1967.
All photographs are by the author, unless noted otherwise.

The real voyage of discovery consists not in seeking new landscapes but in having new eyes.

–Marcel Proust

But to the sugar house, the crop has just been gathered; and by a thousand wings of commerce, it has been scattered over the world; the engines of the sugar house, therefore, are lifeless; its kettles are cold, its store-rooms are empty, and the key that opens to its interior hangs up in the master's house, where it will remain until the harvesting and manufacturing of the new crop.

–T. B. Thorpe, *Harper's New Monthly Magazine,* 1853

Contents

Preface

Louisiana's sugarcane plantations are threatened with extinction. For more than two centuries, sugar plantations dominated much of the southern Louisiana landscape. Since 1969, 44 sugar factories have been reduced to 19 working factories. Farms have declined from 1,687 to 690. In the late 1960s, I synthesized 202 sugarcane plantations and analyzed 6 in depth for their landscape morphology and culture traits. In 1989, I returned to Louisiana to take up where my earlier research had left off and discovered alarming devastation on the contemporary landscape. Two case-study plantations had disappeared, three were in ruins; only one remained reasonably intact. Archaeologists were digging sites that I had analyzed as intact plantations just twenty years before. The time seemed right for an illustrated scholarly book on Louisiana's sugar plantations, and it had to be done before the landscape disappeared beneath the blades of bulldozers.

Delta Sugar describes the creation of Louisiana's sugarcane plantation landscape and explains its recent demise. This book is not a reprise of Heitmann's *The Modernization of the Louisiana Sugar Industry, 1830–1910* or

Stein's *The French Sugar Business in the Eighteenth Century,* both from Louisiana State University Press, or of Sitterson's *Sugar Country,* now out of print from the University of Kentucky Press. *Delta Sugar* resembles the Sitterson book in its historical perspective, but it differs by having more visual materials in historic and contemporary photographs and maps that convey the textures on which visual and cultural geography rely.

The content of *Delta Sugar* focuses on the landscape during two periods. One is the recent past, the last thirty years, in which many of Louisiana's plantations have disappeared. There are no thorough examinations of this part of the Louisiana sugar plantation story. Deeper areas of content probe the long-term evolution of the Louisiana sugar plantation to its origins. The use of historical settlement succession as explanatory description will enable the reader to experience plantation development vicariously through specific French and Anglo sugar plantations. In this regard, the book is a scholarly treatment of change at several levels from both historical and cultural geographic perspectives.

As a cultural geographer, I interpret the panorama of Louisiana's sugarcane plantation landscape from a culturogeographic perspective. In chapter 1, "Plantation Evolution," I introduce the plantation concept, develop the delta environment as a physical geographic foundation, trace historical perspectives from the origin of sugarcane and its industry in the Old World to a Caribbean plantation legacy, and describe Louisiana's sugar plantation distributions over a temporospatial milieu. The text and maps will lead the reader to infer a degree of spatial and temporal stability for sugarcane and the plantations on which it grows.

Chapter 2, "Culture and Form," focuses first on specific trait signatures for cultural identity. Mansion types and settlement patterns are diagnostic traits for French and Anglo-American plantations. Here we examine architectural traits, origins, and differences between French Creole plantation mansions and Anglo-American mansion types. Anglo architectural traits trace to the English-settled Tidewater region on the Atlantic coast and to Scotch-Irish-, Scottish-, and English-settled areas of the South. Settlement patterns also prove to be diagnostically important for identifying French and Anglo-American plantation landscapes imprinted from the first effective settlement. We examine dwellings of all kinds and types, including mansions, for their cultural identity. Chapter 2 also features other dwelling types, such as quarter houses and overseer's houses found "back of the big house." Such buildings carry no lesser importance to the settlement geographer and provide a glimpse of the past in this most rapidly vanishing part of the landscape.

Chapter 3, "The Morphology of the Functional Plantation Landscape," examines those elements in plantation morphology that have functional meaning to the working plantation. The chapter covers form and function in a historical context for sugar factories, barns and outbuildings, stores and churches, fields and fences, agricultural implements, landings, and levees. In essence, we look at those things in the material culture, other than dwellings, that have landscape expression and functional meaning.

Chapter 4, "A Prescription for Landscape Decline," examines the events of the past thirty years that have shaped the changing, especially declining, plantation landscape. Dramatic changes in the structure of plantations are explained through the medium of evolutionary occupancy models that follow phases of sequent occupance. From the traditional plantation, the models evolve into a Louisiana plantation corporation model, a multiplantation corporation that rises from the dust of the Great Depression. The evolution proceeds to the CALA model, a 1970s lamination of a California agribusiness model to the Louisiana plantation corporation model. The result is a bulldozed, sanitized landscape that has lost the old qualities of plantation management and settlement. Finally, chapter 4 examines the small plantation and cooperative sugar factory model, which seems to maintain stability in the unstable business of producing sugar in a marginal environment.

A full understanding of the sugar plantation requires a focused perspective. A plantation is much like a person, exhibiting a personality through its sequence of ownership and landscape expression. A historic photographic record and a settlement succession story is told in chapters 5 through 10 for six case-study plantations–Armant, Ashland, Oaklawn, Cedar Grove, Whitney, and Madewood–which represent a cross section of the cultural heritage and morphology of the Louisiana sugarcane plantation landscape. Spatially, temporally, culturally, and economically, these plantations selectively best represent the Louisiana landscape for two focused time periods–the 1960s and the 1990s. I have arranged the plantations according to degrees of decline and types of ownership. We witness each plantation's development through a succession of owners and their influence on the landscape. In this way, one may travel vicariously through time and space to each plantation. Illustrated with maps and photographs, these chapters demonstrate that the past is the key to the present.

Chapter 11, "Agents of Change," briefly examines changing regional and national patterns to show that Louisiana's sugarcane plantations are not isolated in American culture. They are interwoven into the fabric of processes and events that focuses on external and internal change, regulations, and personal choices affecting the plantation and its landscape. Factors such as sugar prices, federal government controls on sugar, the energy crisis, industrial expansion, environmental protection regulations, the introduction of soybeans, meteorological disasters, sugarcane expansions, economies of size in sugar factories, cane operators and labor, and entrepreneurs all played roles in the dynamic demise of the contemporary plantation landscape.

Acknowledgments

I acknowledge my wife, Judy, for putting up with me in Louisiana's sticky heat and my now-grown children, Karen and Ken, who wondered why I made so many trips to the delta. Special recognition goes to my mentors–Fritz Gritzner, for introducing me to the excitement of geography, and Fred Kniffen, for wisely directing my eagerness into plantation research. I am especially grateful to Don Ralston, Will Fontanez, Brian Dibartolo, and Russell Peeler for their cartography; to colleague Ron Foresta and the peer reviewers for their encouraging advice; and to Linda Forlifer at the Johns Hopkins University Press and George F. Thompson, president of the Center for American Places, for their editorial expertise and patience. It is the planter folk, past and present, whom I should finally acknowledge. Without their vision, perseverance, and cultural traditions, the Louisiana landscape would have had an entirely different appearance.

CHAPTER 1

Plantation Evolution

Southern Louisiana's cultural landscape displays a multiplicity of contemporary surface features. Consider unique urban scenes in New Orleans, Thibodaux, Houma, or Morgan City. Experience the less than pleasant smells, sights, and sounds of more than 138 chemical industrial plants in the Chemical Corridor, also called "Cancer Alley," bordering both banks of the Mississippi River between Baton Rouge and New Orleans. Observe swamp and marsh communities strung along the banks of sluggish bayous that reach across the deltaic plain. Inexorably, you will see sugarcane plantations, with their deep history but questionable future, struggling for space in the scene. The plantations that first produced granulated crystalline sugar were etched into the Louisiana landscape beginning in 1795. But their roots ran deeper to a plantation landscape legacy initially based on tobacco and indigo in the early 1720s and to experiments in sugarcane cultivation in the 1740s.[1] Sugar plantations persistently dominated the rural landscape from 1795 until about 1970. But today's plantations are vanishing, and their dominance as cultural traditions is in danger of disappearing.

FIG. 1.1. *Sugar factories are among the most visible plantation landscape features but today appear on only one of every four plantations. Saint James Sugar Cooperative, Incorporated, Saint James Parish.* Photograph, 1995

This chapter defines the plantation concept, builds the natural environmental setting and ethnic backgrounds, explores deep historical perspectives, and explains spatial distributions over time.

Delta sugar is not a fantasy. It is a complete and genuine plantation landscape that evolved over two and a half centuries, a quarter of a millennium, on a flat, damp surface called the Lower Mississippi River floodplain in southern Louisiana. It is also the last place in the continental United States where a functioning plantation landscape exists. *Delta Sugar* embodies traditional plantations dating from the eighteenth and nineteenth centuries as well as residual forms on contemporary landscapes. Plantation buildings, many in ruins, with weather-beaten walls and missing tin roofs exposing skeletal rafters, reflect past landscapes and patterns of human folk cultural contributions. Few plantations survive, but some of the working ones allow us to view once more, but probably not for very long, a landscape of the past.

Over the past thirty years, about one hundred twenty sugar plantations have disappeared; along with them have gone twenty-five sugar factories that once dominated the landscape. Only eighty-two plantations and nineteen sugar factories remain, and many of them are declining. Ironically, a landscape morphology that once nurtured the crop that initiated it is vanishing. Stranger still, sugarcane production and acreage have reached record levels over the past several years despite damaging frosts and hurricanes. As expanding cane lands and productivity reflect unprecedented growth, plantation morphology experiences continued deterioration.

The Plantation Defined

The southern plantation is a widely popularized but little understood property on the North American cultural landscape. Almost everyone has a concept or a mental picture of a plantation. Some view the plantation mansion as the perfect, definitive symbol, much like fictional Tara and Twelve Oaks in *Gone with the Wind.*[2] Some scholars use functional definitions related to single-cash-crop economics, large landholdings, or a form of measurement that separates this type of farming unit from the family farm.[3] Others treat the subject from a historical perspective dealing with slavery and with antebellum and postbellum materials.[4]

Plantation definitions vary over time and space and reflect different points of view. In seventeenth-century Britain, the term *plantation* was synonymous with *colony* and functionally meant the process of settling. King

FIG. 1.2. *Abandoned late-nineteenth-century plantation quarter houses have become rare landscape relics. Iberville Parish.* Photograph, 1993

James I established plantations of lowland Scots in northern Ireland's Ulster Province in 1610.[5] During the early English settlement of the eastern seaboard of North America, colonizing efforts at Plymouth, Providence, and Jamestown were called plantations.

Definitions expressed social conditions. U. B. Phillips, American historian, linked the plantation definition to slavery (Phillips 1929, 21). Kenneth Stampp argued that the plantation concept was in existence before the peculiar institution of slavery, and he was correct in saying that plantations continued long after slavery was abolished (Stampp 1956, 5). Agricultural historian Lewis C. Gray said that "the plantation was a capitalistic type of agricultural organization in which a considerable number of unfree laborers were employed under unified direction and control in the production of a staple crop" (Gray [1932] 1958, 302). By omitting the word *unfree,* Gray's definition would apply to contemporary plantations.

Location and economic parameters did much to define plantations. According to George McCutchen McBride, "Plantations accordingly are a form of great landed estate, usually in colonial or semicolonial countries, which raise such tropical or semitropical products as cotton, sugar, rubber, coffee, tea, rice, pineapples, and bananas, with a laboring class kept in economic if not political servitude" (McBride 1934, 148). Leo Waibel, geographer, demonstrated a tropical and economic definition: "A plantation is . . . a large agricultural and industrial enterprise managed as a rule by Europeans, which at a great expense of labor and capital, raises highly valuable agricultural products for the world market." Waibel further stated that the plantation was industrial because it processed or preprocessed agricultural products. He placed the plantation exclusively in the tropics or subtropics because these areas "have long and in parts uninterrupted growing periods for vegetation, during which they [plantations] produce certain agricultural products that are lacking in the temperate zones" (Waibel 1941, 157).

Edgar Thompson, an authority on plantations for decades, argued that the plantation is a cultural phenomenon and does not wholly depend upon climate for its location or its definition. In deference to McBride and Waibel, Thompson argued that plantations are so located "not because of climate, but because in the present world community, tropical regions constitute a highly important and accessible frontier . . . Plantations have developed along non-tropical frontiers in the past and conceivably may in the future" (Thompson 1941, 54). Thompson defines the plantation in economic and social terms. "The plantation . . . is a large landed estate, locked in an area of open resources (i.e., land-rich and labor-poor), in which social relations between diverse racial and cultural groups are based upon authority, involving the subordination of resident laborers to a planter for the purpose of producing an agricultural staple which is sold to a world market" (Thompson 1935, 5).

Are plantations social or agricultural institutions? Sidney Mintz, undisputed dean of Caribbean social research, said in 1964 that the plantation is "an absolutely unprecedented social, economic, and political institution, and by no means simply an innovation in the organization of agriculture" (Guerra 1964, xiv). Roland Chardon at Louisiana State University defined the sugar plantation in the Dominican Republic in 1984 in terms of the organization, technology, processes, and spatial relationships surrounding the *ingenio,* or sugar mill. "It [the plantation] includes not only the surrounding area from which the supply of sugarcane comes but also the modern organization, labor force, and equipment for production, storage, and shipment from a central mill" (Chardon 1984, 449). Here the sugar factory came to mean something more than a processing center; it served as a surrogate for the entire plantation operation and landscape. In another plantation inquiry, Chardon examined plantation traits in the New World with topics covering Jesuit missions in Brazil, the diffusion of African slaves to Peru, the role of tobacco in Virginia's urban growth, the diffusion of the watermelon, the peopling of the Danish West Indies, the role of the plantation system in Guatemalan political unrest, William Walker and slavery in Nicaragua, East Indian ethnicity in Trinidad, and plantation slave subsistence in the Old South and Louisiana (Chardon 1983). None of the papers provided a plantation definition, but the themes of the project offer a glimpse of the kinds of material included in plantation research.

Merle C. Prunty, one the South's leading rural geographers, expressed the need for a modern definition.

> The "plantation," as the term is used in the South today, comprises six elements: a landholding large enough to distinguish it from a family farm; a distinct division of labor and management functions, with management customarily in the hands of the owner; specialized agricultural production, usually with two or three specialties per proprietorship; location in some area in the South with a plantation tradition; distinctive settlement forms and

spatial organization reflecting to a high degree, centralized control of cultivating power; and a relatively large input of cultivating power per unit of area (Prunty 1955, 460).

Form and function still play important roles in plantation definitions, but the functional component has a very different manifestation in the contemporary scene. Anyone touring the grand old antebellum plantation mansions around New Orleans will see *plantation homes without plantations.* Homes like Ormond, Destrehan, and San Francisco no longer function as plantations; in fact, many mansions on the touring circuit, such as Oak Alley, Madewood, Oaklawn, and Nottoway, are owned and operated separately from the sugar lands nearby. Today's "sugar planter" is either an individual or a corporation owning some land but leasing much more, ranging from a few hundred to as many as eighty thousand acres, on which to produce sugarcane. Sugar operators manage impressive sugar operations with numerous properties no longer functioning as plantations or even as family farms. As land tenure changes from ownership to leaseholding, the landscape radically changes as well.

Definitions apparently are as prevalent as experts on the subject, but plantations generically demonstrate a relative body of common attributes. The important generic criteria defining a sugarcane plantation are

- large-scale agriculture
- raw sugar as the primary product
- capital-intensive business
- initially a European enterprise in its establishment and management
- distant, extraregional markets
- intensive labor ranging from indentured laborers to slaves to wage laborers over time
- resident labor on all initial plantations, changing in some areas to local, nonresident labor
- special settlement patterns, largely compact agglomerated clusters of buildings
- hierarchical social order

Louisiana's sugarcane plantations possess these characteristics and much more, but a fundamental question in my mind has been how plantations show a landscape expression. For three decades I have pondered the plantation definition, and from the very beginning I sought a working definition that was meaningful to the cultural landscape (Rehder 1971, 5–10). In Louisiana, a complete model sugar plantation appeared to me as an agricultural enterprise with distinctive landscape features. Towering over the cane fields, tall chimneys punctuated the location of a sugar factory–an agricultural factory-in-the-field. One of every four plantations had a sugar factory (also called a sugar mill, sugarhouse, or *central* because of the centralization of sugar factories and milling functions), and that ratio continues to the present. A cluster of barns, sheds, warehouses, molasses tanks, and assorted functioning structures surrounded the sugar factory to form a centralized outbuilding complex. The *quarters* was a village of nearly identical laborers' dwellings aligned along a single road in a linear pattern or grouped in a block pattern based on a grid of streets. An *overseer's house* was nearby but not directly in the quarters. A prominent mansion, colloquially called *the big house* and set amid moss-draped oaks, occupied a site exclusive of the other buildings. At the mansion site, some plantations had an entourage of small structures consisting of large cypress water cisterns attached to the mansion, a detached kitchen, a guest house, a garçonnière, pigeonniers, a privy, and perhaps an office and other small buildings. On larger plantations a company store, a company office, and a church were located on or near the plantation holdings. Extensive fields covering hundreds, even thousands, of acres unbroken by fences stretched long and narrow from stream banks to backswamps. Long, straight ditches divided the fields to give them a characteristic linear appearance. Such was the plantation in the 1960s and early 1970s, when many enterprises matched the model, but today's landscape contains few properties that still fit this description.

My working field definition focuses on a minimum of four quarter houses on site. The presence of quarter houses means that the agricultural unit now or at an earlier time required more on-site, live-in laborers than are required on a family farm. Four houses are the minimum because they can be arranged geometrically into either a line or a block settlement. Although three houses can make a line, three cannot form a block or square. Geometry becomes culturally important because linear settlements are identified with original French plantations and block-shaped settlements have been traced to Anglo plantation sources. If four quarter houses signify a plantation, then what prevents four mobile homes or any four modern subdivision houses from being called a plantation? The presence of four

FIG. 1.3. *A model sugarcane plantation of the 1960s has the manager's house at the* left, *the quarters for agricultural workers in the foreground, and the sugar factory workers quarters behind the sugar factory. Caldwell Sugar Cooperative, formerly Laurel Grove Plantation. Lafourche Parish.* Photograph, 1967

quarter houses serves only as an initial signature that I seek in the field for the working field definition. In the search for four or more quarter houses, I also look for other diagnostic signatures, such as an owner's mansion or manager's house, an outbuilding complex of barns and sheds with perhaps a sugar factory or the ruins of one, large fields either presently or formerly used for sugarcane on large landholdings laced with ditches and served by roads, and other landscape features that together set the quarters into a plantation context and considerably add to the definition.

In my initial search for Louisiana's sugarcane plantations, I wore out a battered blue '51 Ford and later drove a newer '67 Chevy Malibu over ten thousand miles following both banks of the Mississippi River and up and down every bayou road in twenty parishes. I searched from Venice, near the mouth of the Mississippi River below New Orleans, northward to Alexandria in Rapides Parish, from Baton Rouge westward beyond Lafayette, and from Alexandria southward below Houma in Terrebonne Parish and beyond the realm of reason across miles of coastal marshes to the Gulf of Mexico. I found 202 sugar plantations in Louisiana in 1969. I repeated the process in 1993, retracing the same routes but this time in a much better field vehicle, a red '89 Chevy Blazer with my passport to the region, a front license plate that read "This is CAJUN country." I investigated the plantation distribution, field checking and remapping every plantation I had mapped in the 1960s. I discovered that 120 plantations and 25 sugar factories were missing, razed, destroyed, or reduced to the point that they no longer had a landscape presence within the parameters of my plantation definition.

Contemporary Sugar Plantation Distributions and Productivity

The freeze-sensitive tropical crop that we call sugarcane has been tenaciously tended on plantations for more than two hundred fifty years in the oldest but most marginal area of commercial sugarcane production in the United States. Continuously cultivated in Louisiana since its introduction in 1742 from the French Caribbean, sugarcane demonstrates remarkable spatial stability.[6] Current cultivations occupy the very same fields and soils that supported sugarcane in the 1840s and in some areas since the 1790s. Large tracts must be planted on farms with two hundred to eight thousand acres for the crop to be economically feasible. But with dry land at a premium, the ecumene for Louisiana's urban and industrial growth places greater

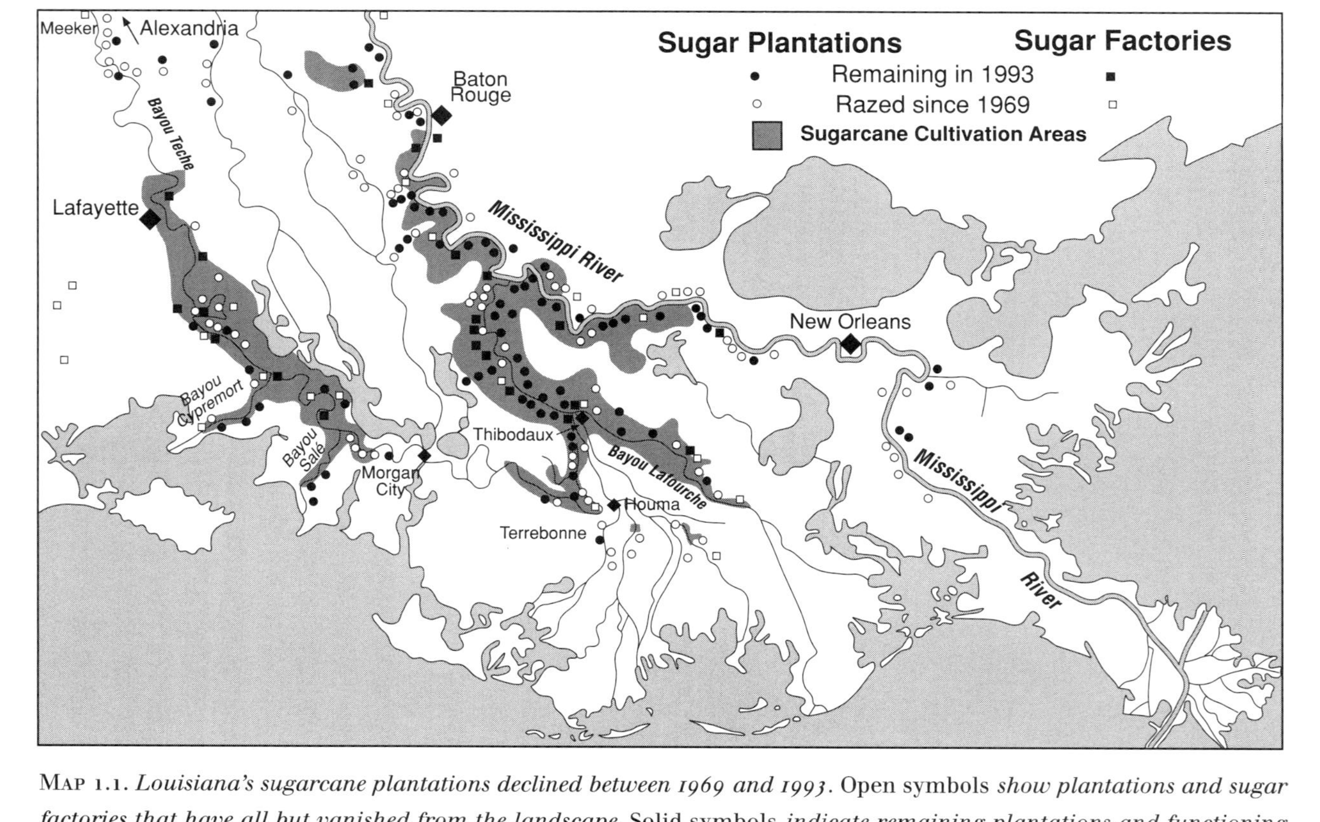

MAP 1.1. *Louisiana's sugarcane plantations declined between 1969 and 1993.* Open symbols *show plantations and sugar factories that have all but vanished from the landscape.* Solid symbols *indicate remaining plantations and functioning sugar factories. Compare sugarcane cultivation areas* (shaded) *with the plantation distribution.*

pressure on large plantation tracts, particularly those along the Mississippi River between Baton Rouge and New Orleans. Historic and contemporary spatial patterns emerge into "dottable data" plotted on maps over space and time.

Louisiana's sugar plantation landscape survives in a well-defined sugarcane growing region concentrated in the Mississippi River floodplain in twenty south-central Louisiana parishes where 364,000 acres are harvested from sugarcane cultivations. Plantations extend northward to Meeker near Alexandria and southward to Houma (see map 1.1). The eastern limit of the plantation area follows the east bank of the Mississippi River from Baton Rouge to a point fifteen miles south of New Orleans. At its widest point, the region extends 120 miles between New Orleans and Lafayette. The western limit follows a line from Meeker near Alexandria southwestward to Lafayette and southeastward to a point near Morgan City on Bayou Teche. A very recent western expansion of cane lands is taking place west of Lafayette toward Lake Charles, where about six thousand acres of new cane lands have opened up. By the year 2000, an estimated thirty thousand additional acres could possibly become sugarcane lands.[7] For the most part, however, sugar plantations have been situated on better-drained portions of natural levees of the Mississippi River, Bayou Lafourche, Bayou Teche, Bayou Salé, Bayou Cypremort, and the bayous of Terrebonne Parish (see map 1.1).

As the plantation landscape vanishes, commercial sugarcane cultivation and productivity flourishes in contemporary Louisiana. Sugarcane's productivity aggressively expands in tonnage and acreage while the number of sugar plantations continues to decline. From a national perspective, Louisiana ranks second behind Florida, slightly ahead of Hawaii, and well above Texas in the four states that produce commercial sugar from sugarcane in the United States. In 1996, Louisiana's zenith year, production was about 30 percent of the U.S. sugarcane crop. The 1995–96 record production of 1,058,000 tons followed a previous record of 1,021,000 tons in 1994–95 (USDA Economic Research Service 1997, 35). Within the state in terms of cash crop rankings, sugarcane was first in 1973, fourth in 1983, and second to cotton in 1993, with the other crops being soybeans and rice. Louisiana's sugar industry regularly suffers from vagaries in weather–frosts and freezes, floods, and hurricanes. For example, in 1990–91 a frost-damaged crop yielded only 438,000 tons (USDA Economic Research Service 1993).

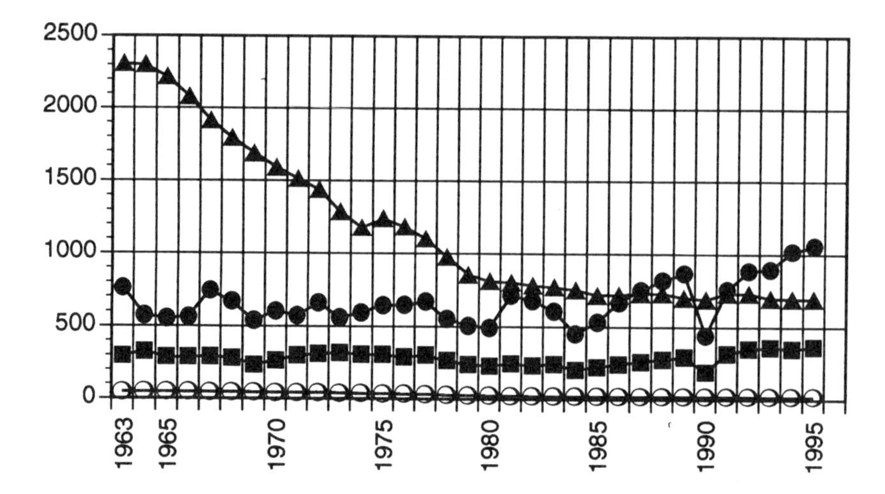

FIG. 1.4. *Trends in Louisiana's sugarcane industry are reflected in raw sugar production* (solid circles, *in thousands of tons), harvested acres* (squares, *in thousands), farms* (triangles), *and sugar factories* (open circles) *in the turbulent years between 1963 and 1995*. Sources: U.S. Department of Agriculture, Economic Research Service 1996, American Sugar Cane League

Despite such problems, the industry recovers quickly, and the crop dominates as a monoculture in more than half of the sugar parishes in the state.

The Delta Environment

The physical requirements of sugarcane for granulated sugar make Louisiana a unique and incomprehensible place for cultivation. Biologically, sugarcane (*Saccharum officinarum*) requires eleven to twelve months to reach maturity (Blume 1985, 44–52). Louisiana's subtropical climate, however, provides only a nine- or ten-month growing season that suffers from unpredictable but devastating freezes. Rainfall is plentiful, but poor soil drainage limits cane cultivation to better drained alluvial natural levees in a land otherwise known for bayous, interlevee basin swamps, and quaking marshes on the margins of terra firma.

The lower Mississippi River floodplain forms one of North America's broadest deltaic river plains. Bounded on the east and west by slightly higher upland Pleistocene terraces, the deltaic plain stretches over one hundred twenty miles wide, a river- and lake-riddled surface where no

TABLE 1.1

Trends in the Louisiana Sugar Industry

Year	*No. of Farms*	*Acreage (× 1,000)*	*Raw Sugar (× 1,000 tons)*	*No. of Factories*
1937	10,260	–	188	92
1959	2,686	–	440	47
1969	1,687	242	669	44
1974	1,180	308	571	37
1978	977	268	550	28
1982	780	247	675	21
1987	725	264	731	20
1989	696	290	844	20
1990	685	201	438	20
1991	725	321	762	20
1992	726	356	868	20
1993	693	360	890	20
1995	690	364	1,058	20

Sources: American Sugarcane League 1991–96; Durbin 1980, 84; U.S. Department of Commerce, Census of Agriculture, Louisiana, 1959–87; USDA Economic Research Service, *Sugar and Sweetener Reports,* June 1987, p. 27; June 1990, p. 25; December 1990, p. 34; September 1992, p. 32; March 1993, p. 29; June 1993, pp. 18–44; September 1994, p. 14.

land is more than thirty-five feet above sea level. Father of Waters and delta land maker, the Mississippi River, tan and swift, surges southward girded by man-made levees on its flanks. Sluggish surface waters shimmer in a myriad of bayous and interlevee basins dotted with large, round lakes filled with cola-colored swamp water. Deep green tree-covered swamps and verdant marshes form a wetlands vegetative cover in fresh to brackish to salt water environments. Rising slightly above the damp delta are ribbons of tan, silty alluvium formed as deposits made by ancient and present courses of the Mississippi River. Land was a gift of the river free and clear; water was a gift, too, but swiftly flooding, brown, silty, and obscure. The people caught in the infamous 1927 flood thought that Louisiana's Mississippi was trying to wash them away (Newman 1974).

The sugar plantation landscape, except for a small area in the western part, lies entirely within the Mississippi River floodplain. Drainage patterns are distributary, and stream courses diverge downstream toward the Gulf of Mexico. Natural levees formed by accumulating layers of alluvium after

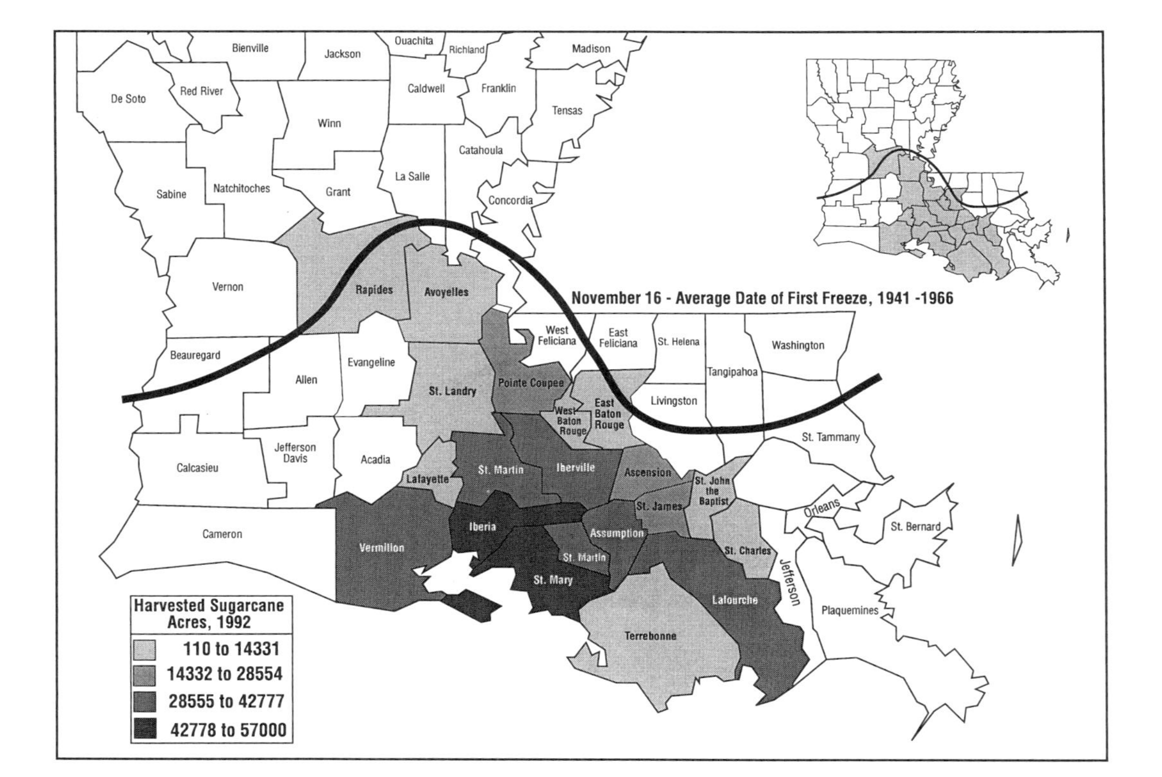

MAP 1.2. *Louisiana's harvested sugarcane acreage by parish in 1992 illustrates the heart of sugar country, where a tropical crop is cultivated in a marginally subtropical region. The isostade for November 16, the average date for first frost, is a crucial climatic boundary.* Sources: U.S. Department of Agriculture 1993; Cry 1968; Buzzanell 1993, 21

overland floods provided the physical foundations upon which most cultural landscapes emerged. In profile, the crest of a natural levee has its highest elevation at the stream bank; the imperceptible backslope descends as little as six inches per mile and extends as much as four or five miles, ultimately terminating at the backswamp. Natural levee crests on the Mississippi vary from thirty-five feet near Baton Rouge to sea level at the Mississippi passes south of New Orleans (Russell 1936, 73–77). Levee crest elevations for other streams are lower and diminish southward toward the Gulf of Mexico. Near the geographic center of the delta, Bayou Lafourche's natural levees were formed by an earlier course of the Mississippi River and stand about fifteen feet above the surrounding floodplain. Levees on Bayou Teche, also the result of a former Mississippi channel and a later Red River course, are ten to fifteen feet high.

Levee crests served as major focal points and sites of choice for human occupance. Relatively safe from damaging floods, settlement successions established stratigraphic sequences of occupation on these natural features (McIntire 1958, 128). From Indian habitations to early French settlements to plantations to current urban and industrial complexes, people competed for land in a limited ecumene. Levee widths, measured from levee crest to backswamp at the normal limit of cultivation, are widest along the present course of the Mississippi River south of Baton Rouge, along the flanks of upper Bayou Lafourche, and along Bayou Teche north of Franklin. Natural levees bordering the Mississippi are about three miles wide south of Baton Rouge for a downstream distance of about forty miles. Maximum widths of six miles occur at points where crevasses have breached the levee at flood time and deposited sediment well beyond the normal backswamp (Russell 1936, 72–77). Lands adjacent to streams with the widest natural levees have been choice sites for agriculture throughout the plantation history of the area. In contrast, where levees are low and narrow, such as along the southernmost courses of the Mississippi and its distributaries, arable land is so limited that plantation-scale agriculture is unlikely. The land limits of a sugar plantation can only be estimated because of the variables in plantation size, owner-builder intentions, and technological capabilities through time. But field evidence taken over the past thirty years indicates an absence of intact sugar plantations on levees that are less than one-quarter mile wide.

Productive floodplain soils possess a wide range of texture and drainage characteristics depending upon their location on natural levees. Surface soils form as a natural result of overland flooding when streams overflow

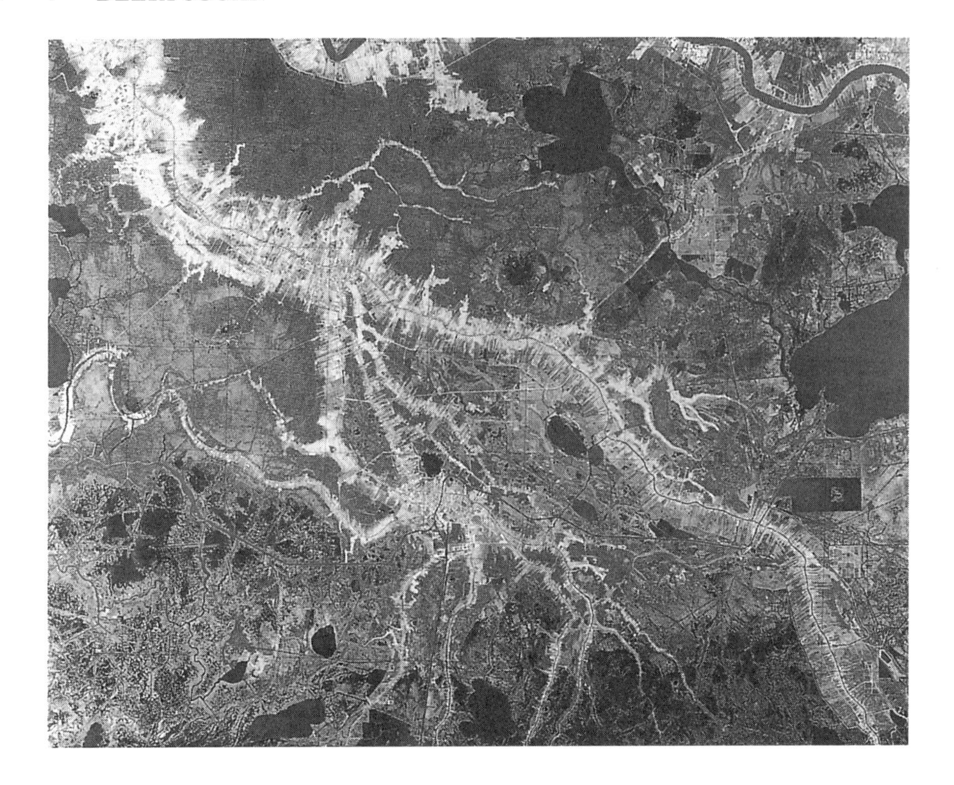

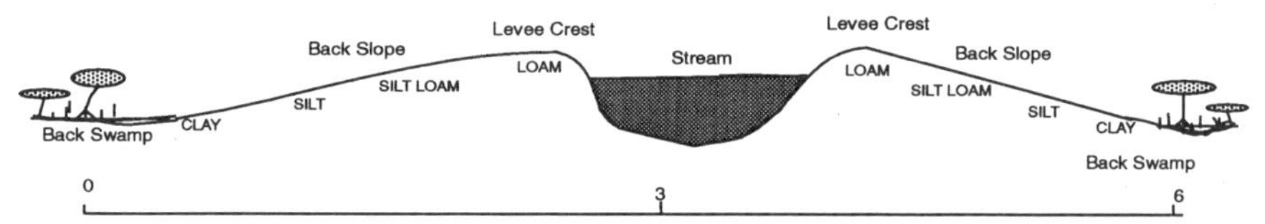

FIG. 1.5. *A satellite image of the delta environment displays natural levees as white fingers of relatively dry arable land with darker areas of open water, swamps, and marshes. New Orleans is in the* upper right. *The profile of a natural levee demonstrates why levee crests are favored settlement sites and illustrates the gradation from levee crest to backswamp in slope and soil types.* Source: U.S. Geological Survey, Landsat mosaic of New Orleans, 1:250,000 scale

their banks, spreading highly turbid water over natural levees. Coarse sands and heavier silts winnow out first and are deposited on frontlands. Farther down the gentle slope toward the backswamp, floodwaters move more slowly, depositing fine silts and clays. A horizontal gradation develops, from coarse, friable, sandy soils nearer the levee crest to fine, thick, stiff clays toward the backswamp.[8] Such natural-levee landscapes served as the physical stage upon which selective crops were cultivated and a plantation landscape was developed. Elsewhere in the floodplain, broadly spaced interlevee basins composed of lakes, swamps, and coastal marshes separate levee systems from other stream courses. Runoff from natural levees collects in the basins, with waters slowly draining southward through interconnected lakes, bayous, bays, and estuaries, ultimately to the Gulf of Mexico.

Sugarcane Physiology, Varieties, and Environmental Requirements

Sugarcane plants are sucrose-laden perennial grasses that reach heights of twelve feet or more. The stalk gets the most attention because it contains a fleshy pith that stores sugar. Joints or nodes divide the cane into segments over the full length of the long stalk, and each segment ranges from three to twelve or more inches in length. Long, sharp leaves radiate out and upward from the nodes. Lateral buds on the nodes are extremely important because planted sugarcane is propagated asexually as a cultigen from cuttings. New canes, called *plant cane*, emerge from planted stalk segments that have two or more buds, or eyes, at the node. Sugarcane also has the unique ability to *ratoon*, that is, reemerge from the stubble or rootstock into a full-grown crop for each of several successive years. Although twenty ratoons are possible, canes diminish in quality because of soil fertility depletion and root problems. Since yields are cut in half by the fourth ratoon, planters allow crops to ratoon for two to four years before replanting with a new crop of plant cane (Blackburn 1984, 158–61). Cane stalks display color (green to yellow to red or purple) depending on variety and stage of growth. Mature canes have stem diameters of one-half to three inches and have feathery tassel seed flowers at the top of the plant. Seeds are important only for cane breeding work and are otherwise undesirable for planting cane crops. A cross section of the cylindrical stalk reveals an outer hard rind and a soft, fleshy fibrous pulp inside. The toughness of the rind depends on variety; soft rinds were easier to chew and to mill with primitive milling technology, but hard rind varieties were more resistant to frost and disease (Earle 1928, 62–63). The root system has shallow, widely branching, superficial roots to take up nutrients and moisture and buttress roots to support the tall stalk, and in dry regions the plant may develop rope system roots that penetrate deep into the soil for moisture. Shallow roots are highly susceptible to root rot and water-borne diseases.[9]

The world's domesticated sugarcane varieties were few and relatively simple from about 8000 B.C. until the late nineteenth century. After 1900, sugarcane varieties became short lived and quite complex through scientific selective breeding and hybridization. The first plant, *Saccharum officinarum,* diffused from its domestic homeland in New Guinea to India, where it hybridized with *Saccharum barberi.* This hybrid became the single most important sugarcane variety to diffuse from India to the Middle East to the Mediterranean region to the Atlantic Islands off Africa and ultimately to the New World. The first sugarcane in the New World was brought to Hispaniola by Christopher Columbus, on his second voyage, in 1493. This old original cane probably had no name as it diffused throughout the Caribbean until the late eighteenth century, when Pacific varieties of *S. officinarum* started to arrive from European exploration voyages in the Indian Ocean and in the South Pacific. These new Pacific varieties were called "Otaheite" and "Bourbon" canes, while the old Caribbean cane came to be called "Creole" (Barnes 1964, 1–5, 30–33). By the late nineteenth century, Dutch experimenters working with Cheribon canes in Java discovered breeding and hybridization methods resulting in numerous new hybrids spreading throughout the sugarcane world.

Louisiana's sugarcane varieties followed much the same pattern of simplicity leading to complexity over time. As few as ten varieties dominated the sugar crops from 1742 until 1940. Creole cane was the first and only cane in Louisiana between 1742 and 1797. Even after Otaheite cane arrived in 1797, Creole still dominated, with more than 75 percent of crops until 1830. These were replaced by Louisiana Purple (Black Cheribon) and Louisiana Striped (Striped Preanger) varieties, which completely dominated cultivation until 1898, when D-74, a variety from the Royal Agricultural Society of Demerara in British Guiana, was introduced. The reliance on a single variety led to disaster in the 1920s, when the sugarcane mosaic disease destroyed D-74 crops and caused production to plummet from 324,000 tons in 1921 to 47,000 tons in 1926. Diversity meant survival as thirty-four more varieties were introduced between 1926 and 1978 (Durbin 1980, 84–85). First came POJ (Proefstation Oost Java) varieties from Java, then CP varieties from the U.S. Department of Agriculture Experiment Station in Canal Point, Florida, and then later ones from the Agricultural Experiment Station in Houma, Louisiana.[10]

In an ideal environment, sugarcane requires a frost-free twelve-month growing season to reach full maturity. Temperatures must remain high throughout the season because cane growth can be retarded at temperatures below seventy degrees Fahrenheit and ceases when soil temperature goes below sixty-two degrees (Barnes 1964, 28). In completely tropical regions of the world, canes are cut at eleven months because the sucrose content has nearly reached its maximum level, beyond which the plant will begin to desiccate and flower. An additional "free" crop can be harvested over a twelve-year period by cutting at eleven months instead of twelve. Water requirements and relationships are particularly important. A mature crop will contain 70 percent water (by volume), with the balance in cellulose, sucrose, and other sugars. Rainfall of more than forty inches relatively evenly distributed throughout the year is required. Irrigation is needed in some areas, such as parts of Hawaii, where one ton of irrigation water is needed to produce one pound of sugar (Barnes 1964, 26–28). In wet, flat regions such as Louisiana, drainage is critical because of root rot problems; as one cane farmer told me, "sugarcane can't stand wet feet."

Louisiana's Immigrant Ethnohistory, 1718–1850

It is important to understand the ethnohistory of any region in which the layers of culture are so deep and so important to the landscape. The immigrant ethnohistory could be viewed in a stratigraphic sequence, but it is perhaps wise to point out that, as Louisiana experienced a long and varied cultural heritage, two primary groups fulfilled the status of sugar planter, French Creoles and Anglo-Americans. Initial European immigrants during the eighteenth century were predominantly French, with minor groups of German and Spanish settlers. Later, in the first half of the nineteenth century, large numbers of Anglo-Americans migrated overland from the Atlantic seaboard of Virginia, the Carolinas, and Georgia and southward from the Upland South into Louisiana after the Louisiana Purchase in 1803 (Writer's Program WPA 1941, 38–44). Africans as slaves were entering Louisiana during both centuries, but African-born slaves came predominantly during the period 1726–43 and sporadically thereafter.[11]

Initial French settlers came directly from European French sources in Nantes, La Rochelle, and Saint Malo and began settling along selected sites on the Gulf Coast. Very early French settlements appeared at Fort Maurepas (1699), Old Mobile (1702), Mobile (1711), and Biloxi (1720). The area of the Mississippi River in the vicinity of New Orleans was initially settled in 1718. Other French settlements began as military forts at Natchez, Vicksburg,

Natchitoches, and Baton Rouge. By 1731, isolated French farming settlements called *concessions* had appeared at various points along the Mississippi River and at Natchitoches.[12]

Acadians, ancestors to today's Cajuns, were people of French ancestry forced by the British to leave established settlements in Acadia (Nova Scotia) in eastern Canada. They began to enter Louisiana in two waves of immigration, between 1755 and 1785. Initial settlement was focused along the Mississippi River in an area that came to be known as the Acadian Coast in Saint James and Ascension Parishes. Later, more Acadians settled much farther west along Bayou Teche in the Attakapas country in the vicinity of the present-day city of Lafayette. Louisiana's Cajuns grew from an initial eighteenth-century population of about 4,000 immigrants to a 1990 population of 432,549 who claim Acadian Cajun ancestry.[13] French West Indian refugees fleeing slave revolts in Sainte Domingue (Haiti) and Martinique entered Louisiana at the turn of the nineteenth century. Composed mostly of planters, this refugee population settled on scattered unclaimed lands in southern Louisiana.[14]

German and Spanish immigrants arrived in smaller numbers during the eighteenth century and for the most part became assimilated by the dominant French culture. Germans arriving as early as the 1720s settled in small numbers in Saint Charles and Saint John-the-Baptist Parishes, which would become known as the German Coast. Other Germans settled the Gulf Coast in the Pascogula and Biloxi areas in the 1720s.[15] Most of the Germans were family farmers contented with small landholdings and farmsteads, and only a few became large plantation owners (Deiler 1909, 1–16). Spanish immigrants came directly from the Iberian Peninsula and the Canary Islands during the period of Spanish political control in Louisiana between 1763 and 1803. They settled in small concentrations along Bayou Terre-aux-Boeufs in Saint Bernard Parish, along both banks of the upper parts of Bayou Lafourche, in Ascension Parish at Galveztown and Gonzales, and at New Iberia in Iberia Parish on Bayou Teche (Martin 1827, 1:43). Most early Spanish settlers established themselves on small subsistence farms, rarely on plantations.

Anglo-Americans and Africans rounded out the eighteenth- and nineteenth-century populations. From 1803 until the 1840s, waves of Anglo-Americans from the Atlantic seaboard and interior South poured into Louisiana, filling the void of habitable lands not already claimed by earlier non-Anglo groups. Most Anglo-Americans settled in northern Louisiana, but some planters moved on to settle in southern Louisiana.[16] African populations arriving as unwilling slaves were widely distributed among all types and sizes of plantations in the region. Ethnic distinctions were made among African-born slaves with such tribal identities as Mandinka, Bambara, Soso, Timne, Kiamba, and Kongo. American-born slaves were called *creole*, meaning born in the Americas.[17]

Planter ethnicity associated with sugar plantations originated in eighteenth-century southern French Louisiana, where the majority of initial planters were from French colonial, Creole (American born of French parents), Haitian, and other French Caribbean backgrounds. Anglo-American planters from the Carolinas, Virginia, Kentucky, Tennessee, Alabama, and Mississippi entered the region in the first half of the nineteenth century and adopted sugarcane agriculture with newly established plantations. The Acadians, Germans, Spaniards, and Africans seldom became large planters, although a few acquired plantations and led a plantation owner's way of life.

Louisiana still preserves two distinct culture areas: *French Louisiana* in the southern part of the state and *Anglo-American Louisiana* in the portions of the state north of Alexandria (Knipmeyer 1956, 3). Even so, the mobility and mixing of ethnic groups and the assimilation and acculturation of older groups create a less than clear picture. In the years between 1699 and 1996, the ethnic pattern, with approximate dates of initial arrival and settlement of the groups, included the following (Brasseaux 1996):

–French (1699)
–French Creole (1700 and 1809)
–African (1719)
–German (1720 and 1850)
–French Acadian "Cajuns" (1755 and 1785)
–Spanish (1765)
–Anglo-American (1803)
–West Indian (1700 and 1809)
–Irish (1803 and 1830)
–Yugoslavian (1820–70)
–Chinese (1865)
–Italian (1850)
–Sicilian (1880)
–Lebanese (1890)
–Hungarian (1900)

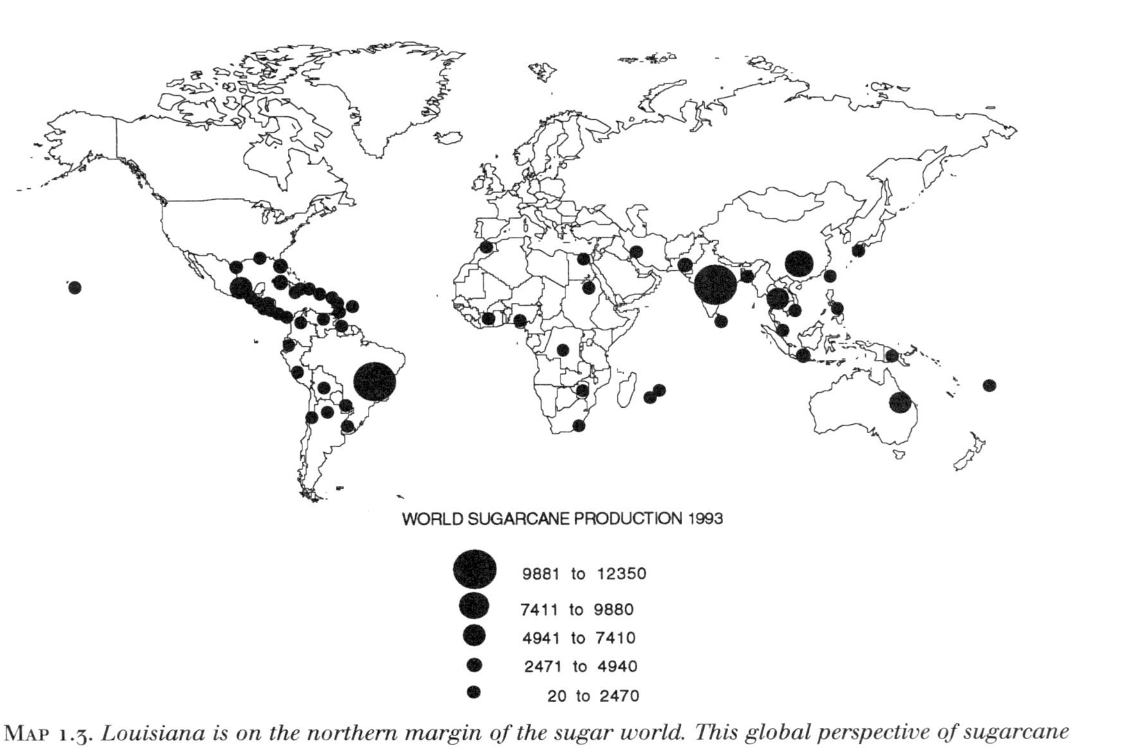

MAP 1.3. *Louisiana is on the northern margin of the sugar world. This global perspective of sugarcane regions in 1993 allows a comparison of spatial patterns and productivity.* Source: U.S. Department of Agriculture, Economic Research Service 1993, Sugar: World Markets and Trade

–Cuban (1959)
–Vietnamese (1975)
–Cambodian (1975)
–Laotian (1975)

In rural southern French Louisiana, except for widespread distributions of Anglo-Americans, a Yugoslavian enclave in Plaquemines Parish, and recent Asian immigrants scattered along the coast, cultural distinctions are difficult to assess, largely because so many were assimilated over time by the French.

Historical Perspectives

Historical perspectives on sugarcane origins and diffusions to the New World and on the Caribbean plantation influence provide time-depth material for a deeper understanding of the foundations leading to the Louisiana sugar industry. The emphasis here is on sugarcane as a cultivated botanical plant first and then as an industry whose primary purpose is to produce granulated sugar crystals, *raw sugar*. It is also important to view the contemporary global distribution of sugarcane cultivations, their relative production, and their tropical spatial limitations to understand the diffusion of sugarcane, its industry, and the position that Louisiana holds in the process. The historical diffusion of sugarcane follows two distinctive patterns: (1) the dispersal of sugarcane plants, which reflects disparate temporal directions and wide-ranging spatial patterns, and (2) a more focused but latent diffusion of sugar technology for manufacturing raw crystalline sugar.

Four related aspects of sugarcane use and manufacture will *not* be treated here in depth: sugarcane for chewing, syrup making, rum production, and sugar refining. Sugarcane for chewing has a much wider world distribution than plantation-grown sugarcane. It takes neither effort nor infrastructure to raise sugarcane for the chewing novelty. Syrup making, too, has a wider spatial and climatic distribution and requires less technical expertise and infrastructure than raw sugar production. Rum making, though largely found in tropical regions, may require more expertise and more infrastructure than syrup making, but both syrup and rum making can still take place on quite small, family-run operations. Neither requires a plantation scale of operation to be successful. Sugar refining requires neither cultivated land nor a tropical location. The majority of the world's sugar re-

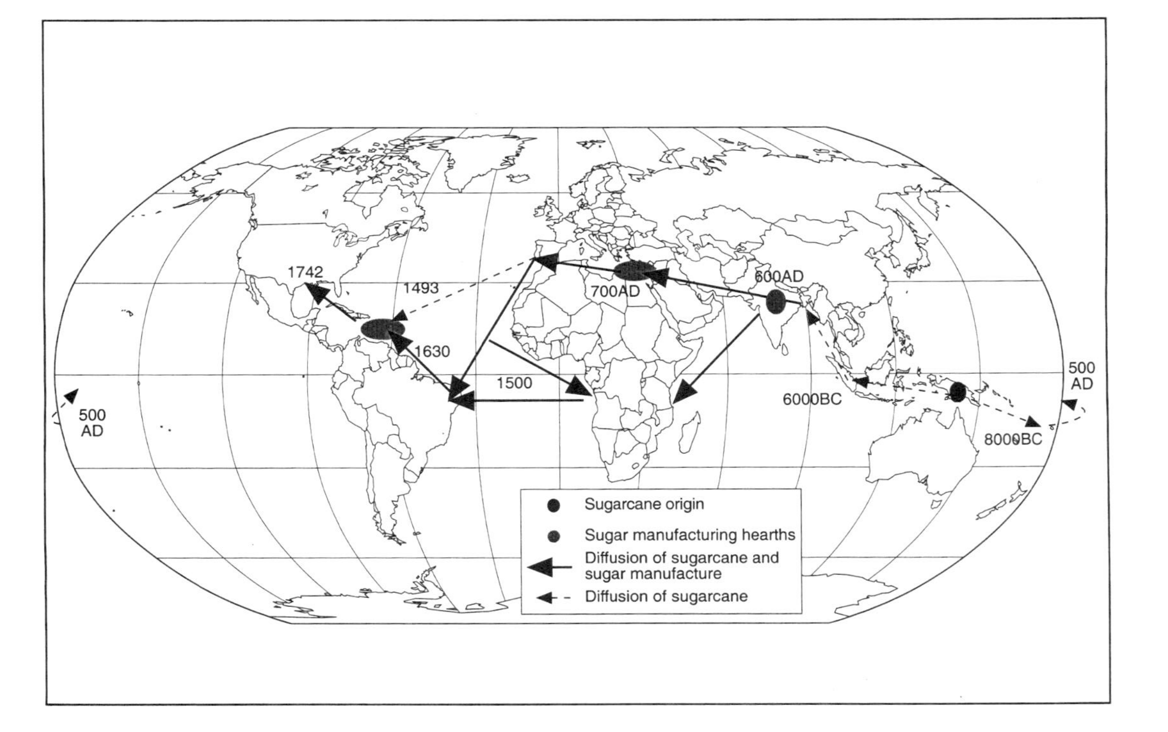

MAP 1.4. *The westward diffusion of sugarcane species began in New Guinea in 6000 B.C. and reached Louisiana by A.D. 1742. The sugarcane industry began in India in A.D. 600 and diffused through a pattern of manufacturing hearths before reaching Louisiana.* Sources: Alexander 1973, 12; Artschwanger and Brandes 1958; Barnes 1964, 1–4; Blackburn 1984, 2; Blume 1985, 22; Brandes and Satoris 1936; Deerr 1949; Galloway 1989, 26, 35, 49

fineries are located in midlatitude industrial cities convenient to port facilities, power, and markets. Sugar refining technology requires great power resources and personnel expertise in chemists, engineers, and management that may exceed those found or needed on plantations. It also follows the tradition of sugar refining in relatively drier, cooler climes because in the wet tropics refined sugar readily absorbs moisture and forms hard lumps when stored and transported in ships. Cane chewing, syrup making, rum making, and sugar refining have their places in the broader scope of the sugar story, but my intent is to focus largely on the biological plant, its production into raw sugar, and its place on plantations.

The Origin and Dispersals of Sugarcane and the Sugar Industry

The sweet, tall grass called sugarcane probably originated in South Asia and was first domesticated in New Guinea. Wild ancestors of sugarcane originating in South Asia spread south and east toward Australia as early as seventeen thousand years ago.[18] This initial dispersal was enhanced by the presence of ancient land bridges linking areas that later would become islands. In New Guinea, a wild species called *Saccharum robustum* probably hybridized with other wild grasses. From this union, *Saccharum officinarum* became the first domesticated species, the original noble canes. The term *noble* refers to two things: (1) the noble, high-yielding quality of the plant and (2) the process of nobilization, a technique used by early twentieth-century Dutch planters in Java who hybridized wild forms to *S. officinarum* to achieve an ennobled cane, one that was disease resistant. Whether it is called noble, official, or native sugarcane, *S. officinarum* is the original domesticated sugarcane species (Alexander 1973, 3–6).

Diffusion routes from the New Guinea domestic homeland took three primary directions. Initial movement, beginning about 8000 B.C., went southeastward from New Guinea to the New Hebrides and New Caledonia areas. The second but most important route, beginning in 6000 B.C., proceeded north and west from New Guinea to the Malay Peninsula, to Burma (Myanmar), and thence to India, where sugarcane would gain a significant foothold for further development.[19] A third, much later, and somewhat less significant route began about A.D. 500–1100 and went east and north from New Guinea to Fiji, Tonga, Samoa, the Cook Islands, the Marquesas, the Society Islands, Easter Island, and ultimately Hawaii by A.D. 750–1000.[20]

In the all-important westward movement, sugarcanes reaching India were to hybridize to the point that some scholars erroneously believed that new and separate varieties had independently developed there (Chaturvedi 1951). *S. officinarum* remained in India for centuries before hybridizing with *S. barberi* and moving westward. The impetus to disperse farther westward was related to the Islamic conquests in A.D. 600–800, when sugarcane cultivation spread from India to Iran, to Syria, and eventually to the shores of the Mediterranean Sea. Still, the sugarcane plant preceded the manufacture of cane sugar by a century or more.[21] Sugarcane arrived at the eastern shore of the Mediterranean and in Egypt by the late seventh century, spread to Cyprus by 700, Morocco about 709, Andalusia in southern Spain about 714, and then Crete about 818 and Sicily about 827 (Blume 1985, 24).

The Mediterranean Hearth, 700–1600

The Mediterranean region became a nurturing hearth for sugar cultivation and manufacture between 700 and 1600. Through the spread of Islam, the diffusion of sugarcane plants and methods of cultivation and irrigation using existing noria water wheels and the underground *qanat* system of canals and wells followed (Watson 1983, 103–19, 190–200). The Mediterranean probably was not a place for significant technical invention. There is controversy over the origin of the three-roller mill, the one major invention in sugar technology popularly attributed to the Mediterranean region. J. H. Galloway refuted Noel Deerr's and Edmund Oscar von Lippmann's statements that the three-roller mill was invented in 1449 in Sicily.[22] According to Galloway, the best documented evidence for the three-roller mill comes from Brazil and Peru in about 1610, and the place of origin may well have been China (1989, 37, 73–75, 206). The early, primitive sugar-making methods that had come from India through the Middle East and thence to the Mediterranean were still being used well into the fifteenth century.

Mediterranean methods of sugar milling were an adaptation of existing animal- and water-powered grain mills and olive presses. Cane stalks were cut into small pieces and pulverized between two horizontal mill stones or milled with an edge runner mill that had a turning vertical wheel. Juice extraction came in two stages; the stone mills first yielded small amounts of juice, and the remaining juice came from pressing the crushed cane pulp in olive presses. The next phase was boiling the sweet but watery juice to reduce water content, skimming the impurities, and allowing the mixture to cool and crystalize. The grainy mixture was poured into clay pottery cones. Each eight- by fourteen-inch cone had a small hole at the narrow end where the uncrystallized syrup, which we know as molasses, dripped out. The remaining crystallized sugar grains inside the cone were raw sugar. Refining was done by simply dissolving raw sugar in water, reboiling, and crystallizing the mixture several times (Galloway 1989, 39–40). Raw sugar produced in Egypt, Cyprus, Sicily, and southern Spain was shipped to Venice to be refined into whiter, finer grade sugars. By the time of the Crusades, sugar products and their techniques of manufacture were being introduced to Western Europe, forming a link between the sugar industry and later colonial Europeanization that would become a world diffusion pattern.[23]

The rise and decline of sugarcane cultivation and manufacture in the Mediterranean region have been attributed to several historic and geographic factors. One explanation focuses on climate. The Mediterranean region, the northernmost for sugarcane cultivation, would not seem to favor it because of cold winter rains, occasional freezes, and fierce summer droughts. However, in A.D. 1000–1200, the Mediterranean experienced warmer than normal weather conditions, perhaps warming by one to two degrees centigrade, which enabled the budding cultivation of a purely tropical crop to flourish.[24] Between 1550 and 1700, conditions returned to much colder winters, especially during the Little Ice Age, thus placing climatic limits on an already declining agricultural activity. The decline of Mediterranean sugar cultivation came as climate cooled, wars destroyed land and irrigation schemes, and plagues like the Black Death in the 1300s caused widespread depopulation. But the most important reason was that sugar interests had already turned to the Atlantic islands off the coast of Africa.[25]

The New World Dispersals, 1450–1800

Spain and Portugal provided much of the impetus to propel sugarcane beyond the Mediterranean region. From 1450 to 1550, the Portuguese dominated the international sugar trade and initiated the dispersal of sugarcane plants and sugar manufacture to the islands of Madeira, São Tomé, and Fernando Po off the African coast, to Angola on the African mainland, and eventually to Brazil (Blackburn 1984, 3). In the New World, the sugarcane hybrid of *S. officinarum* and *S. barberi* was the foundation stock on which the early sugarcane industry in the western hemisphere would be solely dependent until the late eighteenth century. It was the same old stock that had come from India, through the Middle East, to the Mediterranean, and it was the very same sugarcane that Columbus introduced to the Caribbean on his second voyage in 1493 (Galloway 1989, 11). This was the cane later called *Creole.*

Dispersal of sugarcane varieties and sugar manufacture to and through the New World took disparate temporal and spatial directions. The first canes to arrive in the New World were brought to the large island of Hispaniola (today's Haiti and the Dominican Republic) as early as 1493 by Christopher Columbus. The first attempts to grow sugarcane met with failure, but by 1506 cane was being cultivated on the western end of the island. A working sugar industry on Hispaniola did not begin until 1515, twenty years after the cane plants first arrived (Barnes 1964, 22). Why so late? This was the typical history of sugar manufacturing. First the canes arrived, and then decades or even centuries later a struggling industry appeared. For Hispaniola, the explanation was that Spanish conquistadors were interested only in searching for gold and other mineral riches and that Spanish missionary movements initially were interested only in the search for lost souls. It would take another 115 years before a sugar industry would be introduced to much of the rest of the Caribbean. Meanwhile, Brazil had received its first sugarcane in 1500, brought there by the Portuguese, and by 1526 raw sugar was being shipped from Pernambuco, Brazil, to Lisbon (2–4).

The Portuguese may have been the first to introduce sugarcane plants to Brazil in about 1500, but, as in the sugar industry elsewhere, there was a lag time of twenty-five or more years before an established sugar industry emerged. This time it was in Pernambuco, the easternmost province on South America's Atlantic shoulder. Almost from the very beginning, a plantation system based on slave labor was the basis of commercial sugarcane in Brazil. As a Portuguese sugar industry continued to develop between 1532 and 1630, Dutch sugar interests emerged in Pernambuco and in Dutch Guiana (Suriname). Between 1630 and 1654, the Dutch gained control of Brazilian coastal areas and São Tomé, the old slave and sugar island just off the African coast. Although Brazil at this time was under Dutch rule, Portuguese Jews contributed to sugar industrial technology and became instrumental in its diffusion. After the Dutch lost control of Brazil, Jewish and Dutch refugees were forced from Brazil and carried the sugar industry northward with them.[26] The Caribbean was waiting for the right technology to produce raw granulated sugar, and an unlikely group from a distant place was to bring it with them, the Dutch from Brazil.

Caribbean Sugarcane Plantations, 1630–1800

In the American tropics in the sixteenth and seventeenth centuries, the dispersal of sugar manufacturing was linked to European colonization and a unique agricultural industry: the plantation system as developed by the Portuguese in Brazil, by the Dutch in the Guianas, and by the French and English in the Caribbean. In spite of its early introduction to the Spanish Caribbean, where for many years sugarcane had been cultivated as a chewing novelty, a bona fide raw sugar industry in the Spanish Caribbean would not become important until the nineteenth century. However, for the French and British Caribbean, a full-scale agricultural-industrial endeavor based entirely on sugarcane grew to major importance in the seventeenth and eighteenth centuries (Blume 1985, 31–32, 171).

Two sugarcane varieties dominated the emerging sugar industry in the Caribbean Basin. Until the late eighteenth century, Creole cane was the only sugarcane variety in the New World. Then, in the late 1700s, Otaheite cane, a new sugarcane variety, arrived in the Americas from an entirely different direction. This second, quite important sugarcane was a Pacific variety of *S. officinarum* that had been taken westward from Tahiti to Mauritius in the Indian Ocean by Bougainville in 1768. From Mauritius, the Otaheite cane, later renamed Bourbon, was introduced to the French West Indies in 1780 and to Saint Vincent in 1793. By 1795, the variety was introduced to Jamaica, this time from Hispaniola (Barnes 1964, 4). In 1797, Otaheite cane was introduced to Louisiana. These events provided the parent stocks for the two most remarkable cane varieties in the New World, canes that would dominate New World cultivations until 1890.

Quite separate from historical trends in sugarcane varieties was the trend in the initial or first effective settlement of Europeans that would eventually lead to plantations. As a concept, the first effective settlement established settlement patterns within a cultural context, and it represented the first cultural imprint of permanent settlement on the landscape. It became an important benchmark for all subsequent European settlements in the Americas (Noble 1992). The imprint might have been a thin, light touch or a heavy, deep, long-lasting one, but it was a mark made by an initial culture that led to an understanding of the subsequent cultural milieu. The Caribbean islands were to be marked for centuries with such a cultural imprint after rival European colonial groups had determined which islands would eventually be permanently settled.

TABLE 1.2
Evolution of Settlement and Sugarcane Plantation Development for Selected Caribbean Islands, 1624–1850

Island	*Date of Permanent European Settlement*	*Sugarcane Plantations*	
		Year	*No.*
Barbados	1627	1683	358[a]
		1731	500[b]
		1773	430[b]
Martinique	1635	1671	12
		1736	447
		1770	286
Guadeloupe	1635	1687	105
		1730	252
		1765	401
St. Kitts	1624	1680	10 ?
		1753	51[b]
		1850	93[c]
Jamaica	1655	1673	57
		1739	429
		1774	775
Ste. Domingue	1670	1681	0
		1739	350
		1775	648

Sources: Merrill 1958, 60, 70, 94; Lasserre 1961, 1:350–51; Blume 1985, 172–73; Dunn 1972, 119; Sheridan 1974.
[a] Cattle mills
[b] Windmills
[c] Sixty windmills and 33 steam mills

In 1624, the first permanent English settlement in the Caribbean was established on the small island of Saint Christopher, better known as Saint Kitts. Saint Kitts was considered the mother island from which colonization to the other British-settled islands in the Caribbean developed. Englishmen from Saint Kitts established settlements at Nevis and Anguilla in 1628, Tortue in 1630, and Antigua and Montserrat in 1632. Even the early, small French enclaves on Saint Kitts sent settlers to Guadeloupe, Martinique, Marie Galante, and other smaller French islands between 1635 and 1642 (Merrill 1958, 59). Fitful starts and stops at English settlement on Saint Kitts were marked by skirmishes with Carib indians, struggles over two parts of the island under temporary French occupation, and assorted military battles. There were feeble attempts at plantation agriculture based on tobacco with indentured servant labor before a stable sugar industry emerged sometime after 1670.[27]

Barbados was permanently settled in 1627 and struggled for a few years with poorly equipped agricultural pursuits. The Dutch introduced sugar technology, and the island straight away settled into a sugar industrial format that would effectively evolve over the next 350 years. Barbados grew to become the mother island for sugar plantations and sugar technology. In the 1640s and 1650s, about three hundred sugar planters were making sugar at least a generation ahead of their counterparts in the Leewards and in Jamaica. By 1680, nineteen Barbadian planters owned 200 slaves each and eighty-nine planters owned 100 slaves each (Dunn 1972, 46). By 1683, there were 358 cattle mills (Blume 1985, 172–73). Sugar was so important that seventeenth-century Barbados would become the number one sugar producer in all of the Caribbean. Technical innovations developed in Barbados spread northward to other islands. Galloway's diffusion models of five innovations from Barbados show the dispersal of (1) windmills for grinding sugarcane stalks; (2) cane holing, a field pattern made of five-foot squares deepened to 2.0- × 3.0- × 0.5-foot holes for planting cane on steep slopes; (3) the "Jamaica Train," which was not a train at all but a new, improved furnace for open kettles in the boiling process during the 1700–1780 period; (4) manure for fertilizer; and (5) a fuel use of bagasse, the dry cane stalk residuals after milling (Galloway 1989, 94–102).

Seventeenth-century English sugar planters were a steady group. They adopted and adapted a successful sugar industry on islands thought to be too small for such enterprises. Before 1700, their only real competition came from the Portuguese in Brazil and to a lesser extent the Dutch in tropical America. Before 1730, the French had not yet entered the sugar industry in earnest. Historian Richard Dunn summed up the early years: "In 1700 the English planters in Barbados, Jamaica, and the Leewards supplied close to half of the sugar consumed in western Europe" (Dunn 1972, 48).

Islands took turns wearing the mantle of importance in sugar production. Zenith sugar production was in Barbados in the seventeenth century, in Martinique in the mid-eighteenth century, and then was passed to Jamaica and Sainte Domingue in the late eighteenth century. The French Caribbean was doing very well by the time sugarcane and sugar manufacturing diffused to Louisiana in 1742–95. By the eighteenth century, the

Caribbean Basin had evolved into the most important hearth for the New World sugar industry, and its landscape revealed a plantation presence that would dominate the islands and later diffuse northward, eventually to Louisiana.

Caribbean Plantation Traits

The Caribbean probably did not have a singular "typical" plantation because of its cultural diversity in English, French, Dutch, and later Spanish sugar plantations and the changing of its plantations over time. However, with help from many sources, I attempt to describe the common traits of eighteenth-century Caribbean sugarcane plantations in terms of land, slaves, dwellings, land use, sugar mills, boiling houses, and the processes therein. I also provide a paraphrased translation of Père Labat's eighteenth-century description a sugar plantation on French Guadeloupe to illustrate a French Caribbean plantation as a model for plantations in Louisiana.

Land was absolutely essential to the scale of operation that a sugar plantation demanded. Caribbean plantations were on the whole smaller than their Brazilian and American counterparts. Acreages ranged from less than a hundred acres to as much as one thousand acres, with typical units of about two hundred acres.[28] Slope has always been an issue in agriculture, but in the Caribbean it became a two-edged sword. Flat or gently sloping terrain was preferred but was more likely to be arid. Islands with elevations over two thousand feet had the best opportunity for orographic rainfall as moisture-laden trade winds were forced aloft on the windward sides of islands to cool adiabatically with elevation, condense moisture, form clouds, and produce dependable rain. If one wanted reliable rain, higher elevation and steeper slopes came with the territory.

Caribbean soils clearly reflect orogeny and terrain. Low, flat, dry islands have old coral reef limestone bedrock on which red or black limestone soils formed. If there was enough rainfall, flat, limestone-based soils were attractive to planters. The Grand Terre half of Guadeloupe and much of Barbados, with their limestone soils, are fortunate to have rainfall barely sufficient to support sugarcane, which has been cultivated there continuously for 250 to 350 years, respectively. Mountainous Caribbean islands with volcanic orogenies have sufficient elevations to provide orographic rain to rich volcanic soils. Islands such as Saint Kitts and Martinique still have successful sugarcane plantation agriculture based on mountain volcanic soils.

Other islands, such as Nevis, Antigua, Montserrat, Saint Vincent, Saint Lucia, Dominica, and even the Virgin Islands once had sugar estates but no longer cultivate sugarcane on a commercial plantation basis.

Slaves constituted the single most important monetary investment on an eighteenth-century Caribbean sugar plantation. Early documented evidence reveals a commitment of about 30 to 40 percent of the total investment for slaves. On a medium-size Jamaican plantation, 37 percent of the value was in slaves, 31 percent in land, and 21 percent in sugar-making equipment. On a small plantation in Cuba, 12 slaves accounted for 33 percent of the value. Bryan Edwards, considered to be the on-site eighteenth-century expert on the British Caribbean, estimated the cost needs for a late 1700s sugar plantation in Jamaica: slaves and livestock, 49 percent; land, 34 percent; and buildings, 17 percent.[29] The number of slaves obviously varied among different scales of operation, but an average plantation with 200 acres in Antigua had 100 slaves, and a typical plantation in Saint Kitts would have fewer than 100 acres in cane worked by fewer than 150 slaves.[30] Overall, the ratio was about one slave for every two acres.

Dwellings sheltered the plantation owner and his family, his servants, overseers, and slaves. Early homes tended to be simple frame structures with thatched roofs. Long, narrow bungalows measuring sixteen by sixty feet with three or four rooms were common houses for planters and some lesser folk alike on Saint Kitts in 1706 (Dunn 1972, 139). The sugar prosperity of the eighteenth century brought with it a planter's way of life, replete with a mansion called the *Great House* on English plantations and the *Maison du Maître* on French ones. Locally quarried volcanic stone blocks formed the foundations and walls for two-story, multiroomed, square houses with hipped roofs in Saint Kitts and Nevis (Merrill 1958, 79). French mansions also had stone foundations, but more wood was used elsewhere throughout the construction. On French islands, a typical Maison du Maître was a two-story house with multiple rooms accessed by multiple outside doors opening onto a *galerie* (porch), with the ensemble sheltered under a wide-hipped roof. The hip roof, believed to have originated from both sides of the English Channel in Europe, was ubiquitous throughout the English and French Caribbean (Doran 1962, 97). Roofing materials varied temporally from thatch to wooden shakes or shingles, to slate, to tiles, and to galvanized steel.

A slave quarters consisting of three to fifty structures was located near the boiling house–sugar mill complex at the functional center of the plan-

FIG. 1.6. *An early-nineteenth-century West Indian house with a hip roof exposes the internal structure of the roof design; note the stone foundation and door and window openings. Saint Kitts, West Indies.* Photograph, 1995

tation. Slave huts originally were small, flimsy stick-and-thatch shelters; others were made with wattle and daub, a framework of sticks daubed with mud. As the West Indian cottage began to take shape, slave houses became single rooms that measured about ten by twelve feet, covered by either a small hip roof or a straight saddle roof. Doors were culturally diagnostic. Dwellings on English plantations had a single front door. On French plantations, all houses from the Maison du Maître down to the smallest slave houses had multiple doors front and back.

Caribbean sugar plantations involved four types of land use. The most important was sugarcane land, which took up the majority of the cultivated surface. Smaller, steeper patches of land supported gardens set aside to furnish food for the plantation population. Vegetable gardens on English islands were called *provisioning grounds* and on French ones were *jardins à nègres,* gardens for the negroes. Pasture lands were extremely important, as livestock were essential to power cattle mills (a common name for animal-powered sugar mills) and as a source of manure for fertilizer. Forest land located on the higher, steeper slopes completed the ensemble. Forest resources were used for building materials and especially for fuelwood in the furnaces of the boiling house, where sugar processing was conducted.

Sugar manufacture required a mill to grind the cane and produce cane juice; a boiling house, or *sucrerie,* to boil off the water in the juice, clarify, and crystallize the mass; and a purgery room to separate the molasses from the brown raw sugar. Slaves cut sugarcane by hand in the fields and transported the canestalks in animal-drawn two-wheeled cane carts to the sugar mill. Throughout the Caribbean, the sugar mill was the ubiquitous three-roller mill. Canes were fed into the mill, which consisted of three vertical rollers, each measuring eighteen to thirty or more inches length and integrated with toothed gears supported by a large wooden frame (see fig. 3.4). One or more long poles were fixed with one end to the top of the center roller and the other end attached to a power source. Power varied over space and time. Cattle mills (mills powered by cattle, oxen, horses, or even people) were the oldest types of mills and were widespread throughout the American sugar world. Sugar mills evolved from cattle mills to windmills and on some islands to water mills. By the late 1600s, windmills appeared first in Barbados and then diffused through the Leewards northward to Guadeloupe, Antigua, Saint Kitts, and as far as the Virgin Islands. Watermills were much more common on the French islands of Martinique and Guadeloupe. All three methods of milling might be used simultaneously. For example, in 1770, Martinique had 184 cattle mills, 116 water mills, and 12 windmills (Blume 1985, 173). On several islands, especially Saint Kitts, remnant windmill stone towers stand today as mute testimony to a sugar landscape of the past. It was not until the nineteenth century that steam-powered sugar mills arrived.

Sugar making took place in the boiling house. Cane juice from the mill ran through pipes or troughs into vats, where it was stored momentarily before being placed into the first of a series of kettles. The signature feature of the boiling house was a row of iron kettles or copper cauldrons called a *battery.* Most boiling houses had just one battery, but some had two, set in a foundation beneath which was a long furnace called the *Jamaica Train,* an improved furnace for open kettle sugar processing. The kettles were named and arranged from largest to smallest, known in French areas as *grande, flambeau, sirop,* and *batterie* in accordance with variations in furnace heat. Cooler temperatures were required at the large kettle, and higher heat was needed at the smallest. Cane juice placed in the first, *grande* kettle was heated over low heat until impurities rose to the surface to be skimmed off

TABLE 1.3
Eighteenth-Century Caribbean Sugar Mills

	Cattle Mills		*Windmills*		*Water Mills*	
Island	*Year*	*No.*	*Year*	*No.*	*Year*	*No.*
Barbados	1683	358	1731	500	–	–
	1773	14	1773	430	–	–
St. Kitts	–	–	1753	51	–	–
			1850	60		
Martinique	1770	184	1770	12	1770	116
Guadeloupe	1738	174	1738	1	1738	80
	1790	228	1790	140	1790	133
	1818	117	1818	222	1818	136

Sources: Blume 1985, 172–73; Dunn 1972, 119; Lasserre 1961, 1:356; Merrill 1958, 60.

by slaves. Next the liquid was ladled into the second kettle, where ashes were added to precipitate more impurities; as the liquid boiled, moisture was driven off. Ladling, boiling, and skimming continued sequentially from kettle to kettle. By the time the liquid reached the fourth kettle, cane syrup was in the making. Once in the final, smallest, hottest kettle, the thickening syrup, now called *massecuite,* was reaching the strike point of crystallization. A skilled sugar maker made the split-second decision of when to "strike" the batch. At the strike, the massecuite was placed into a cooling vat to cool and crystallize for several hours.

In the next stage, the grainy, damp slurry mixture of sugar crystals and molasses was placed into barrels or clay cones in a separate room called the *purgery.* Molasses dripped by gravity from holes in the barrels and cones and collected in vats below. The purging process was "as slow as molasses," taking several weeks to complete. Raw sugar took different names according to how it was purged. *Muscavado sugar,* purged through barrels, was a darker, raw brown sugar with considerable residual molasses. *Clayed sugar* was a much cleaner, whiter sugar purged with the use of water rinses and clay filtration through clay cones. After purging, the remaining raw sugar was barreled and made ready for shipment to sugar refiners in Europe.[31] Molasses was also barreled and prepared for either overseas shipment or local rum distilling. Molasses in the English Caribbean became important as a trade item in the triangular trade between the Caribbean, New England, and Africa; as rum, it was important because the

British Navy issued a daily rum ration to each of its seamen from 1655 until 1970.[32]

An Eighteenth-Century French Plantation on Guadeloupe

The search for a plantation in the French Caribbean to be used as a model for French plantations in Louisiana leads us to one ideally described by Father Jean-Baptiste Labat, a Jesuit priest who visited Guadeloupe in 1696. The *habitation*, or plantation, was an elongated landholding with a 1,000-yard river frontage and a 3,000-yard depth that extended upslope to a forested area. A thin line of trees bordered the river and protected livestock and canes from the wind. Beyond the windbreak of trees, broad grassy savannas used for pasture extended another 300 yards inland. On an elevated site beyond the savanna was the Maison du Maître, a large and well-ventilated mansion surrounded by a garden. The garden separated the mansion site from offices and *magasins* (warehouses) located immediately behind the mansion. The sugar mill was located at a reasonable distance behind the warehouses, where the noise from the mill would not disturb the master. Downwind from the mansion was an alignment of slave houses very close together to form a quarters settlement along one or two roads. Cattle pens were located next to the slave houses so that slaves could watch over the animals. All of the buildings at the mansion site and at the slave quarters, along with the cattle pens, mill, outbuildings, and gardens, covered an area measuring three hundred yards square.

A total of sixty-one fields were in sugarcane cultivation. The best cane fields covered fifty-two acres, beyond which and up slope were an additional one hundred acres. The mill was located at the center of the cane fields to simplify the transport of cane, which weighs sixteen to twenty tons per acre. Carts would fetch canes from the fields along eighteen-foot-wide roads that separated fields at 100-yard intervals. This grid pattern of roads and fields set at 100 by 100 yards had many advantages: it limited the danger of fire, it enabled the cane carts to penetrate the fields, and it facilitated the surveillance of the work done over plantation lands.

Above the cane fields were *jardins à nègres* and the plantation provisioning grounds where potatoes, manioc, corn, bananas, and other food plants were cultivated. In accordance with the Code Noir of 1685, slaves were allowed one day a week, usually Saturday, to tend to their own gardens. The

Map 1.5. *A seventeenth-century plantation plan became the model for eighteenth-century plantations in the Caribbean and Louisiana.* Sources: Labat 1742; Lasserre 1961; Rehder 1978, 146. Used with permission from the Louisiana State University School of Geoscience

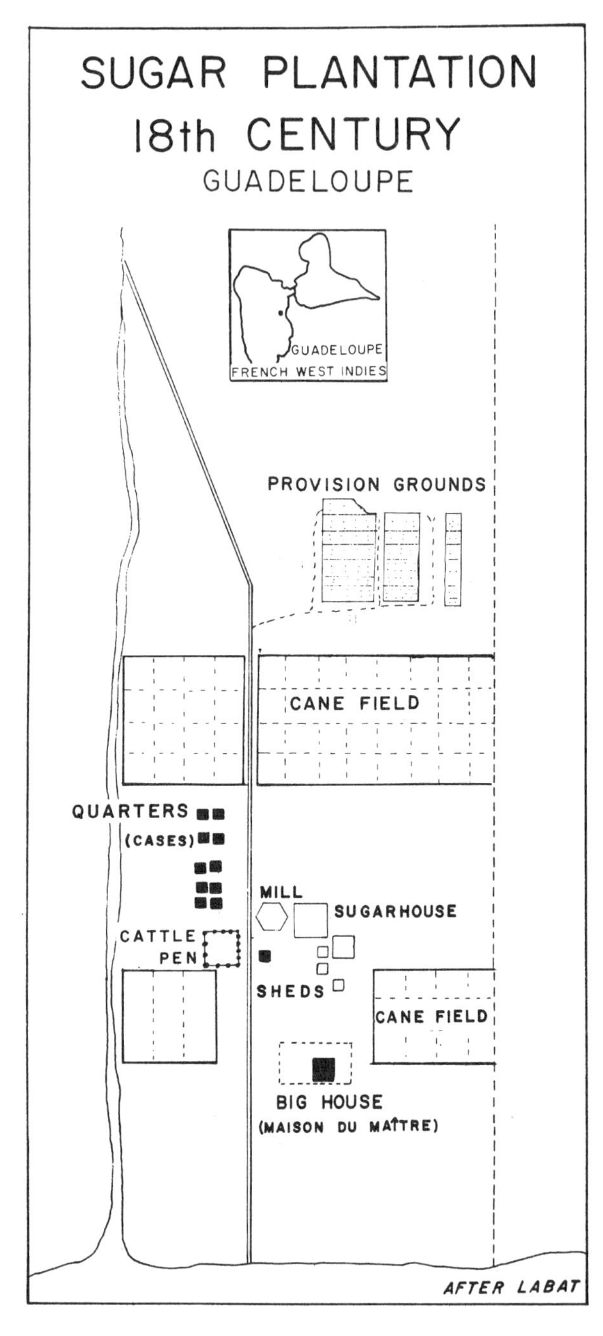

master was to furnish clothes, salt meat or cod, and occasionally medicine to his slaves. Labat mentioned a cacao grove with several hundred trees at the edge of the forest, which would occupy about fifteen slaves. Additionally, these slaves would be in charge of manioc cultivation and other provisioning on the plantation. Finally, above the cultivated lands were large woodland tracts that extended up the mountain slope.[33]

The plantation had a labor force of 120 slaves, 38 oxen, and 12 horses. The 120 slaves were organized into the following divisions of labor:[34]

–6 at the sugar factory
–5 at the mill
–1 to wash and bleach
–1 making vinegar
–8 to run the cane carts
–2 coopers
–2 at the forge
–3 at the purgery
–3 carpenters
–2 masons
–1 woodworker
–1 wheelwright
–1 livestock guard
–1 nurse
–25 cane cutters
–6 woodcutters
–2 to process food grains
–1 commander
–4 domestic servants at the mansion
–7 temporarily sick slaves
–25 infants, and
–10 invalids and elderly slaves.

Labat's plantation description was a model that would be repeated over time and space but in different dimensions throughout much of the region.

The Caribbean Basin that had become the New World sugarcane plantation hearth in the seventeenth and eighteenth centuries would continue to produce sugar in subsequent centuries. On old sugar islands like Saint Kitts, Barbados, Martinique, Guadeloupe, Trinidad, Jamaica, Cuba, and the Dominican Republic on Hispanola, the industry has continued on a plantation basis right to the present. On other islands it ceased with the emancipation of slave labor in the late eighteenth and early nineteenth centuries, and on others it ended with political independence and nationalization in the twentieth century. In the 1700s, however, Louisiana would receive traits from a French Caribbean sugar plantation legacy, particularly from Sainte Domingue. Sugarcane plants, methods of cultivation, methods of sugar making, French planters, skilled sugar makers, slaves, and French architectural traits would ultimately diffuse from this French Caribbean hearth to Louisiana.

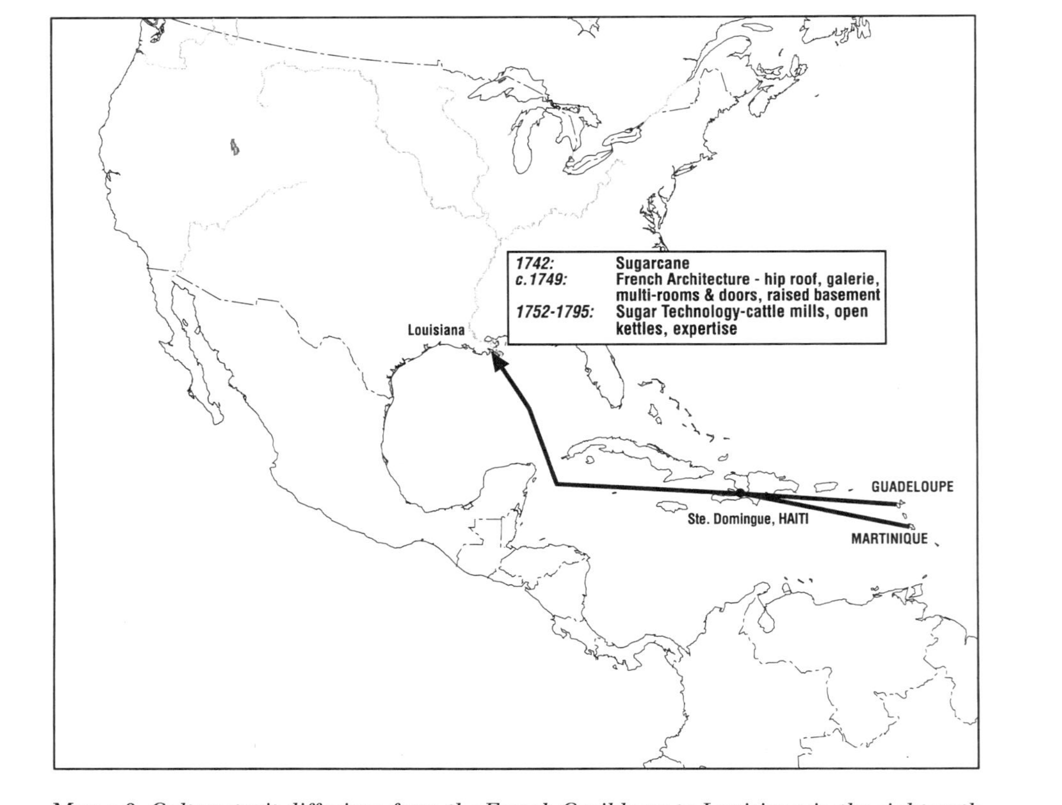

MAP 1.6. *Culture trait diffusions from the French Caribbean to Louisiana in the eighteenth century provided Louisiana with significant components of sugar technology and French architecture.* Sources: Brasseaux, Conrad, and Cheramie 1992; Dart 1935; Oszuscik 1992a, 1992c; Rehder 1971; Wilson 1987

Sugarcane in Louisiana, 1742–1795

The story of sugarcane diffusion to Louisiana is not simple, especially considering the questionable dates of introduction. The year 1751 is the most popular date, but other dates given are 1725, 1733, and 1742.[35] The popular account is that Jesuit priests brought the first sugarcane stalks to Louisiana from Sainte Domingue in 1751. According to Avequin, indigo planters and local produce vendors obtained cane cuttings from the Jesuits and began growing sugarcane stalks to be chewed like candy and to be sold in New Orleans (Avequin 1857, 615–19). Local planters obtained cane stalks for experimentation, and some began to make a rumlike beverage called *tafia*. Joseph Villars Dubreuil, whose plantation was on the east bank of the Mississippi River near New Orleans, wrote about his experience with sugar cultivation in Louisiana in a letter dated September 10, 1752 (Dart 1935, 286).

> I am working now Monseigneur, in an effort to establish the sugar industry in this country, as sugarcane can successfully be grown below here and 12 leagues up the river. I sincerely believe that this culture will not cause any serious injury to the islands [French West Indies] from what I can see, we can make sugar only during two months; but as cattle are plentiful and cheap we can increase the number of mills. The sugar industry is still a good branch of trade.
>
> I have proof that sugarcane does not freeze in the ground during the winter if we cover the stalks that have been cut four inches from the ground; in the spring at the beginning of February, the cane sprouts and is ready to be cut in October. I hope Monseigneur, that Your Highness will not disapprove of this culture, while I am actually undergoing all necessary expense to establish same, since I am certain the stalks will not die in winter if properly covered.

These statements were made in 1752, just one year after the popularized date of sugarcane introduction to Louisiana. Dubreuil reported that sugarcane could be successfully grown below New Orleans and twelve leagues above it. Is this a known geographic fact, or was he speculating about the spatial potential for cultivation? He made a startling discovery that sugar could be made only during two months. Furthermore, he had proof that cane does not freeze in the ground, that cane sprouts in February, and that October is the best time for harvest. One year is certainly not enough time for anyone in Louisiana to have gained such agronomic knowledge. Years of experimentation would have been needed to know Louisiana's climatic limitations, the annual round of sugarcane in this most unnatural setting, and the range of spatial probability for sugarcane cultivation. Based on Dubreuil's information, we must conclude that sugarcane arrived in Louisiana before the popularized date of 1751, perhaps in 1742 or even earlier.

The fifty-three years between the introduction of sugarcane in 1742 and the first commercial production of granulated raw sugar in 1795 were years of agricultural experimentation and political and economic upheaval in Louisiana. Throughout the history of sugarcane, a lag time separated the introduction of the plant and the emergence of raw sugar manufacture, and Louisiana was no exception. From 1742 until 1763, French Louisiana was testing the waters for commercial agricultural pursuits. Cultivations in tobacco, indigo, sugarcane, mulberries for silkworms, citrus, grapes for winemaking, and a host of other crops were being tried, many without success. By 1756 Joseph Villars Dubreuil, the same Dubreuil mentioned earlier, had a working sugar mill that produced a thick cane syrup but failed to make a granulated raw sugar. In 1757, Dubreuil died and a man known only as Mazan purchased his mill. Mazan maintained the mill and in 1763 sold a half-interest to Louis de Boré, father of Étienne de Boré (Conrad and Lucas 1995, 3–7).

In the years between 1763 and 1803, Louisiana reluctantly became a colony of Spain. There was political and economic turmoil for the French as Spanish authorities set about to change government, transportation, commerce, and many of the elements to which colonial French Louisiana had become accustomed (Arena 1955). Under Spanish control, cash crops like tobacco and especially indigo were strongly encouraged, but sugarcane was not promoted for several reasons. Spain already had sugar interests elsewhere and saw no need to invest in a crop that was expensive to capitalize and would be marginal at best. Rebellions and general unrest in Spanish Louisiana kept the area off balance politically. Furthermore, the subtropical climate had experienced several years of bad weather, enough to discourage attempts at cultivating sugarcane. Indigo was thus championed, and some planters staked their careers on its production, which lasted until the 1790s, when indigo became less productive. Indigo was a tropical crop struggling to grow in subtropical Louisiana, and indigo-producing areas in the Carolinas and in much of Latin America competed successfully against the Louisiana crop. Finally, diseases and pests claimed much of the crop in the waning years and after 1794 virtually destroyed indigo forever in Louisiana (Gray [1932] 1958, 73–74). As indigo's cultivation declined, tobacco and other commercial crops struggled to continue on concessions and plantations.

It became abundantly clear that local experimentation and expertise from the Caribbean were needed before a bona fide sugar industry could emerge in Louisiana. Enter Joseph Solis, perhaps the crucial link between uncertain experimentation and granulation in sugar production. Solis had sugarcane cultivations from which he made molasses, which he distilled into a rumlike liquor, tafia. The operation struggled for several years, and in 1794 Solis sold his land, canes, sugar mill, and boiling equipment to Antonio Mendez. By this time, desperate planters, including Mendez, had decided to hire outside expert help in the form of experienced sugar makers from Sainte Domingue. The man hired by Mendez in 1794 was Antoine Morin. After his 1794 crop failed, Mendez sold all of his sugarcane as seed cane to surrounding planters. Among the purchasers was Étienne de Boré, who planted some of the Mendez cane but, more importantly, hired Antoine Morin as his sugar maker for the 1795 crop. Granulation was achieved, 100,000 pounds of raw sugar were produced, a $5,000 profit was made, and, thanks to Étienne de Boré and Antoine Morin, Louisiana's future and success at sugar making was under way.[36]

The Origins and Dispersals of Louisiana's Sugarcane Plantation

The New Orleans Hearth

Louisiana's sugarcane plantations originated in the vicinity of New Orleans in the period 1742–95 (Avequin 1857, 616). From 1718 to 1742 agriculture had focused on indigo and tobacco plantations, which served as an economic foundation where plantation-based sugarcane would eventually emerge.[37] The earliest large agricultural areas originated at various granted concessions along the Mississippi River, especially in the vicinity of New Orleans. By the mid-1720s approximately fifteen concessions were already operating on a plantation basis, cultivating tobacco and indigo and experimenting with citrus and mulberry trees for a new silk industry (French 1851, 78). These early concessions became the genesis for a plantation system that subsequently led to sugar plantations dispersed over the Louisiana landscape.

New Orleans became the political, economic, and cultural center of colonial Louisiana and was the port of call for incoming ships from France and the French West Indies. The French Caribbean helped initiate the sugar industry by supplying New Orleans with sugarcane stalks, sugar technology, and sugar makers. Jesuit priests from Sainte Domingue probably introduced the initial sugarcane plants and cultivation techniques to form an ethnobotanical hearth in New Orleans.[38] Louisiana sugar planter J. V. Dubreuil experimented with sugarcane cultivation in the 1750s near New Orleans but could not achieve a lasting success at producing granulated raw sugar (Dart 1935, 286). Eventually, in 1795, Étienne de Boré and West Indian sugar maker Antoine Morin received credit for producing the first granulated sugar on a commercial plantation scale (Gayarre 1889, 607). As these events were unfolding, other plantation crops (indigo and tobacco) declined, enabling sugarcane to rise as the primary plantation crop of southern Louisiana after 1795 (Avequin 1857, 615–19).

Until the climatic limits of sugarcane were well understood, political and economic factors drove the development and spatial distribution of the initial sugarcane plantations in the budding industry. While Louisiana was under French and Spanish political control, colonial sugar-producing areas in Central America and the Caribbean dominated the industry and left Louisiana at a competitive disadvantage. But when the United States acquired Louisiana in 1803, sugarcane became a unique, attractive product that could be marketed to a broader U.S. market.[39] Almost immediately the promising new sugar industry came under the protective revenue duty of 2½¢ per pound on brown sugar. By 1812 the duty was raised to 5¢ as a war measure, but in 1816 it fell to 3¢. Sugar selling for 8½¢ in 1815 included a duty of 3¢ to 5¢ per pound. With increasing sugar prices, potential planters were drawn into the sugar industry.[40] Besides a national market in the United States, Louisiana sugar planters had the advantage of a large local market and an international port in New Orleans, the primary outlet for products throughout the Mississippi Valley.[41]

Sugar Plantation Dispersals, 1795–1860

From a modest beginning, the industry experimented and succeeded with sugarcane cultivation, but the planters' successes or failures relied on important location factors (Rehder 1973). Little is known of the distribution of plantations before 1803, but by that year seventy-five enterprises of various sizes shared both banks of the Mississippi River north and south of New Orleans. The northward expansion of contiguous sugar plantations by 1806 led probably no farther than Saint James Parish, about forty river miles above New Orleans. Even there, sugar was being planted less than cotton. Sugar cultivation intermittently extended northward to Pointe Coupee Parish, where an outpost of former tobacco enterprises was described by William Darby in 1806 as "one of the most and best cultivated settlements on the Mississippi." Here, commercially produced commodities were "cotton, lumber and sugar; the latter yet in very small quantity" (Darby 1817, 143). Whereas the frontier of cane cultivation had reached Pointe Coupee by 1806, sugar plantations had not yet become a part of the landscape in the western bayou regions of Lafourche, Terrebonne, and Teche.

The Western Expansion of the Sugar Region

Sugar plantation landscapes developed in areas west of the Mississippi after 1812, when Anglo planters, largely from the English Tidewater and Lowland South, began to arrive.[42] Writers traveling the area described a land of opportunity. William Darby's far-reaching account declared Louisiana's lands "the most extensive unbroken continuous body of productive soil on the globe" (Darby 1817, 45). Attracted by the cheap public lands of Terrebonne Parish, the lower Bayou Teche regions, and the backlands of upper Bayou Lafourche, the Anglos came as land speculators and as would-be planters. Between 1812 and 1850 they entered southern Louisiana by water, traveling on the Mississippi and on the western bayous. The rich, arable lands on natural levees formed by these bayou waterways were the initial plantation sites for Anglo planters.

The migrations of Anglo-Americans were important to both the distributions and the internal character of plantations on the landscape. Anglo planters learned techniques of sugar culture by borrowing from French planters along the Mississippi (Pugh 1888, 143–67). Anglo-Americans introduced unique traits of material culture in settlement patterns, buildings, dwelling types, and agricultural practices. By 1844, a date by which most initial-occupance plantation patterns had been established, significant cultural patterns were evident. French-owned plantations were then concentrated along the Mississippi River; Anglo plantations came later at the extremities of the distributional pattern.

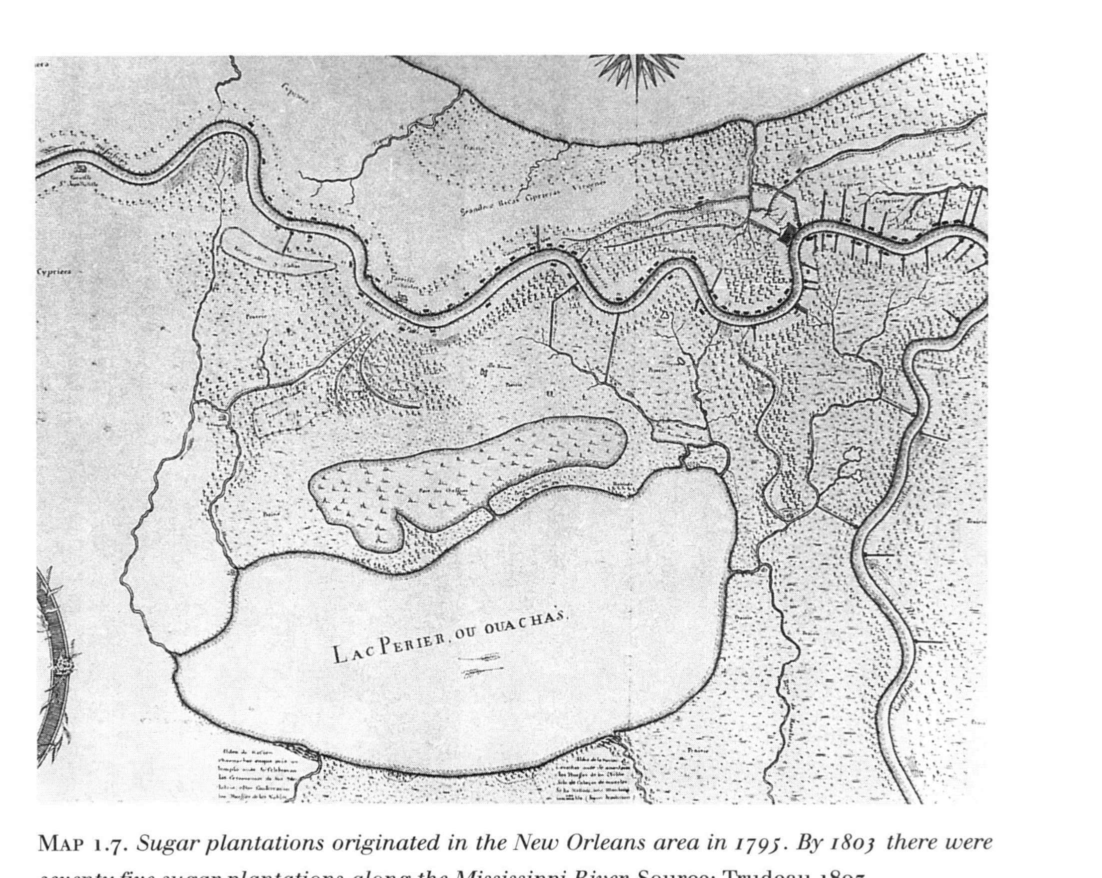

Map 1.7. *Sugar plantations originated in the New Orleans area in 1795. By 1803 there were seventy-five sugar plantations along the Mississippi River.* Source: Trudeau 1803

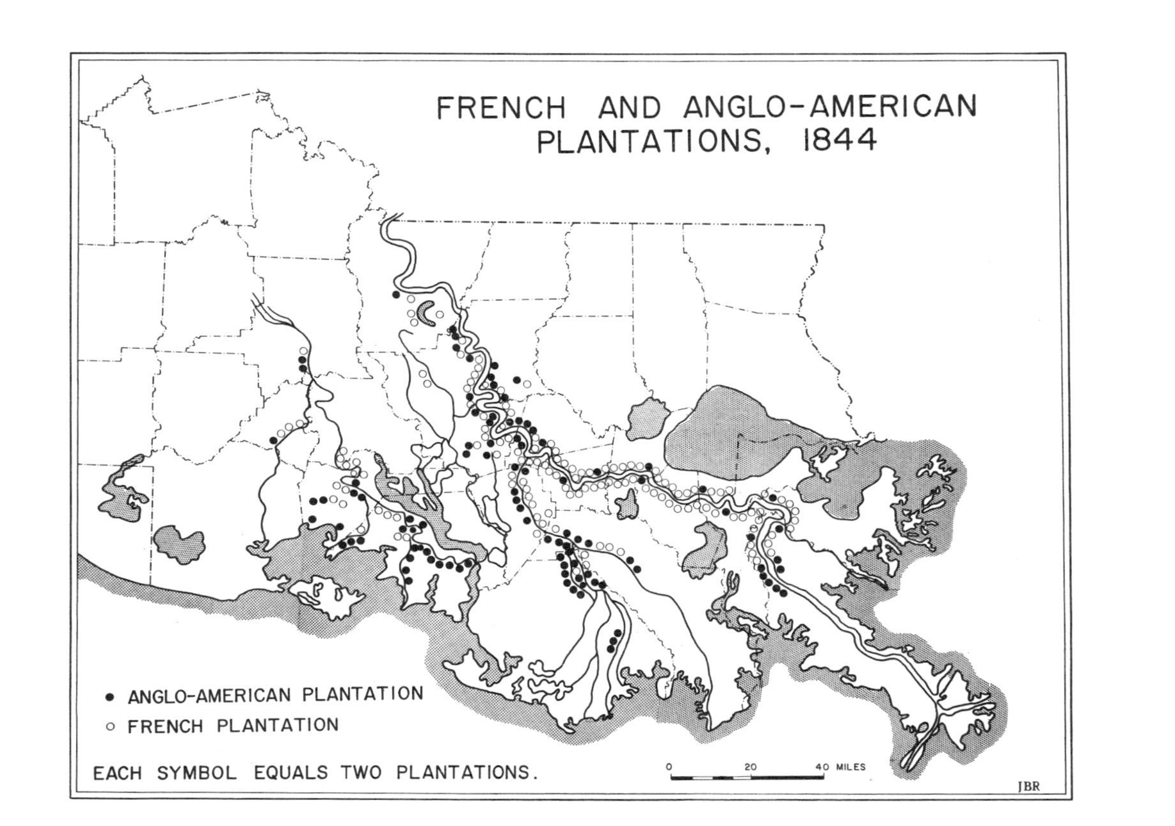

MAP 1.8. *By 1844, the spatial pattern of French and Anglo-American plantations was firmly in place.* Sources: Champomier 1844; Rehder 1978, 137. Used with permission from the Louisiana State University School of Geoscience

The westward expansion of the sugar plantation landscape reached no farther than the natural levees of Bayou Teche and the two lower distributary bayous of the Teche, Bayou Salé and Bayou Cypremort. A further westward expansion was discontinued because lands west of the Teche were not included in the Louisiana Purchase. Still in the possession of Spain until 1819, the area remained a no-man's-land buffer between U.S. territory and Spanish holdings farther west.[43] Moreover, southwest Louisiana's soils were on Pleistocene terraces that were unsuited for sugarcane because the soils were deficient in organic composition and became hardpan in dry weather (Darby 1817, 10–11). Limited navigability on small bayous, the scarcity of well-drained floodplain soils, and the distance from the New Orleans market and sphere of influence were limiting factors, even if the territory had been included in the Louisiana Purchase. Regions west of Bayou Teche were not effectively settled until 1880, and even then the settlers were midwestern grain farmers who adopted rice, the present dominant cultivation west of the Teche.[44]

Southern Dispersals

The southern extent of effective sugar plantation settlement in the Lafourche, Terrebonne, and Teche areas was restricted by drainable lands, all of which are characterized by a progressive narrowing of natural levees. At points of extreme levee narrowing where levees were a quarter mile wide or less, sugar cultivation terminated. In the Bayou Teche area south of New Iberia, plantation settlement was achieved in the first half of the nineteenth century by Anglo-American planters (Richardson 1886, 593). After 1812, planters found available public lands in the sparsely settled Lower Teche, Bayou Cypremort, and Bayou Salé areas (*American State Papers* 1843, 119–21). By 1844, sugar plantations had extended southward to Berwick, four miles from Morgan City. Today, however, the lower limit of sugar on Bayou Teche is near Calumet, ten miles upstream from where the Teche empties into brackish water estuaries at Morgan City. Lands north of Saint Martinville and Lafayette had been settled in the late eighteenth century by French Acadian and Spanish small-farming groups; consequently, large plantations were rarely ever numerous there.

Northern Expansions and Limitations

Sugar and cotton were especially competitive in nineteenth-century Louisiana. In 1806, sugar plantations had expanded as far north as Pointe Coupee, where they were meager secondary enterprises. Cotton had become a primary staple and was cultivated as far south as Saint John-the-Baptist Parish in 1806 and even westward on small farms in the Lafourche and Teche region because cotton prices were good and sugar was still a rather new and developing crop. Besides, few individuals among the *petits habitants* (German, Spanish, and Acadian small farmers) possessed the necessary capital of forty thousand dollars for a fifty-hand sugar plantation.[45]

Sugar plantation expansion north of Pointe Coupee to the Red River area and the conversion from cotton to sugar plantations between Ascension Parish and Pointe Coupee Parish came in the 1830s and 1840s with the fall of cotton prices.[46] The northernmost sugar parishes and the lower portion of the cotton region evolved into a transition zone, where fluctuations in prices could cause one crop to advance and the other to retreat. Before 1825, cotton prices had remained relatively high; consequently, sugar was held back by the "high-price tide" of cotton. Between 1826 and 1832, cotton prices began to fall, giving way to the slight northward expansion of sugarcane. Further expansion of the sugarcane industry came in the 1840s, when cotton prices fell from fifteen to five cents per pound between 1835 and 1842 (Gray [1932] 1958, 697, 715, 1027). A strong protective tariff for sugar encouraged a greater expansion of sugarcane (Homans and Danna 1861, 511). Former cotton plantations converted to sugarcane in the parishes of East Feliciana, West Feliciana, Avoyelles, and Rapides. Cane became so widely accepted that sixty new sugar plantations emerged on the northern margins of the sugar region in 1845 (Champomier 1844, 10). The agricultural writer Solon Robinson said of the area in 1848, "Short crops and low prices of cotton combined with the fact of several planters in the hill lands between Woodville and Bayou Sara [Wilkinson County, Miss., and West Feliciana Parish], having been very successful in the cultivation of cane the past season or two, is creating considerable excitement about making sugar in a region that it would have been considered only a few years since, madness to talk about" (Kellar 1936, 146).

As new and converted sugar plantations developed northward, deep into the cotton parishes of Rapides, Avoyelles, and even Concordia, Catahoula, and East and West Feliciana, the inevitable was to happen, a northern limit to the expansion. In the mid-nineteenth century, little was known of the climatic limits of sugarcane, against which Louisiana planters were ever so swiftly pushing. Sugarcane, a tropical crop, required a frost-free, twelve-month growing season to reach full maturity; Louisiana had only a nine- or ten-month frost-free season in its southernmost parishes. Naturally, canes in the northern areas of the region were extremely vulnerable to freeze damage, so the area was and still is marginal for sugar production (Hebert 1964, 2–4). The first freeze date was critical; if too early, it brought devastation to the cane crop. Freezing causes the sucrose content to diminish, even sour, while an unseasonably warm period after a freeze produces a sour, useless product. Molasses could be recovered from some frost-damaged cane, but granulated raw sugar was always the desired product (Coleman 1952, 342–43, 377–80).

In the early nineteenth century, planters had not ascertained the northern limits for successful production; consequently, sugar cultivation extended beyond critical freeze limits (DeBow 1846, 442). A northern expansion of sugar plantation distributions ended abruptly in 1856, when a severe early frost devastated the entire sugar region. Plantations along the northern fringes later reverted to cotton, a safer, more reliable crop (Champomier 1856). The northern climatic boundary today is effectively delimited by the average first-frost date of November 16. Severe freezes in 1984 and especially in 1989 reminded us that sugar is sensitive to vagaries in the weather, and Louisiana continues to be a surprising place for sugarcane cultivation at a commercial plantation scale.

Soils were important to the northern distribution. In the vicinity of Meeker, alluvial soils were rich, fertile, and well drained on wide, almost level natural levees (Lytle 1968, 17–18). But beyond the Red River Valley, lesser soils that formed on Pleistocene terraces and Tertiary deposits lacked the drainage and fertility requirements for sugarcane. Phosphorus levels in terrace soils did not match those in the Red River and Mississippi floodplain alluvium.[47] Such soil variables contributed to a concentration of plantations in the Red River Valley on Bayou Boeuf around Meeker near Alexandria.

Eastern Limits

Very little expansion, dispersal, or otherwise significant movement of the sugar plantation culture occurred east of the Mississippi River. In the nineteenth century, sugar expanded into only one small area in East Baton

Rouge and East and West Feliciana Parishes; even then the movement was short lived. Plantations on the east bank of the Mississippi north of Bayou Manchac were limited by poor Pleistocene terrace soils, early frosts, an economy long based on cotton, and an early cultural separation from southern sugar areas. Deficient soils were the most significant and continuous limitations to the eastern margins of the sugar region. Pleistocene terrace soils had inadequate drainage, and oxidation weakened the soils to the point of unsuitability for sugarcane (Lytle 1968, 17–18).

The crucial isostade of November 16 for the first freeze passes southward and across the center of West Feliciana Parish and then southward through the eastern part of East Baton Rouge Parish. This climatic boundary places the parishes northeast of the line beyond the limits for effective sugar cultivation. Culturally and historically, during Louisiana's experimental period of sugar cultivation from 1742 to 1795 the eastern region north of Bayou Manchac (called the *Florida Parishes*) was under British and later Anglo-American domination.[48] The area was already successful in cotton, and experienced Anglo cotton planters were reluctant to attempt sugar cultivation, particularly when cotton prices were high (Berquin-Duvallon 1806, 166). Moreover, when sugarcane cultivation began to spread northward after 1812, the new crop was not widely adopted because it was anomalous to the area and expensive to process. A common belief was that sugarcane could not be successfully cultivated beyond Bayou Manchac (DeBow 1846, 442).

With the collapse of cotton prices in the 1840s, commercial sugarcane cultivation temporarily expanded to terrace soils north and east of Baton Rouge (Gray [1932] 1958, 687, 748, 1027). Here cotton planters attempted to cultivate and process sugarcane with the view that sugar was to be their financial salvation. Mild success was achieved in 1855, when the Feliciana Parishes together produced 10,793 hogsheads of sugar, accounting for 3 percent of the total Louisiana crop.[49] The years after 1855 were marked by damaging frosts, the Civil War, the planters' realization of inadequate soils, and an agricultural reversion to cotton, all of which destroyed the sugar industry northeast of Baton Rouge.

The Plantation South in 1860

The Louisiana sugar plantation by 1860 had become particularly meaningful within the spatial and temporal context of other Southern plantations. The South in 1860 had come under the direct influence of a system of

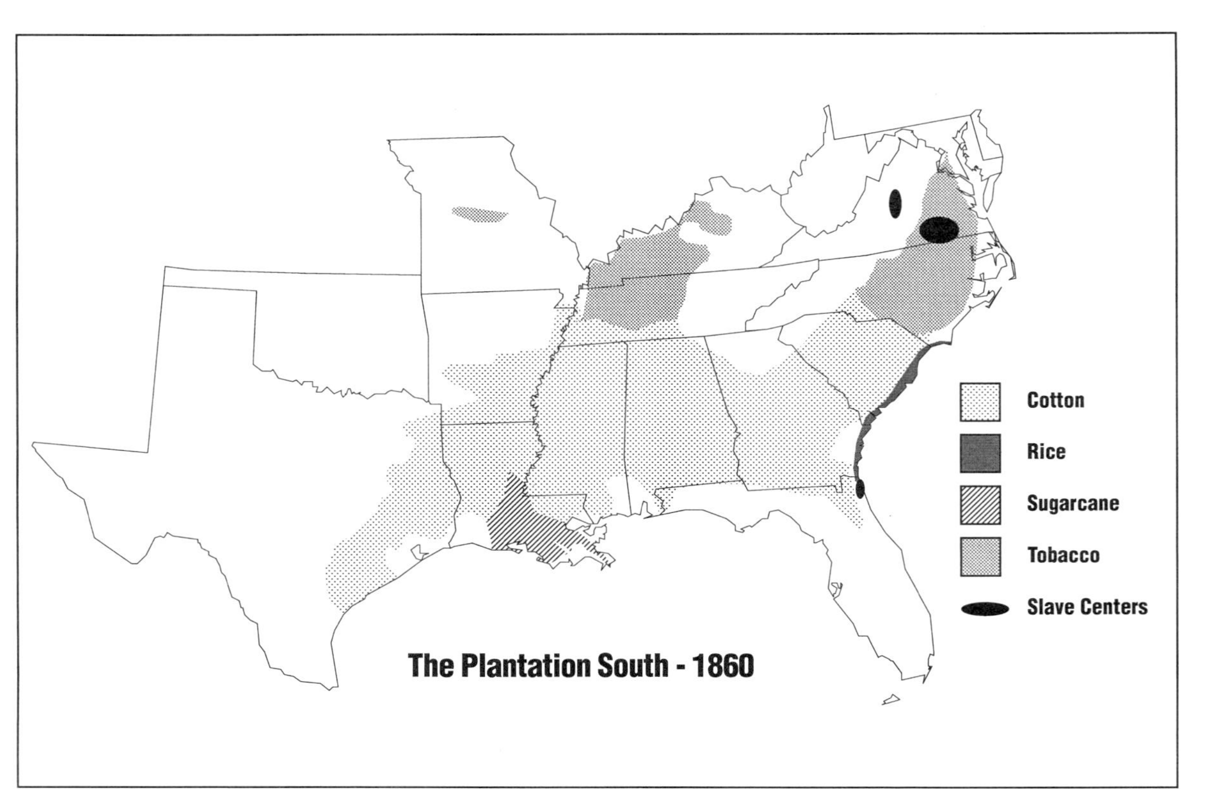

MAP 1.9. *The plantation South in 1860 was producing tobacco, cotton, rice, sugar, and slaves.* Sources: Hilliard 1984; National Geographic Society 1988, 133

thousands of plantations distributed from Maryland to Texas. At their zenith in 1860, plantations numbered 46,274, based on a minimum of twenty or more slaves per farm unit. Nearly half, or 20,789, were small plantations with twenty to thirty slaves each. Larger plantations with a hundred or more slaves numbered only about 2,300. Our image of a plantation probably is one of the larger ones, but in reality the majority of slave holders were rather small operators.[50] Plantation types were identified by the products they produced, with each creating a different landscape. Four plantation types based on tobacco, rice, cotton, and sugarcane dominated the South in 1860. But other products also came from plantations, ranging from eighteenth-century experimental crops like indigo, citrus, and mulberries to early-nineteenth-century supply products such as draft animals, foodstuffs, and, strangely enough, slaves.

Tobacco began as the earliest plantation crop in Virginia and Maryland in the seventeenth century and spread through the Carolinas during the eighteenth century. By 1850, there were 15,745 tobacco farms and plantations producing at least 3,000 pounds of tobacco each in the South (Gray [1932] 1958, 213, 233, 529). By 1860, tobacco as a plantation crop still covered parts of eastern Virginia and the Carolinas and had diffused westward to central and western Kentucky and into central Missouri. Tobacco plantations were best identified by unique tobacco barns, tall, two-story flue curing barns measuring sixteen by sixteen feet at the base in the eastern states and much larger, square, air-cured and fire-cured tobacco barns that developed in Kentucky in the West. Tobacco fields were irregularly shaped and quite small when compared to sugarcane fields. A typical plantation might have a few hundred acres with numerous scattered tobacco fields, appearing much like quilt patches, with each field measuring about two to ten acres. But tobacco was labor intensive. A typical slave in the East could work two acres of tobacco and produce 1,600 pounds, while a similar slave could work eight to ten acres of cotton in Mississippi and average 250 pounds per acre (Gray [1932] 1958, 912).

Rice was a very special coastal crop that entered Charleston, South Carolina, from Madagascar in the early 1660s. Plantation rice emerged in a Charleston hearth between 1690 and 1720. By the mid-1700s, rice cultivation had expanded along the Atlantic coastline in a narrow band from the Charleston hearth northward to the lower Cape Fear River in southeastern North Carolina at Wilmington and southward along the Georgia coast beyond Brunswick.[51] Rice plantations had threshing platforms and winnowing houses before the nineteenth-century development of steam-driven rice mills. Rice fields were unique in all of southern agriculture, with their tidal gates, canals, and hand-built dikes etched into fresh-water marshes in the Carolina and Georgia estuaries (Hilliard 1978, 94–112).

Cotton, like tobacco, was native to the Americas. Cotton cultivation developed slowly on colonial farms in the American South and emerged as a large-scale plantation crop beginning in the late eighteenth century with the invention of the cotton gin by Eli Whitney in 1793. Plantation-grown cotton centered in several southern hearths in the Piedmont of South Carolina and Georgia, central Alabama, and southwestern Mississippi by 1820 and then rapidly expanded across the South in a broad Cotton Belt from Virginia to Texas by 1860 (Gray [1932] 1958, 680–84). A cotton plantation was best marked by a gin house and bale press facility located at the center of the plantation in an outbuilding complex along with mule barns and sheds. The block of slave quarters was very near or adjacent to the gin-house complex. Fields were relatively small, irregularly shaped, and scattered as cotton patches over the landscape (Aiken 1978, 151–65).

Indigo, along with some other experimental crops, was one of the early plantation crops in the eighteenth-century South that did not survive to 1860. Indigo reached plantation status in South Carolina and Georgia but only between 1739 and 1800. Indigo declined rapidly, as evidenced by South Carolina's exports, which dropped from 839,666 pounds in 1792 to 96,000 pounds in 1797 to 3,400 pounds just three years later in 1800 (Gray [1932] 1958, 610–11). In Louisiana, indigo cultivation began in 1725, citrus was attempted between 1748 and 1773, and mulberry trees intended for a potential but failed silk industry were tried on plantations in the 1720s (69, 73, 75). All failed as long-range plantation crops.

A lesser known fact in the plantation history of the South is the existence of plantations whose primary purpose was to produce slaves. As commercial breeding centers, plantations in Virginia, the border states, and even Florida produced a growing population of slaves to be sold to other agricultural plantations. George Ross, a slave born in 1817 and raised in Maryland, testified at the age of forty-six in 1863 (Blassingame 1977, 406):

> I don't know as I have known any particular instances where slaves were raised for the purpose of selling them; but I have been on a farm where they had 30 or 40 colored people, & as the younger growed up, they sold off the older; so I rather think by that it was done pretty near for that purpose. Some-

> times, when they could get good bargains, they would sell the young just the same. I have often heard of slaves being kept for the purposes of breeding, but I have never seen it. That may be done down in Virginia and the other foreign States, perhaps.

Much like southern suppliers of hog meat and hoecake for foodstuffs, mules and horses for draft animals, and other materials desperately needed by plantations, there were plantations producing and nurturing humans for the southern slave market.[52] One of the best documented examples is the Kingsley Plantation on Fort George Island in Duval County, north of present-day Jacksonville, Florida. Between 1813 and 1843, Zephaniah Kingsley raised cotton, sugarcane, and proportionally a greater number of slaves than were needed to work the plantation. It was quite evident that Kingsley was operating a slave-breeding plantation. The importation of slaves was made illegal in the United States after 1808, but the Florida territory imported slaves legally until 1821. For many years thereafter, Kingsley raised, trained, and smuggled slaves to Georgia and Carolina rice and sea-island cotton plantations.[53]

Sugarcane first arrived in Louisiana in 1742, became a genuine commercial plantation crop in 1795 in the New Orleans area, and diffused northward to its climatic limits in Louisiana and east and west to its soil limits by 1860. Perhaps the signature landscape feature of an 1860s sugar plantation was the sugar factory, consisting of a steam-powered sugar mill, a boiling house (also known as a sugarhouse) for extracting raw sugar and molasses, and chimneys and vents spewing smoke and steam in the cool crisp autumn air during grinding season. In 1844, there were 762 sugar mills, of which 408 were steam-powered sugar mills and 354 were cattle mills (Gray [1932] 1958, 741). By 1849, Louisiana sugarcane cultivation had reached its zenith, with 1,536 sugar mills of various types (Champomier 1850, 43). But by 1860, the number of Louisiana's sugar mills had dropped to 1,308, of which 992 were powered by steam (Degalos 1892, 65–68).

Sugarcane fields covered hundreds to thousands of acres and were laced with drainage ditches and cart paths at regular intervals. For much of southern Louisiana, landholdings were very long and narrow, resulting in "long lots," a misnomer when defined as a survey system. In truth, the Louisiana landscape was surveyed by the French arpent survey system, which resulted in long, narrow landholdings measuring two to twenty arpents wide along stream courses and having a standard depth of forty ar-

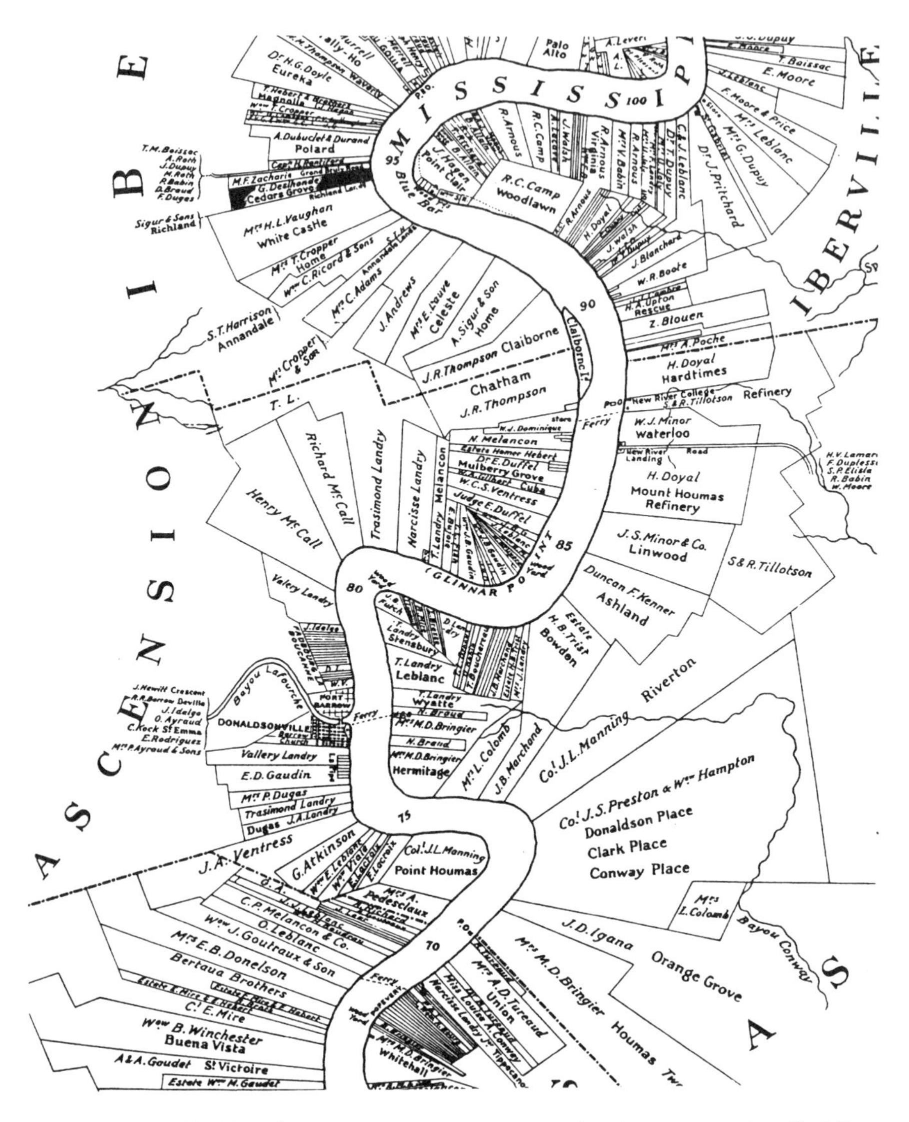

Map 1.10. *The French arpent survey system created unique narrow landholdings, as evidenced by plantations in the Donaldsonville, Louisiana, area about forty-five river miles south of Baton Rouge.* Sources: Persac's map, or Norman's chart [1858] 1931; Oszuscik 1992a, 143. Used by permission

pents. Each arpent equaled 192 linear feet and variably measured 0.84 to 1.28 acres. This meant that all initial landholdings had frontage on the river or bayou, the principal transport route. In the 1860s, a typical modest plantation landholding of five by forty arpents would have dimensions of 960 by 7,680 feet. Settlement features would include the planter's mansion, overseer's house, perhaps a dozen or more small quarter houses in the slave quarters, a sugar factory surrounded by barns and sheds in an outbuilding complex, and extensive fields covered with tall stalks of sugarcane. Despite major setbacks during the Civil War period and from cane disease in the 1920s, sugarcane survived and continues as a plantation crop to the present.

Developmental Factors for Louisiana's Sugar Plantations, 1742–1998

The perpetuation of the sugar industry and its plantation system in Louisiana relied on historic events and responsible factors. Technological factors in sugarcane varieties, sugar mills, sugar-processing boiling equipment, and mechanical harvesting equipment ensured growth and a level of stability for the industry. Other factors, such as the Civil War, organized leadership, and government controls, affected the industry in negative and positive ways. At one point the industry suffered a nearly complete collapse during the Civil War, but by the late nineteenth century, with improved regionwide leadership and government controls, the industry was saved (Rehder 1973).

Four technological improvements perpetuated the sugar plantation and its industry: sugarcane varieties, sugar mills, boiling apparatus, and mechanical harvesting. Creole cane was the earliest known sugarcane variety grown in Louisiana. Its cultivation continued from the time of introduction in the 1740s well into the 1830s. The soft stalk's susceptibility to frost and insects caused Creole cane to decline.[54] Otaheite cane was introduced to Louisiana in 1797 as a companion cane to the Creole, but it, too, diminished because of its susceptibility to frost and failure to regenerate annually, or ratoon. Both canes were soft and easily milled by animal-powered mills but did little toward the northern expansion of sugar plantations because both varieties lacked cold-weather resistance (Earle 1928, 62–63). In 1824, the entry of the Purple Cheribon and Striped Ribbon or Striped Preanger varieties was important because their hard outer rinds enabled cane cultivation farther north (Avequin 1857, 619). The protective rinds were detrimental at first because they were difficult to mill with animal-powered equipment. However, in 1825 the milling problem was overcome when steam-powered sugar mills entered Louisiana (DeBow 1856, 275). Steam engines were far faster than cattle- and horse-powered mills, and speed at grinding time was especially important for stabilizing the industry and its plantations (DeBow 1849, 157–58). Planters could allow canes to mature further by delaying harvest because they knew that they could grind quickly with steam-powered mills.

Refinements came to the plantation in the 1840s and 1850s when crude, open-kettle boiling methods changed to sealed vacuum pan processing.[55] Plantations with vacuum pan equipment in the sugarhouse could process more cane juice faster and with better quality control. This technical improvement not only accelerated sugar processing but also contributed to the stability of the industry. In the 1920s, the old ribbon and purple varieties and the D-74 cane variety imported in 1898 from British Guiana suffered severe losses to the mosaic cane disease and were replaced by mosaic-resistant hybrids imported from Java (Durbin 1980, 84–85). After the Proefstation Oost Java (POJ) varieties saved the sugar industry, other varieties were developed in U.S. Department of Agriculture Experiment Stations at Canal Point, Florida, and at Houma, Louisiana.[56] In 1935, the mechanical cane harvester became one of the most important technical innovations in the sugarcane industry (Hebert 1964, 8). The harvester provided greater speed and efficiency at harvest time than had formerly been achieved with hand labor. With this improvement, planters delayed harvest and benefited from greater sucrose in a more mature crop.

Another factor was the temporary but severe economic effects of the Civil War on the Louisiana sugar industry (Roland 1957). After 1860, the sugar region diminished but did not shrink any smaller than its 1844 spatial pattern. By 1870 the industry was approaching recovery (Prichard 1939, 318–32). Postbellum plantations retained the same landscape morphology that they had had antebellum. They had the same settlement patterns, identical field patterns, and undoubtedly many of the same plantation buildings. The compact settlement pattern in the quarters continued in the post–Civil War period because sugarcane required gang labor, with teams of men working and housed together as wage-earning residential laborers. Conversely, on Southern cotton plantations, quarter houses were dispersed when freed slaves and other laborers became sharecroppers and tenants who worked individually on small cotton fields.[57]

The founding members of the Louisiana Sugar Planter's Association in the 1870s should be credited with fostering scientific technological development through continued improvements in sugar factories, agricultural experimentation, and mechanized harvesters and other technological elements to enhance the sugar industry and foster the perpetuation of plantations. Moreover, the association's planters should be lauded for their persistent persuasion for federal tariffs, subsidies, and other government assistance throughout the twentieth century (Heitmann 1987, 71–97). The results of their efforts can be seen in the perpetuation of the industry today.

The origins and diffusions of sugarcane, sugar manufacturing, and the plantation concept reflect global tropical spatial patterns and deep historical perspectives. The Caribbean's European and African cultural legacies contributed substantially to the early material culture, the plantation process, and landscape personality. Louisiana sugar plantations and their distinctive landscapes originated with French Creole planters in the New Orleans vicinity and dispersed northward along the banks of the Mississippi River. Waves of Anglo-Americans later stimulated the dispersal into the western bayous of the Lafourche, Terrebonne, and Teche regions. By the mid–nineteenth century, sugar plantations had expanded beyond the limits of successful sugarcane cultivation. Throughout the period of initial sugar plantation origin and dispersal, routes followed the lines of current travel via streams and bayous. Specific plantation sites focused on the banks of streams, the crests of natural levees, and backslopes.

Location factors influenced sugar plantation spatial distributions. Physical factors set particular limits on sugarcane cultivation and its accompanying plantations. The climatic first-freezing temperatures determined the northern extent of sugarcane grown for granulated sugar. Soil drainage and fertility characteristics determined plantation location on the better drained and richer soils in the Mississippi floodplain as opposed to the weaker and harder-to-drain Pleistocene terrace soils on the eastern and western upland boundaries.

Particularly significant are historical factors such as the demise of other plantation crops. Early contacts between the French colony of Louisiana and the French West Indies sugar industry resulted in the arrival of sugar mills, sugar makers, immigrants from Sainte Domingue, dwelling traits, and even the sugarcane plant itself from the French Caribbean. Cultural factors were instrumental to the origin, dispersal, and distributions of plantations. Initial sugar plantations near New Orleans were followed by a dispersal along both banks of the Mississippi precipitated by French planters. Anglo-Americans, with their former experience with cotton, carried the plantation system into the western and northernmost margins of the sugar region. Acadian, German, and Spanish peasant farming groups hindered the expansion of sugar plantations by their prior occupation of potential plantation lands, thus influencing new arrivals to seek lands elsewhere.

Political factors also played a role. The Louisiana Purchase opened the Louisiana Territory to waves of incoming Anglo-American planters. Conversely, territorial boundary disputes hindered land settlement and plantation expansion west of the Teche region. The differing policies of Louisiana's colonial possessors affected the spread of sugar enterprises. France and Spain discouraged the development of commercial sugar crops in Louisiana because they had more lucrative production areas in the Caribbean. After 1803, Louisiana's early sugar industry was encouraged by U.S. government support. The sugar industry for much of its history, except for a few gaps such as the 1974–84 period of deregulation and the 1996 Farm Bill, has been politically and economically supported and controlled by U.S. government subsidies and tariff systems that effectively perpetuated the industry.[58]

Economic factors also played an especially pivotal role in nineteenth-century sugar plantations in Louisiana. The initial establishment of an American market for Louisiana sugar and protective tariffs as a result of political policies encouraged the spread of sugarcane plantations. Fluctuations in sugar and cotton prices were responsible for temporarily dispersing the sugar plantation northward into the cotton kingdom. Perhaps the silent factor of inertia is the real reason that sugarcane plantations have persisted in Louisiana. With all economics and politics aside, inertia fostered a traditional plantation landscape from 1795 until the 1970s and its sugarcane products to the present.

CHAPTER 2

Culture and Form

The settlement geographer observes and interprets the forms and patterns of material culture that give character to the landscape. Not all scholars have the same view of the content and purpose of settlement geography. Kirk Stone believed that buildings were the only important objects of study because he saw them as surrogates for population distribution (Stone 1965, 346–55), an opinion with which I do not fully agree. Other scholars have seen the subject focused on those forms on the landscape that are derived from the process of settling, the results of first effective permanent settlement and subsequent layers of settlement over time. Terry Jordan, the leading cultural geographer in America today; Otto Schlüter, the leading settlement geographer in nineteenth-century Germany; Carl Sauer, the founder of American cultural geography; Fred Kniffen, the father of American settlement and folk geography; and E. Estyn Evans, the father of Irish folk geography, all acknowledged that the study of settlement belongs to those things that are inscribed into the landscape and give to it characteristic expression.[1] We seek those things in the material culture that may possess diagnostic qualities to be observed and interpreted within a cultural, historic, and geographic context.

Sugar plantations in southern Louisiana once lined the banks of streams like beads on a string, sometimes in an uninterrupted succession, and their peculiar arrangement and function created a distinct agricultural landscape. As a settlement form, the sugar plantation invites study, and in this chapter I describe plantation landscape features and offer an explanation of plantation morphology. Diagnostic culture traits are best represented by plantation house types and settlement patterns and provide particular evidence for a cultural interpretation. The landscape signatures of mansions and settlement patterns best identify the initial culture groups who created them. French Creole mansions and Anglo-American Tidewater and Lowland South house types play a major role in the diagnostic process. Furthermore, settlement patterns may be linear and identified with French origins, or they may be block, as in nodal-block or bayou-block patterns with an Anglo identity. Other landscape elements may not be as culturally diagnostic but are still important to the plantation ensemble. These include quarter houses, overseers' houses, sugar factories, outbuildings, stores, fields, and other features found on a plantation.

Louisiana's sugar plantations are examined here for their landscape morphology, but their quantitative history evolved out of 12 tobacco and indigo concessions in the 1740s, grew to 75 genuine sugar plantations by 1801, then to 762 plantations by 1844, and reached a maximum of 1,536 enterprises in 1849 (Champomier 1844, 10; 1850, 43). The Civil War, cane diseases, and the consolidation of enterprises brought the number of plantations to 202 by 1969, with forty-four sugar factories (Rehder 1971, 4). One in 4 plantations maintained a sugar factory because it was more feasible to process sugarcane at large, centralized factories. Economic, political, and other cultural factors by 1993 reduced the number of morphologic units that I defined as sugar plantations to 82 plantations and twenty factories. As of 1997, nineteen factories survive; ten are cooperatively owned and operated by cane growers, and the others are privately owned (USDA ERS 1997, 35). Louisiana remains the last place in the United States to have working plantations that have occupied the landscape for more than two centuries.

The myriad of buildings, fields, and other features in the landscape morphology of plantations can be condensed into a single morphologic landscape model. Degrees of cultural identity and association proceed in descending order from mansions, with the best diagnostic traits, to linear and

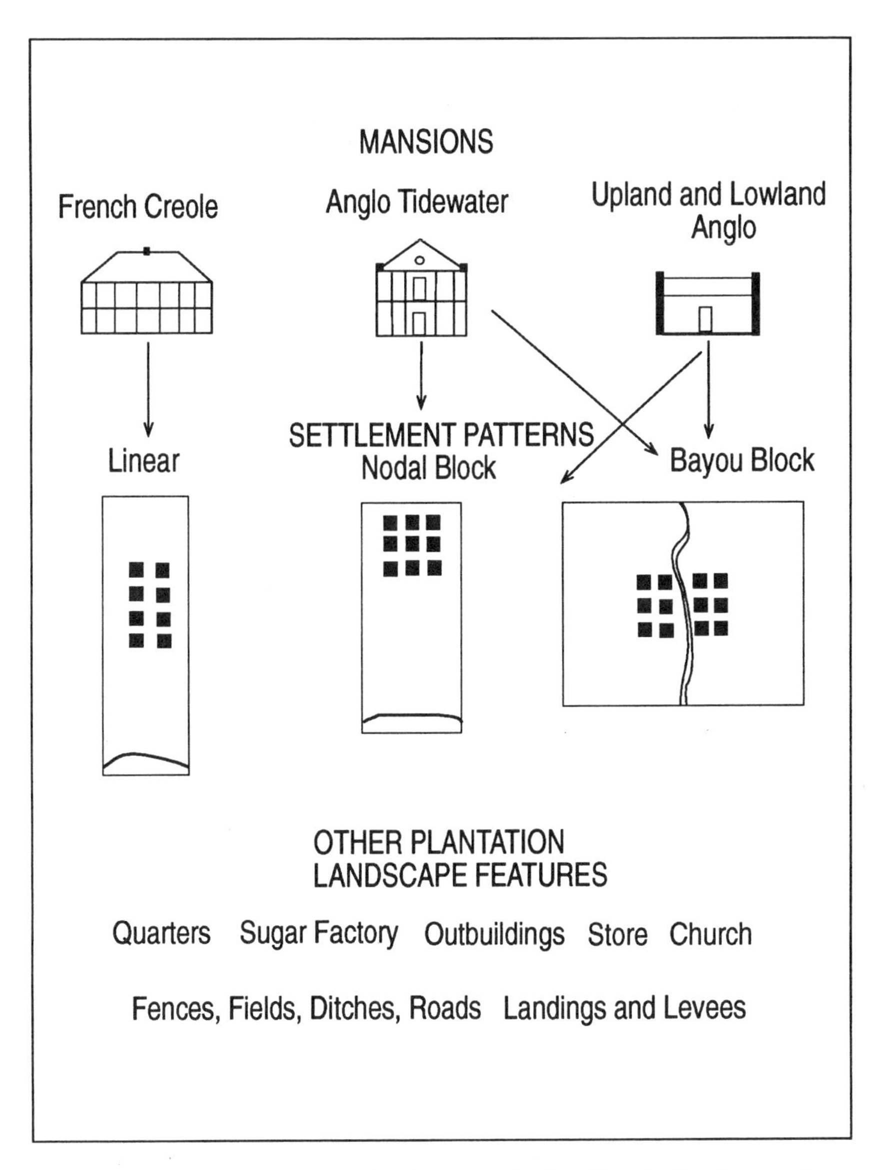

FIG. 2.1. *A model of diagnostic landscape traits has French Creole mansions linked to linear settlement patterns. Anglo Tidewater and Upland and Lowland Anglo mansions are linked to either nodal- or bayou-block settlements. The remaining features (quarters, sugar factory, outbuildings, store, church, fields, ditches, roads, landings, and levees) may be found with any of these mansion-settlement patterns.* Source: Rehder 1978, 138

block settlement patterns, with the next level of identifiable traits. Plantation elements with progressively much less diagnostic cultural identity are quarter houses, outbuildings, stores, churches, fields, landings, and levees.

Diagnosticity is anchored to two settlement features: house types and settlement patterns. Three mansion types embody architectural diagnosticity: Creole mansions with a French identity, Tidewater mansions with an Anglo identity, and Lowland South houses that also have an Anglo identity. Linear settlement patterns are alignments of plantation buildings set perpendicular, or at right angles, to streams and are identified with initial French occupance, the first effective settlement. Block-shaped plantations with squares gridded by streets have an Anglo-American identity. A combination of mansion type with settlement pattern reinforces the diagnosticity so that a plantation with a Creole mansion and a linear settlement pattern just *might* have a French cultural identity. As much as I would like for culture traits to be completely predictable, humans within any cultural context and at any given time can be unpredictable. In nineteenth-century southern Louisiana, someone in a well-informed group of aristocratic planters might have the eccentricity to go against his cultural traditions. Fortunately, that did not happen often, and my research bears out the recurrent validity of the diagnostic culture traits used in this study. The remaining landscape elements–in barns, sheds, stores, roads, fields–contain insufficient diagnosticity for a reliable cultural association. Such features certainly are important to the understanding of a plantation's function and morphology, but historic cultural affinity cannot be accurately assigned to them.

Mansions

The mansion attracts and even demands attention, but it has been perhaps the least understood landscape component from a cultural, settlement viewpoint. We must make a clear distinction between architectural period, style, and plan. *Period* refers to time period and can be especially confusing when dealing with variations between temporal periods in Europe and North America and trying to apply them to southern Louisiana. *Style* refers to fashionable decorative characteristics, such as millwork trim, that would not be culturally diagnostic in a folk sense. *Plan* in the floor plan or layout of rooms is a more desirable element of architectural history to the cultural geographer because it is a reliable and enduring diagnostic trait in structural typology.

Misnomers in architectural terminology are used repeatedly in reference to plantation mansions. The period term *Southern Colonial* serves little use because the house is neither universally Southern in the folk-house tradition nor entirely colonial in the temporal sense. Many extant plantation houses were built after 1803, which in Louisiana follows the colonial period. *Antebellum,* another period meaning "before the war," only dates houses prior to 1860, before the Civil War. The term *Georgian* refers to a time period associated with the English kings George I, II, III, and IV and with the styles and plans of the late seventeenth to early eighteenth centuries in England.[2] The classic Georgian plan, consisting of a central hall flanked by two rooms on each side, was introduced to eastern North America between 1735 and 1760, was assimilated in 1760–75, and was molded into an American pattern between 1775 and 1790 (Rifkin 1980, 18–19). Although *Georgian* is an accepted floor plan term in American architecture, it should not be used to indicate architectural periods, styles, and decorations introduced to the South by Anglo-American planters and hired architects after 1825. However, architectural plans with Georgian symmetrical order–four rooms to a floor, central halls, paired chimneys, paired wings–began to appear in the American colonies between 1735 and 1790 but came to southern Louisiana's landscape after 1825.[3] Terms like *Greek Revival* and *Classic* characterize only the façade and pillars of certain mansions, but not the floor plans of folk dwelling types (Hamlin 1944, xv–xvii). The architecture of a plantation mansion is determined through various means, but among the most important are the ethnic and folk traditions of the initial owners. In keeping with cultural tradition as a basis for typology, I prefer the terms *Creole plantation mansion* for French planters' homes and *Anglo plantation mansion* for Anglo-American planters' houses.

Site and Setting

In the riparian landscape of southern Louisiana, the higher elevations offered by natural levee crests were always the choice sites for human habitation dating from prehistoric times to the present. For plantation mansions, these elevated places on the landscape offered a commanding view of the river, cool breezes, slight relief from overland flooding (first to flood but first to dry out), and close proximity to the plantation landing. Mansion sites were important in setting the planter's residence apart from the remainder of the plantation settlement. Common denominators of all sugarcane plan-

FIG. 2.2. *The tree-lined entrance to Oak Alley, formerly known as Bon Secours Plantation, is one of the most photographed mansion sites in Louisiana. The French Creole plantation mansion was built between 1837 and 1839 by Jacques Telespore Roman III. Saint James Parish.* Photograph, 1993

tation mansions in Louisiana were (1) location on or near a levee crest, the highest, closest point on a stream; (2) a setting distinguished by yard size, entrances, gates, and fences, with *pigeonniers* and *garçonnières,* especially at Creole mansion sites; (3) distinctive ornamental foliage; and (4) distance from the remainder of the plantation. Distances separating the mansion site from the plantation quarters ranged from as few as fifty yards to as much as two miles. The mansion site for Pellico plantation illustrates the nature of large yards with special trees for shade, fragrance, fruit, and borders that set the mansion into a landscape frame much like a framed painting (see frontispiece).

Mansion sites were overwhelming with ornamental biota. Nineteenth-century plantations had live oak trees, magnolias, mulberry trees, orange trees, crepe myrtle shrubs, oleander shrubs, gardenias, azaleas, osage orange, and many other plants. Some even had imported tropical and subtropical plants, as at Valcour Aime's Petit Versailles plantation in Saint

Fig. 2.3. *The Anglo-American plantation mansion and site at Ashland-Belle Helene was built for Duncan Kenner in 1841. Ascension Parish.* Photograph, 1967

James Parish (Aime 1878). The mansion was destroyed about 1920, but the rich botanical collection on site survives in a junglelike thicket conveniently owned by a landscape architect. Oak Alley, with its long avenue of ancient live oaks and clearly one of the most photographed plantation mansions in America, is a showplace with an extraordinary landscape presence. A smaller avenue of oaks leads to Houmas House. Magnolias and live oaks in no particular geometric pattern surround a plantation appropriately named Bocage. Moss-draped oaks surround Ashland-Belle Helene. And the aptly named Mound Plantation mansion, in a most remarkable and logical site, sits atop an ancient aboriginal Indian mound in the flood-prone Bayou Maringouin area. Mansion sites varied in size. For example, both Armant and Cedar Grove mansion sites covered less than an acre each, whereas Madewood's six-acre yard and Oaklawn's seventy-six-acre site dominated their respective plantations. Throughout the delta region a plantation's mansion site still comprises a rich botanical, social, and cultural island set quite apart from the remainder of the plantation complex.

FIG. 2.4. *The French Creole plantation mansion Keller-Homeplace was built about 1790. Significant diagnostic traits are the hip roof, central chimney, gallery, multiple front doors, half-timbered main floor, and raised basement bricked ground floor. Exceptions are the new corrugated metal red roof covering and front steps. Saint Charles Parish.* Photograph, 1993

French Creole Plantation Mansions

French Creole plantation mansions are the oldest and perhaps most widespread plantation mansions in southern Louisiana. Despite the deterioration and destruction of many mansions in the region, a few Creole plantation houses survive. One of the finest examples is Keller-Homeplace in Saint Charles Parish. The plantation once had been called Concession and Pelican, but when the Keller family later acquired the plantation in the 1880s they named it "Home Place." The house was built for the Louis Edmond Fortier family about 1790, possibly by the mulatto builder known only as Charles. With eight rooms on the main floor, no two rooms are the same size.[4] The mansion exhibits the diagnostic traits expected of a Creole plantation mansion.

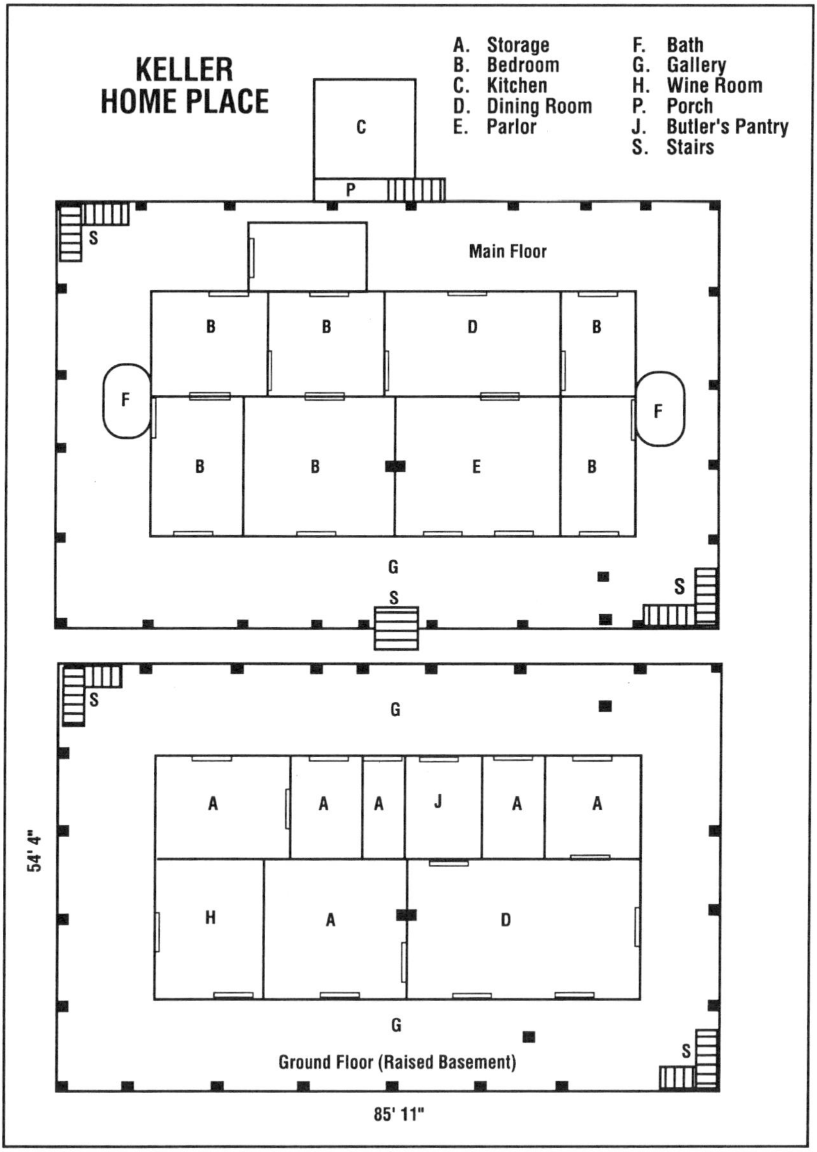

FIG. 2.5. *The floor plan of the Keller-Homeplace mansion illustrates room sizes, their relationship to the all-important gallery, the external gallery staircases, multiple doors, and central chimney.* Source: Daspit 1996, 155

French Creole plantation mansions are best diagnosed by five traits. (1) Interior chimneys are always located on the inside walls, particularly at the center of the roof line; they are never placed on the exterior gabled ends of the house. (2) All front room doors open onto a gallery or porch, so that multiple doors characterize the front of the structure. (3) The floor plan may be several rooms long and one to three rooms deep, and it distinctly lacks a central hallway inside. (4) All stairs are located on the exterior. (5) The hip roof, gallery, and raised one-and-one-half to two-story height are inherently French traits that diffused from France and the French Caribbean.

Floor Plans

Creole plantation mansion floor plans are important diagnostic traits. The house plan is rectangular, with external dimensions ranging from 60 to 90 feet long by 35 to 50 feet wide. The arrangement of rooms is irregular, with two to six rooms along the long axis and one to two rooms deep. All rooms open onto the gallery, but not all rooms are connected internally. Some rooms, especially bedrooms, can be entered only from the gallery, analogous to rows of rooms on the second floor of a motel. The gallery functions as an open-air hall, allowing one to pass from one room to the next. Inside the Creole house some rooms, of course, are internally connected (e.g., the parlor and the dining room). In houses that are two rooms deep, door alignment allows through-ventilation between interconnected parallel rooms. Chimney placement is always at the center, near the center, or at some internal point inside the house. Chimneys are never on the outside walls, as is common with Anglo house types. Front rooms function as the parlor and bedrooms; back rooms serve as dining room, kitchen, pantry, dressing rooms, and additional bedrooms.[5]

The Raised Creole House and Construction

The raised position or raised basement feature of some Creole houses gives cause to wonder why and how this part of the house was used. Three possible interpretations apply. First, the raised trait may have originated from a practical necessity to elevate the living quarters of houses above the threat of floodwaters in the delta region. However, the raised feature was also found in France and in the French Caribbean, where hill sites had no flooding problems. It may have been traditional, locked into the culture, with no practical explanation. Third, the raised trait may have had a social hierarchical meaning, indicating the elevated status of the plantation owner. Other dwellings on the plantation occupied by lesser folk had no such raised basements, even though they were more likely to experience flooding at lower elevation sites (Oszuscik 1992b, 153).

Initial mansion foundations were thick cypress posts that elevated houses two to four feet and higher (Porteous 1926, 259). Later the underside of the structure was bricked in to provide storage space for garden tools. As Creole houses evolved into larger houses and eventually into plantation mansions, the height of the underpinning space between the house floor and the ground grew to a half story and ultimately into a full story. This increased height provided greater protection from floods and increased the utility of the storage space. Functionally, the space became a raised basement for not only the storage of implements and provisions but also a wine room, a kitchen, or, in elaborate mansions, a dining room and ballroom. The space was rarely used for bedrooms or long-term occupation because of dampness, mildew, darkness, insects, rodents, and snakes. Building materials and construction techniques were vastly different between the two levels. The ground floor was built of brick filling interstices between pillars in the underpinning of the structure. Since Louisiana lacked building stone, brick became the material of choice for foundations and the raised basement feature. The upper floor was the living space for the planter and his family and in early homes was constructed using half-timbering techniques.

Construction methods for Creole mansions took two forms: *colombage*, a French building trait of half-timbering, popular in the eighteenth and early nineteenth centuries, and all-wood construction, used in the nineteenth century. With colombage, heavy cypress timbers were hewn, numbered, and brought from the backswamp *cyprière* (cypress timber lot) to the construction site, where they were mortised, tenoned, and pegged together to form the *carré*, or frame (Du Pratz 1758, 21). Interstices between the timbers of the carré were filled with brick or mud nogging called *bousillage*. A tempering of Spanish moss, shells, or lime was necessary to stiffen the mud. Small rods, or *barreaux* (also called *rabbits*) were placed between the heavy-timbered studs and braces to hold the mud nogging in place. Brick nogging, or *briquette entre poteaux*, required only mortar to be self-supporting in the interstices between the timbers. Exposed walls on the gables

FIG. 2.6. *Details of half-timbered construction display heavy cypress timbers in the frame and horizontal rabbits, called* barreaux, *that hold a matrix of* bousillage *(mud and Spanish moss nogging) at the Landeche Plantation mansion. Saint Charles Parish.* Photograph, 1969

and rear were weatherboarded and sometimes painted white. The front wall, protected by the gallery overhang, was traditionally whitewashed but not weatherboarded. All-wood construction began to replace half-timbering as early as 1800; however, some houses continued to be half-timbered until about 1850.

The Façade

Creole plantation houses have unique but rather simple, unpretentious façades composed of the gallery (porch), front doors that also function as shuttered windows, and exterior stairs. The gallery serves as a shade-producing porch and is integrated into the roof line in all Creole houses, those with ridged roofs and straight-sided gables as well as ones with hipped roofs. The gallery is an extension of the roof and is not an attachment (Oszuscik 1992b, 136–56). On one-and-one-half and two-story raised Creole mansions, galleries are doubled where the roof overhang is the true roof gallery and the porch floor of the main level forms a secondary gallery sheltering the raised basement below. The use of multiple front doors leading to separate rooms on the gallery is an extremely important diagnostic trait of southern Louisiana's French Creole plantation architecture. Each door opens into a separate room. Furthermore, door and window forms and functions are integrated so that shuttered French doors and shuttered French windows are frequently one and the same. In contrast, Anglo-American plantation mansions have but a single front door that leads to a central hall and windows are in quite separate, smaller fenestrations. Seldom are columns or pillars very large on Creole plantation mansions. On raised Creoles, pillars on the ground floor are plastered brick columns, and smaller wooden colonnettes support the section from the second-floor porch to the gallery overhang.

All major stairs are on the outside of a French Creole plantation mansion; there are no interior stairs or halls. Many raised Creole houses have stairs built into a well in a corner of the porch, where they are protected by the roof gallery. These were functionally better stairs, allowing discrete movement from one floor to the next without one having to brave the elements entirely. According to Jay Edwards, the popular legend that Spanish tax laws levied taxes on interior stairs and not exterior ones is unfounded (Edwards 1988, 18–21). The most logical explanations for exterior stairs are that human movement in and around the house focused on the gallery porch area, and the natural location for the staircase was under the protection of the gallery roof. On some Creole plantation houses, a single flight of stairs leads from the ground to the gallery porch, just as with ordinary front steps. With smaller folk houses, a set of stairs leads to the attic from the front porch and up into the gallery overhang, a trait with a regional spatial pattern in Louisiana, where houses with these outside stairs are associated closely with Cajun areas in the southwestern part of the state (21).

The Roof

Creole plantation mansions have two types of roof design: the ridge roof with sideward facing gables and the all-important hip roof (see fig. 2.4). The hip roof is the most attractive type from an architectural point of view, and it is especially meaningful from the perspective of cultural diffusion. When viewed from the ground, a hip roof slopes upward on all four sides, meeting at a common horizontal ridge at the apex of the roof. Unlike the pyramidal roof, whose steep slopes meet at a point, a hip roof has slopes that rarely exceed forty-five degrees, and the pinnacle of the roof is a horizontal roof ridge. The hip roof is a key diffusion trait, which traces directly to the Caribbean and to France. According to Edwin Doran, the hip roof originated in the English Channel area (the English as well as the French side). The trait diffused to the Caribbean in the seventeenth and eighteenth centuries and became widespread throughout the region (Doran 1962, 97). From the French Caribbean the hip roof diffused to Louisiana, becoming one of the premiere diagnostic culture traits that trace to the Caribbean (see fig. 1.6). To place a spanner in the works, as it were, the hip roof also has been connected to aboriginal Middle American traits, as in the *bohio* house with a high hip roof (Vlach 1976, 8[pt. 2]:66–69). Other arguments suggest that the hip roof came with slaves from West African source areas, such as the Yoruba homeland.[6]

Construction materials for roofs changed over the 275 or more years of planter occupance in Louisiana. Original roof coverings for plantation buildings were hand-riven shakes made from split cypress logs. Shakes were produced by riving with a froe, using a wedge and a mallet. Thin wedge-shaped slabs, which measured six to eight inches wide by two to three feet long, became the roofing material. Eventually, slate became the material of choice, especially for nineteenth-century urban architecture and plantation mansions after the 1860s (Hamlin 1944, 214). Galvanized, corrugated tin came to the plantation in the first quarter of the twentieth century, and some mansion owners resisted putting tin roofs on their homes. Since cypress shakes and expensive slate had fallen out of favor, plantation mansion owners resorted to asbestos shingles in the past and to asphalt tar shingles more recently, except for the Keller-Homeplace mansion, which now sports a red-painted tin roof installed about 1990.

The dormer window built into the roof is another feature that can be seen from the front façade of a Creole plantation mansion. The summer heat was far too intense for anyone to consider living in the attic. For decorative reasons and to provide light and ventilation to the attic, however, plantation mansion owners installed dormer windows.

Traveler's descriptions from Frederick Law Olmsted and C. C. Robin offer a glimpse of nineteenth-century Creole plantation mansions and their surroundings. Olmsted, the famous landscape designer of New York's Central Park, was traveling just north of New Orleans in 1860 when he described a plantation on the right bank of the Mississippi River (Olmsted 1861, 319).

> Fronting upon the river, and but six or eight rods from the public road which everywhere runs close along the shore inside the levee, was the mansion of the proprietor: an old Creole house, the lower story of brick and the second of wood, with a broad gallery, shaded by the extended roof, running all around it; the roof steep, and shedding water on four sides [hip roof], with ornaments of turned wood where lines met, and broken by several dormer windows. The gallery was supported by round brick columns, and arches. The parlours, library, and sleeping rooms of the white family were all on the second floor. Between the house and the street was a yard, planted formally with orange trees and other evergreens. A little to one side of the house stood a large two-story, square dovecot[e], which is a universal appendage of a sugar-planter's house.

C. C. Robin, an obscure scientist and writer during the French Revolution, made his way to the French Caribbean and to Louisiana in 1803, before Anglo-Americans and outside architectural influences had arrived in the region. His descriptions of dwelling traits and construction techniques for Creole houses have become invaluable documentary evidence (Robin [1807] 1966, 122–24).

> Some of the houses are of brick with columns, but the usual construction is of timber with the interstices filled with earth, the whole plastered over with lime. The houses have ordinarily only two or three large rooms, but the heat of the climate makes galleries around the house a necessity. All of them have one, some all around all four sides of the house, others on two sides only, and rarely, on only one side. These galleries are formed by a prolongation of the roof beyond the walls, but the prolongation forms a break in the angle in the plane of the roof . . . The roofs of the houses are covered with bark, with

pieux (a sort of plank) or with shingles . . . Houses are built either on the ground or elevated on blocks.

The Evolution of the Creole Mansion

In the analysis of Creole houses in Louisiana, we see an evolutionary pattern in which small Creole houses are conceptually enlarged and modified over time. Antecedents to the Creole plantation mansion originated with a basic double-room Creole dwelling characterized by a central chimney, double front doors, sideward-facing gables, and a front gallery incorporated into the roof design. All other forms of Creole dwellings evolved from this basic structure by the simple addition of room units, particularly to the sides, to form three- or four-room Creole houses, one to three rooms deep. Throughout the evolution of the Creole plantation house, each front room opened onto the gallery. The front gallery also formed a continuum from the simplest Creole to the largest multiroom plantation house.

The antecedent Creole house blossomed into the Creole plantation mansion with the addition of a hip roof, a raised position with a bricked ground floor, additional rooms to the sides and rear, green shuttered French doors and windows, and exterior weatherboarding and white paint. Some mansions retained sideward-facing gables and a straight roof ridge, much like that of the antecedent Creole house. Examples of Creole plantation mansions are Destrehan, built by the mullato freeman Charles in 1787, and Keller-Homeplace, also believed to have been built by Charles for the Fortier family in 1790 in Saint Charles Parish (see fig. 2.4). Other excellent examples include the Hermitage, built by Marius Pons Bringier in 1812–14; Oak Alley, built in 1837–39 (see fig. 2.2); Saint Joseph, a half-timbered mansion built about 1820 by C. B. Mericq in Saint James Parish; Armant, built in 1795 by Jean-Baptiste Armant (see chap. 5); and Whitney, built about 1790 for Jean-Jacques Haydel (see chap. 9).[7] Remarkably, the Creole house type has had a resurgence and is appearing in small modern forms across the delta region.

Anglo Plantation Mansions

After 1812, the lower Mississippi Valley attracted scores of Anglo-American planters from the Atlantic coastal Tidewater region and the interior Lowland South (De Laussat 1831, 95–97). Accompanying the planters were not only folk house plans, but also architects who introduced styles and

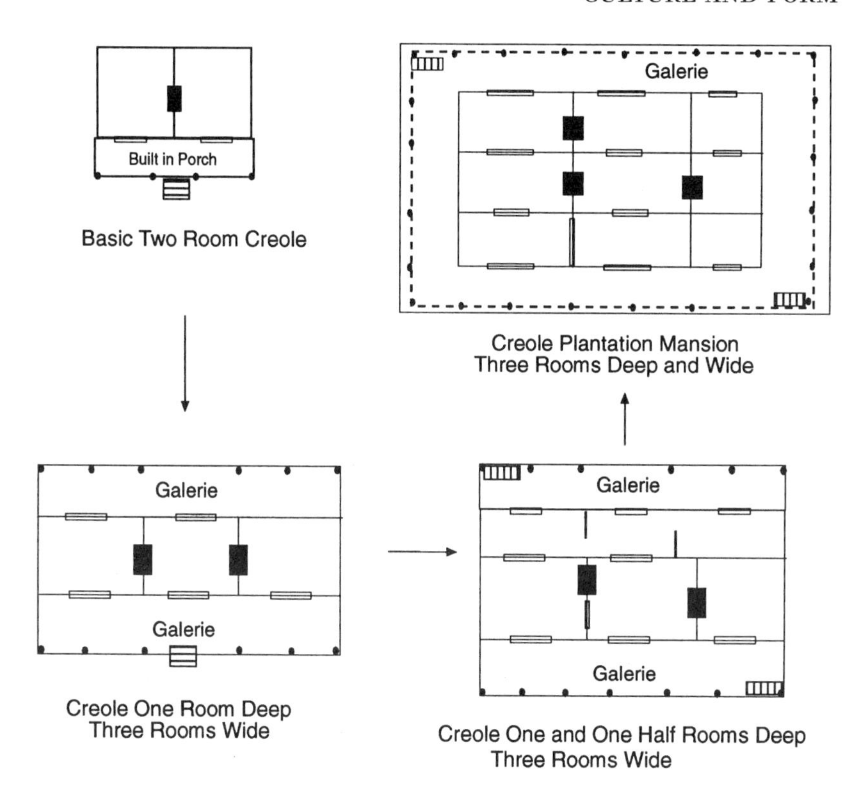

FIG. 2.7. *The evolution of the Creole house can be demonstrated by representative floor plans from the basic Creole two-room house to Creole plantation mansions.* Sources: Smith 1941, 214; Edwards 1988, 18–21; Oszuscik 1992a, 147

decorations from the Old World and popular styles from the Atlantic seaboard.[8] Anglo plantation mansions share diagnostic traits: (1) one and one-half to two stories with both floors used as living quarters; (2) usually one to two rooms deep; (3) no more than two rooms wide; (4) a central hall or passage that extends the full width of the house from the front door to the back; (5) interior stairs; (6) end chimneys that can be either inside-end or outside-end; and (7) a single front door. The central hallway serves especially well as a diagnostic feature for Tidewater and other Anglo houses (Glassie 1968, 54, 64–69, 80–83, 94–99, 109–11). There are no central hallways in any Creole house type. Chimney placement at the ends, particularly at the outside ends of gables, also strongly indicates Anglo influence.

Gables on the Anglo houses may be either side facing or front facing. Mansions with front-facing gables are especially identified with the Tide-

Fig. 2.8. *The Creole house type formed the basis for other larger and smaller subtypes that appear on plantations. Diagnostic traits are the built-in porch that serves as a gallery, multiple front doors, half-timbered construction, and a central chimney (now missing from this mid-nineteenth-century house, which was converted to gas heat). Modeste Point, Ascension Parish.* Photograph, 1993

water region (Morrison 1952, 316–17, 368, 416–17). The front-facing gable was the result of the architectural influence popular in Tidewater Maryland, Virginia, and the Carolinas during the eighteenth and early nineteenth centuries (Forman 1967, 62–63). The trait reflected a Greek Revival decorative style with columns, a portico, and pediments. Sideward-facing gables on Anglo plantation houses are generally associated with both the Lowland and the Upland South traditions; however, the side gable trait is unreliable in house type identification because it appears on other Tidewater regional dwellings and obviously on French Creole dwellings.

The façade of the Anglo plantation house displays more variations in detail than any other part. It is a pegboard upon which are placed numerous nonfolk architectural decorations, such as Greek Revival pediments, massive white pillars, and porticos. The single front door on the façade is a useful identifying trait because it indicates the existence of a central hall. Vari-

FIG. 2.9. *This gable-end view of an early-nineteenth-century Creole house reveals internal features not often seen. From* right *to* left, *notice the built-in porch gallery, interior ground floor rooms, loft rooms above, central chimney, and* colombage *(half-timbered) construction that used bricks between the posts, called* briquette entre poteaux. *Ascension Parish.* Photograph, 1969; source: Rehder 1978, 139. Used with permission from the Louisiana State University School of Geoscience

ety also prevails on the porch. The porch on a Tidewater-Anglo house with front-facing gables takes the appearance of a portico. On other Anglo types, the medium-width porch is either attached to the façade and roof or built in.

The Materials and Construction of Anglo Mansions

Techniques and materials used in the construction of Anglo dwellings differed from those used in Creole mansions. In lowland Louisiana, building materials were brick, plaster, and cypress wood. Materials did not vary between the first and second floors. Walls were either entirely brick or entirely wood, but never half-frame timbered with mud or brick nogging as on earlier Creole mansions. Interior walls were plastered, even in brick-walled homes, and on wooden structures, lath and plaster covered the inte-

FIG. 2.10. *The Hermitage, a French Creole plantation mansion, was built by Marius Pons Bringier between 1812 and 1814. Bringier was impressed by his friend Andrew Jackson and so named his plantation after Jackson's Tennessee home. Ascension Parish.* Photograph, 1967

rior wall surfaces. Brick pillars formed the foundation. However, the raised position of Creole houses was uncommon among Anglo houses. Underground basements or cellars could not be constructed because of the high water table in the delta region.

All-wood frame construction, an inherent building trait of the English-settled Tidewater since the 1620s, was introduced from the Atlantic seaboard (Shurtleff 1939, 101–58). Frame construction and weatherboarding were building traits later identified with Tidewater-Anglo architecture. Southern planters also contributed wood frame construction to the building traits of many southern Louisiana plantations. The few Anglo planters to arrive from the Upland South came from an area steeped in log construction techniques. However, most planters viewed log housing as beneath their dignity, and the availability of sawn cypress planks precluded the need for log construction. Exterior brickwork was a common construction technique for dwellings in the English Tidewater (Maryland, Virginia, and

FIG. 2.11. *The return of a traditional folk house type can be seen in this new Creole house under construction on the left bank of Bayou Lafourche. Lafourche Parish.* Photograph, 1991

the Carolinas) so it, too, accompanied some of the Anglo planting class of residents and architects to southern Louisiana.[9]

Tidewater Traits

The Tidewater culture region is situated along a narrow coastal zone from the lower Chesapeake Bay area of Maryland and Virginia to southeastern Georgia and no more than fifty miles inland. The region was settled primarily by English people in the seventeenth and eighteenth centuries to form an early southern component of British colonial America. The English Tidewater evolved into an extremely important culture region that supplied subsequent English culture traits to interior culture regions of the Lowland South and, to a lesser extent, the Upland South. It was from the English Tidewater hearth and subsequent areas in the Lowland South, then, that Anglo traits diffused to the Gulf Coastal region including southern and, yes, French Louisiana.[10]

During the time of Anglo planter influence in Louisiana, which was to continue until the late nineteenth century, Greek and Classic architectural revivals were appearing in the American East. Almost simultaneously, non-folk architectural styles and decorative features represented by large pillars, entablatures, pediments, and porticos made their appearance in Louisiana with the arrival of architects and Tidewater-Anglo planters. A typical southern Louisiana plantation mansion with Tidewater origin exhibited the diagnostic traits of a front-facing gable, two full stories, large white pillars, a pediment, a portico, inside-end chimneys, and, in some cases, side pavilions, or wings, so common with planned symmetry. The Tidewater dwelling had a central passage or hallway. This central passage evolved from the change from a single central chimney to inside-end chimneys from the Atlantic coastal regions back East.[11]

The best examples of Tidewater mansions in Louisiana are Madewood and Oaklawn. Madewood, on the left bank of Bayou Lafourche, measures sixty feet by sixty-eight feet and displays a front-facing gable, portico, columns, and wing pavilions as characteristic Greek Revival traits. It has a central hall and a single front door, inside-end chimneys, and interior stairs. Constructed of brick covered with plaster, Madewood was built in 1840–48 for Thomas Pugh, a pioneering Anglo planter who came from a Tidewater area, Bertie County, North Carolina (Lathrop 1945, 11–15). Oaklawn Manor, on Bayou Teche in Saint Mary Parish, was built in 1837 by Alexander Porter, an Irishman who first settled in Nashville, Tennessee, before moving to Louisiana (Stephenson 1934, 8–11).

Lowland South Traits

Based in part upon southern folk architecture, the Anglo plantation house was more Southern in form, materials, and plan than it was Greek Revival. The Anglo plantation house had a symmetrical, balanced plan that was one to two stories tall, one room deep, and two rooms wide, separated by a wide hallway. Brick chimneys were usually on the outside of the gabled ends or at the very least on the inside ends of gables. The second floor usually had the same floor plan as the first. Exterior doorways were to the front and rear of the dwelling at each end of the central hall. An Anglo example is Ashland-Belle Helene, built for Duncan Kenner in 1841 and located in Ascension Parish on the east bank of the Mississippi (see fig. 2.3). Look beyond the columns that surround the house, and you will see that

FIG. 2.12. *A good example of an Anglo plantation mansion is Crescent Plantation. William A. Shaffer began developing the plantation in 1827 and built the mansion in 1847. Terrebonne Parish.* Photograph, 1993

the structure has basic Anglo features of inside-end chimneys and a central hall. Furthermore, with its symmetrical floor plan of four rooms to each floor, the mansion's first floor represents a Georgian cottage plan.[12]

After arriving in southern Louisiana, some of the Anglo dwelling plans were altered by modifications resulting from locally borrowed traits. A syncretism or creolization in a blending of traits began to occur as Anglos introduced Lowland South traits and adopted French Creole ones.[13] The gallery's built-in porch was perhaps the most suggestive borrowed Creole folk trait. To the unwary observer, an Anglo plantation dwelling with a gallery in Terrebonne or Saint Mary Parish would appear to be a Creole house type. However, further scrutiny would reveal chimneys at the gabled ends, a single front door, a central hallway, and a symmetrical floor plan two rooms wide and one to two rooms deep, crucial diagnostic traits for Anglo-American houses. Examples can be found at the Schaffer mansion on Crescent Plantation near Houma and at other plantations in Terrebonne and Saint Mary's Parishes.

Origins

The origin of the Anglo plantation mansion followed an evolutionary track that began with houses of the English pen tradition. A pen was a singular room unit that measured approximately sixteen by sixteen feet (Addy 1898, 17, 66–69). Subsequent dwelling types were enlarged by simply adding pens. American houses of the pen tradition evolved from the single-pen house to double-pen houses (dogtrot, saddlebag, and Cumberland house); to the almost ubiquitous I house, a two-story house that is two rooms wide and one room deep; and ultimately to the four-pen house, which has four rooms over four rooms. In the Tidewater and Lowland South, the classic Georgian plan followed a pattern of four rooms to a floor, especially with two rooms on either side of a central hall and with interior or inside-end chimneys.[14] Multipen houses with paired rooms separated by a central passage or hall have the clearest connections with the Anglo plantation mansion.

Two house plans are important to determining the source of the Anglo mansion's configuration: the double-pen dogtrot house and the Georgian plan. A dogtrot house is a two-pen folk house with an open-air passage between the two pens. Chimney placement is on the outside gable ends of the house. Add to this a front porch and a rear porch that may become enclosed shed rooms at the rear. Construction of the two pens would have been originally log, but later dogtrots were built of sawn wood beginning in the late nineteenth century. The dogtrot house has an uncertain origin and a curious spatial distribution, with significant geographic concentrations from middle Tennessee westward to Arkansas and the Ozarks in Missouri and southward through Alabama, Mississippi, northern Anglo Louisiana, and into central Texas.[15]

The classic Georgian plan had a central hall flanked by two rooms. Chimney placement was either on the inside walls in a paired pattern or as massive outside chimneys. The Georgian plan diffused from England to the Atlantic seaboard and was well represented in the eighteenth-century southern colonial Tidewater region. The symmetrical Georgian plan diffused well into the southeast's interior Lowland South region; from there, the plan ultimately reached Louisiana in the early nineteenth century.[16] As Anglo planters arrived in Louisiana, architectural traits accompanied them, only to be modified as climate, style, costs, family size, and status dictated. The basic symmetrical floor plan, central hall, and outside chimney place-

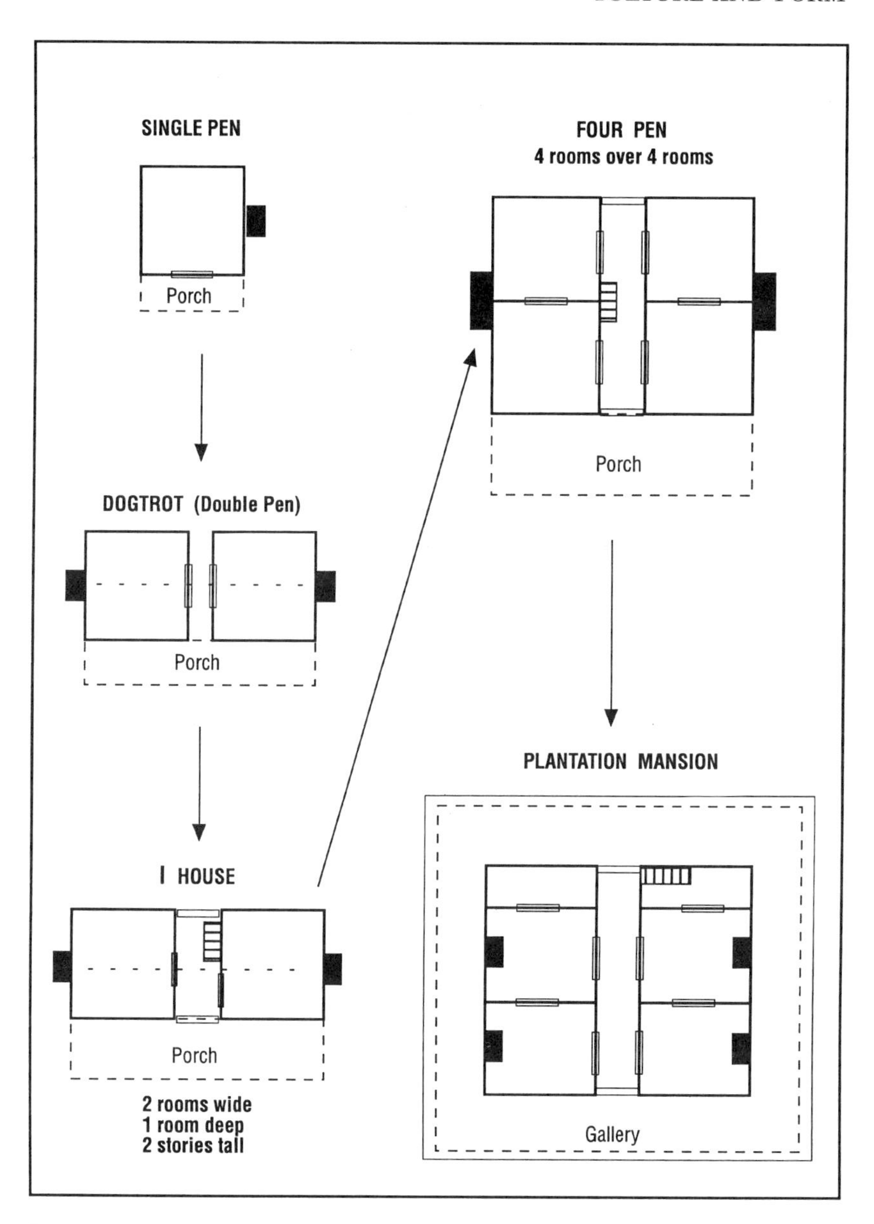

FIG. 2.13. *The evolution of the Anglo-American dwelling in floor plans proceeds from a single-pen house to double-pen houses, such as the dogtrot, to the I house, to a four-pen house, and ultimately to a plantation mansion.* Sources: Smith 1941; Kniffen 1965; Rehder 1992

ment remained the diagnostic constants in the formula. Local modifications were built-in porches that became galleries and a raised height on brick foundations to avoid floodwaters. Construction of all-wood frames and weatherboarding with the widespread use of interior plaster became customary building traits as Anglo plantation mansions dispersed across the delta.

Nineteenth-century French Creole and Anglo planters alike borrowed generously from Greek Revival and other decorative styles, but closer inspection reveals basic dwelling types to be fundamentally identified with inherent folk characteristics in floor plans and chimney positions. Among Creole and Anglo planters, a syncretism, creolization, or simply blending of traits occurred, whereby Creoles borrowed Anglo traits and Anglos borrowed Creole ones. Nineteenth-century Anglo planters, particularly, adopted the built-in gallery porch so distinctive in French Creole architecture. Anglo Tidewater mansions, however, bore architectural styles, decorations, and front-facing gables that, as a complex matrix, became identifying traits.

Growing Anglo Influences in Mansions

The nineteenth-century sugar planter was usually predictable based on his cultural heritage and tradition. However, sometimes a planter chose to have a house type that ran counter to his cultural habits. The Shadows on the Teche, David Weeks's mansion built between 1831 and 1835, has architectural features that blend a Creole plan, to take advantage of the climate, with Anglo traits, as evidenced in its Greek Revival style and use of American millwork.[17] Nottoway, named for an eastern Virginia county, was built in 1857–59 for John H. Randolph and was designed by New Orleans architect Henry Howard, who matched the then-current East Coast fashion. Southdown in Terrebonne Parish was started in 1858 by William Minor and had major renovations, such as a new second floor, added by his son Henry C. Minor in 1893. Bocage was built in 1801 for the daughter of Marius Pons Bringier. A later renovation of Bocage in the 1840s used a Minard Lafever handbook design for the Greek Revival details (Oszuscik, personal communication, 1997). The plantation home San Francisco has the appearance of a steamboat, with so many ornate stylistic details that it has been described as "Steamboat Gothic." San Francisco was begun in 1856 by Edmond Bozonier Marmillion and was completed by his son Valsin. While both *père et fils* were French planters, the remarkably ornate outward ap-

FIG. 2.14. *Nottoway plantation mansion, named for an eastern Virginia river and county, was designed by New Orleans architect Henry Howard for John H. Randolph in 1859. It is the largest plantation mansion in Louisiana, with sixty-four rooms. The refurbished mansion and cane land are now separately owned. In the 1850s the plantation covered approximately seven thousand acres. Iberville Parish.* Photograph, 1993

pearance of the house "speaks little or no French" to the casual observer (Doré 1989, 20).

Between 1825 and 1860, French and Anglo planters hired architects, acquired fashionable design books, and introduced a confusing array of styles in structures that were not entirely of the folk tradition. As more and more Anglo planters came to Louisiana from the Tidewater eastern seaboard and Lowland South, they brought with them increased wealth, a broader perspective of the world, and a greater complexity in mansion styles and types.

Pigeonniers, Garçonnières, and Other Structures at the Mansion Site

Every plantation had small structures back of the Big House that served the needs of the planter and his family at the residence. Associated with the

French Creole plantation mansion were small outbuildings called *pigeonniers* and *garçonnières* that were particularly important because they reflected French plantations and occupance patterns in the region. A pigeonnier was a pigeon house, or dove cote, used as a roost for pigeons (Vlach 1993, 98, 222). Squab, or young pigeons, are considered a delicacy in French cuisine, so a pigeonnier had a respectable role to play in provisioning the planter's family. Unlike other outbuildings for livestock, which were kept hidden in back, pigeonniers frequently came in pairs and decorated the entrance to the mansion (see fig. 9.6). With a form like a turret, they measured ten to twenty feet tall with a six- to eight-foot diameter. Shapes could be round, square, or octagonal with a pointed roof. Most were built of brick at the lower half and of wood or half-timbers above; however, others were entirely brick. The story is told that, whenever the planter's sons misbehaved, the boys would be sent to be locked in the hot, foul, malodorous pigeonnier for several hours. Knowing what pigeons do to statues, you can imagine the severity of the punishment.

The garçonnière was a small cottage located behind or flanking the mansion, usually on larger Creole plantations. The structure was used as a dwelling for the planter's sons or as a guest house to accommodate travelers for short stays. The cottage was sometimes patterned after the mansion but with much smaller dimensions of about twenty by thirty feet. Garçonnières were far less common than pigeonniers, and both were considered decorative augmentations to the Creole plantation mansion site.

All plantation mansions had cisterns attached to the house to collect rainwater. Despite Louisiana's drenching propensity for water in the air, under and on the ground, and in swamps, bayous, lakes, and rivers throughout the delta, potable water required a cistern system. Rainwater collected from roof tops and directed by gutters ran into enormous cisterns. Some cisterns were constructed entirely of brick; some had only brick foundations; others were made entirely of cypress in the form of a huge barrel with staves and metal bands.[18] At smaller sites, hogsheads, the wooden barrels used for storing and shipping raw sugar and molasses, became rain barrels that collected the precious liquid.

The all-important kitchen appendage located to the rear of the mansion was either completely detached as a separate, stand-alone structure or attached by additional roof covering and flooring from the gallery. Separation between mansion and kitchen was said to keep cooking heat out of the mansion and to protect the mansion in the event of a kitchen fire (Vlach 1993, 43–62). Kitchens varied in size and shape, but most were square and rarely measured more than thirty feet on a side. The structure on the former Armant plantation mansion aptly validated the term *kitchen appendage* (see fig. 5.2), and the raised kitchen at Keller-Homeplace in Saint Charles Parish was once completely detached but was moved closer and attached to the house in the late 1880s (Daspit 1996, 154–56). Another excellent extant example is the kitchen attached to the Creole mansion at Laura Plantation in Vacherie (Saint James Parish). The hipped roof, raised kitchen has a small connecting hall attached at the mansion's roof and at the main floor. The plantation cook usually had a living quarters either in a room adjacent to the kitchen or in separate quarters out back. Quarter houses for house servants could be found on most large plantations throughout the South (Vlach 1993, 18–32). In Louisiana, house servants' dwelling types were not unlike the quarter houses found in the quarters proper for field laborers. The only differences were the location in back but at the mansion site and the generally better quality and upkeep of the structures.

Nineteenth-century plantation mansion sites had other separate, small buildings that functioned as the carriage house or livery stable, perhaps an office, a gardener's shed, and, of course, a privy.[19] Carriage houses and livery stables were located in back or to the side of the mansion and functioned much like our modern garages for cars today. Except for large carriage houses, the other structures were small, square to rectangular buildings ranging in size from four feet to sixteen feet square or larger. Most plantation mansion sites had one or more small privies (also called the necessary house, outhouse, crapper, or other colorful names) located out back. In the mansion, people also made use of chamber pots that were kept under the bed. Unlike plantations elsewhere in the South, where numerous outbuildings were common (Vlach 1993, 63–106), southern Louisiana plantations did not have ice houses and few had dairies; they kept food and especially wine cool in cellar rooms in the raised basement. Moreover, smokehouses were not as common, especially on French plantations. Over time, the logical functions for small outbuildings at the mansion began to diminish. By the early twentieth century, plantations had begun to lose these almost-forgotten surface features. By the time I was in the field to observe the plantation of the 1960s, many small, obsolete outbuildings already had disappeared from mansion sites.

Settlement Patterns

In rural settlement geography, the pattern of settlement is an important fundamental part of understanding the cultural imprint on a landscape. Whether the pattern is described in geometric terms or in spatial diffusions and dispersals, settlement patterns are a landscape expression of culture. August Meitzen, a nineteenth-century German scholar believed that the geometric patterns in rural villages and scattered settlements in Europe could be correlated to ethnic groups and that relic forms on the landscape possessed a degree of diagnosticity (Meitzen 1895). Even though his work was only partially correct, Meitzen's ideas influenced Fred B. Kniffen, who in turn introduced to me the idea of correlating settlement patterns to culture groups. The following discussion celebrates the memory and influence of Fred Kniffen and the works of August Meitzen.

Sugarcane plantations all maintain a fundamental settlement trait called *agglomeration.* Buildings are clustered in such a way that small, discrete villages appear on the landscape. Along the natural levees of the Mississippi River, plantation buildings are arranged as linear settlements along lines perpendicular to the stream. The linear pattern here is not to be confused with linear *marschhufendorf*-like settlements, which produce strips of continuous settlement parallel to streams and bayous for twenty to thirty miles in the delta (Dickinson 1949, 256–57). Linear here means an alignment of buildings facing a central road that runs at right angles to the course of the river. A block pattern designated *nodal-block* characterizes the arrangement of plantation buildings in the Bayou Lafourche area. Elsewhere, to the south in Terrebonne Parish, to the west in the Bayou Teche area, and to the north in the Red River area, settlements also form a block pattern but are adjacent to the streams and are called *bayou-block plantations.*

Linear Plantations

Linear sugar plantations appear in an irregular distribution along the Mississippi River from twenty miles north of Baton Rouge to a point just below New Orleans. Linearity is achieved from the alignment of quarter houses in a row of laborer's dwellings along a centralized road that lies perpendicular to the Mississippi River. The sugar factory and outbuilding complex are located at the end of the road, usually equidistant between the levee crest and the backswamp. Thorpe in 1853 explained that the sugar

FIG. 2.15. *Poplar Grove Plantation, with its tightly agglomerated linear settlement pattern, was a robust working plantation in 1966. West Baton Rouge Parish.* Photograph, 1966

factory and outbuildings were so located as "to divide up as much as possible the distance that must be traversed in hauling the wood from the 'swamp,' the cane from the fields, and the crop to the River for shipment" (746–67).

My surveys in the late 1960s showed that the average linear plantation measured 0.3 mile in length and contained approximately twenty-five dwellings. The largest plantation settlement had sixty-seven houses; the smallest had four (Rehder 1971, 86). Barns and sheds usually numbered fewer than ten per plantation. In addition to these structures, a model plantation would have a mansion and a sugar factory. With the overall decline of plantation buildings, even the largest plantations today have fewer than twenty quarter houses, and many of them are in ruins.

Two historical factors are important in understanding the present linear plantation settlement: (1) the initial parceling of land in accordance with the French arpent system of land surveys and (2) the early existence of linear plantation settlements in source areas in the French West Indies. The

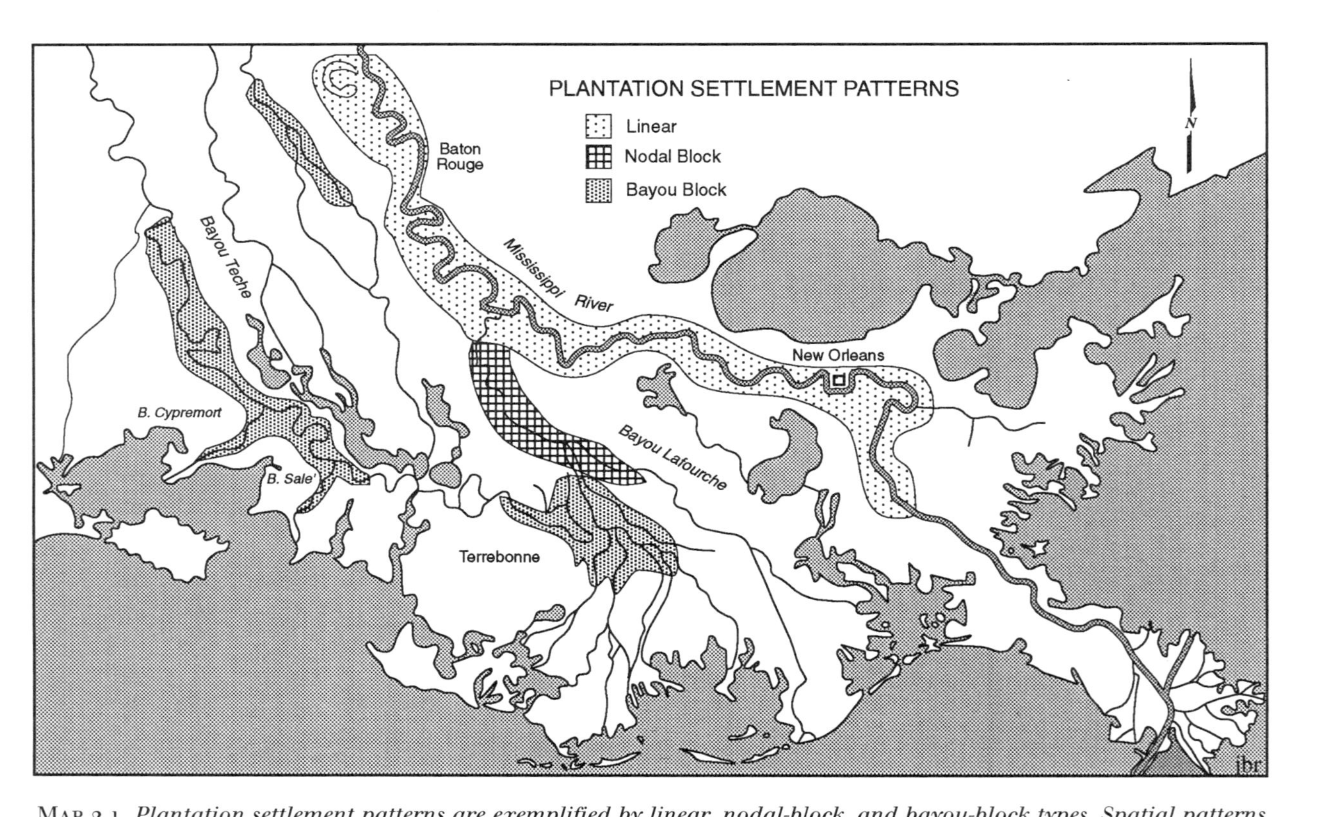

MAP 2.1. *Plantation settlement patterns are exemplified by linear, nodal-block, and bayou-block types. Spatial patterns focus along the Mississippi River, Bayou Lafourche, and in the Bayou Teche, Terrebonne Parish, and Red River areas.*

French arpent system was based on the *arpent*–a linear unit of measure equaling 192 feet. Lands along streams were surveyed from stream bank to backswamp at a standard depth of forty arpents. Widths were variable, but because initial settlements focused on the favored frontlands and levee crests of streams, frontages were narrow, with most plantations less than ten arpents wide. The resulting landholdings were long, narrow parcels that produced a slender frame in which the linear settlement pattern developed.[20] I cannot prove a direct cause-and-effect relationship, but the association suggests that the slender landholdings may have contributed to the linearity of plantation settlements in those areas where early arpent surveys were made. Map evidence supports the antiquity of linear plantation settlements and provides clues to a broader distribution of the pattern. An 1815 map of the New Orleans area illustrated twenty-three plantations with a linear arrangement of buildings placed within long, narrow landholdings (Reed 1915). Even after the Civil War, when plantations were breaking up elsewhere in the South, the Louisiana sugar plantation retained its tight, agglomerated pattern in the quarters.

Evidence of early linear plantation settlements can be found in descriptions by early writers. In 1861, Howard Russell, an English traveler, saw Louisiana sugar plantations along the Mississippi River in this way: "The sugar plantations are bounded by lines at right angles to the banks of the river, and extending through the forest . . . and in the vicinity of each are rows of white washed cabins, which are the slave quarters" (Russell 1863, 98). At another plantation in Saint James Parish, "rising up in the midst of the verdure are the white lines of the negro's cottages and plantation offices and sugar houses, which look like large public edifices in the distance" (103). Frederick Law Olmsted in the 1850s observed the arrangement of buildings. "From a corner of the court [mansion yard] a road ran to the sugar-works and the negro settlement, which are five to six hundred yards from the house" (Olmsted 1861, 320).

Much earlier, in 1803, C. C. Robin described the French Louisiana sugar plantation in these terms:

> They [Negroes] are housed not far from the master's house in a little house or cabin perhaps a dozen feet square . . . On most plantations where the masters take care of things, all cabins are aligned and spaced regularly. It looks like a little village and is usually called a camp. The master's house at some distance apart dominating these humble cabins by its greater size and eleva-

FIG. 2.16. *The linear settlement at Landeche Plantation had shotgun houses in the quarters and a closed plantation store with a faded root beer sign. Saint Charles Parish.* Photograph, 1967

> tion brings to mind feudal times, when the haughty chateaux of the seigneurs looked out over the miserable huts of the serfs (Robin [1807] 1966, 237).

Extraregional evidence supports a connection with the French West Indies as a source area from which many plantation culture traits diffused to Louisiana. A linear sugar plantation in Guadeloupe was described by J.-B. Labat, a Jesuit priest who visited the island in 1696 (see map 1.5) (Labat 1742, 412–18). The landholding and the arrangement of buildings were distinctly linear. Historic maps of plantations in Martinique as early as 1671 also indicated the presence of linear settlements (Revert 1949, 247–60). At various times between 1718 and 1809, Louisiana and the French Caribbean were in very close contact. Such significant traits as sugar technology, sugar makers, slaves, and even the sugarcane plant itself diffused from the French West Indies to Louisiana. The existence of a linear plantation settlement pattern in both areas at a time when French plantation occupance was taking place in Louisiana not only suggests a French connection, but also points to a possible place of origin for the linear pattern.

Examples of linear plantations with French ancestry are Landeche, with its shotgun quarters, Armant, and Cedar Grove. Armant Plantation, named for Jean-Baptiste Armant, the initial owner, was established in 1796. For two centuries, through a succession of early French and later Anglo owners, the plantation retained its linear landholding and settlement pattern until the period 1974–81, when all but three buildings were leveled in response to new corporate ownership. Cedar Grove was an excellent linear plantation with more than 110 years of French ownership. Originally established by Haitian planter George Deslondes, Cedar Grove was maintained by French planters until 1939 (Rehder 1971, 295–303). The plantation fell into ruins in the 1980s, and by 1995 all landscape evidence of the plantation's dwellings had vanished. By comparing map 1.8, which illustrates the distribution of French and Anglo plantation owners in 1844, with map 2.1, showing settlement patterns, you can confirm a map correlation that demonstrates concentrations of linear settlements along the Mississippi River where French planters were concentrated in 1844. Furthermore, by 1858 French planters and their narrow plantation landholdings were still clearly evident along the Mississippi River.

Nodal-Block Plantations

Plantations with a block-shaped pattern of buildings display two different types of locations relative to streams. Along Bayou Lafourche, nodal-block plantations form nuclear clusters of buildings located at a distance of one-half to one mile from the stream banks, near the centers of the landholdings. On other bayous, bayou-block plantations are situated directly on the natural levee crests.

Nodal-block plantations are numerous in the upper portion of Bayou Lafourche but decrease in number downstream. The block pattern is established by four to twenty quarter houses arranged in a grid pattern formed by three to seven streets. The sugar factory and its surrounding cluster of outbuildings constitute the nucleus of the settlement. The mansion is usually located nearer the levee crest, about one-quarter mile to one mile from the quarters. A centrally placed road connects the remotely located settlement with the levee crest and bayou.

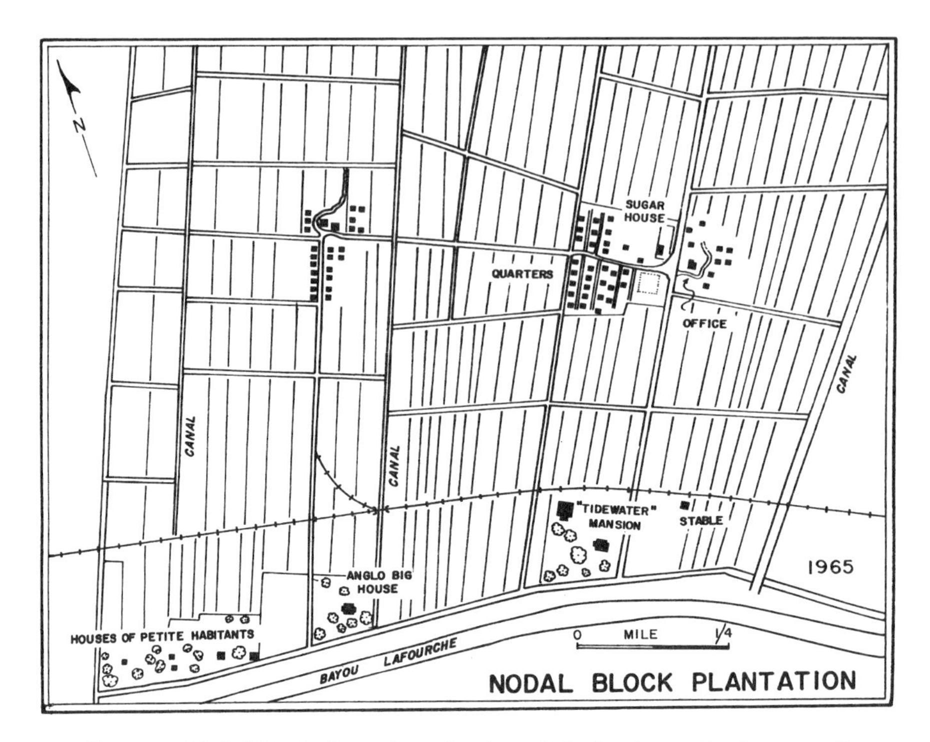

FIG. 2.17. *Nodal-block plantations dominated the landscape in the upper Bayou Lafourche area in Assumption Parish.*

These unusual sites for plantations located far from the levee crest can be explained by examining initial settlement. As the successions of plantation settlement swept over southern Louisiana between 1720 and 1860, the widest natural levees along the Mississippi River were occupied first by French Creole planters. By the latter part of the eighteenth century, small Acadian and Spanish settlers began to occupy the frontlands on the large, natural levees of upper Bayou Lafourche. By the time Anglo-American planters ultimately reached southern Louisiana in the years after 1803, they found the only suitable unoccupied plantation lands limited to the backlands of Bayou Lafourche and along the narrow levees of smaller bayous to the south and west.

Along upper Bayou Lafourche, the purchase of wide parcels of frontage from small farmers occurred so infrequently that plantation holdings became large backland parcels with very narrow access parcels to the bayou. Madewood Plantation in 1830 best illustrates the shape of landholdings that

Anglo planters such as Thomas Pugh were establishing during the initial occupation of the landscape (see chap. 10). As subsequent land acquisitions took place, landholdings began to fill out. However, the initial site locations in the early nineteenth century established the pattern of nodal block plantations so well that today's patterns are simply vestiges of the past.

Bayou-Block Plantations

Plantations with the *bayou block* label dominate the Bayou Teche and Terrebonne Parish plantation areas and appear in a remnant stage in the northern portions of the sugar region. Plantation buildings here are located close to levee crests and often directly on them. On many such sites, the levee crests on both banks of the stream are occupied, with half of the plantation on each side of the bayou, joined by a bridge. Excellent examples of the pattern can be seen on Ashland and Oaklawn Plantations featured in chapters 6 and 7.

The settling process of the initial Anglo-American planters, land surveys, and resulting landholdings help to explain the bayou-block pattern. The Bayou Teche and Terrebonne Parish areas were among the last sections of the sugar region in southern Louisiana to be claimed. Anglo planters, finding the Mississippi and upper Bayou Lafourche levees already occupied, pushed farther west and south to settle finally along the narrowing bayous in the area (Pierce 1851, 606). Moreover, as an incentive to attract settlers, land in 1823 was selling for as little as six dollars per acre (Rehder 1971, 333). After 1811, lands in Louisiana were surveyed according to the General Land Office survey system. American surveyors were instructed to resurvey all landholdings into 160-acre units. Based on township and range delineations, the surveys produced grid patterns elsewhere in southwestern Louisiana. However, public lands on arable natural levees along the smaller bayous between Bayou Lafourche and the lower part of Bayou Teche were meandered, that is, measured in accordance with stream bank locations on both sides of the stream to conform to the earlier arpent surveys of riparian lands.[21] Depths followed the standard forty arpents, but frontage widths were unusually wide. Resulting landholdings had wide frontages measuring thirty-five to sixty arpents to compensate for the extremely narrow depths of arable land between levee crests and backswamps. As landholdings became established on both banks of the streams

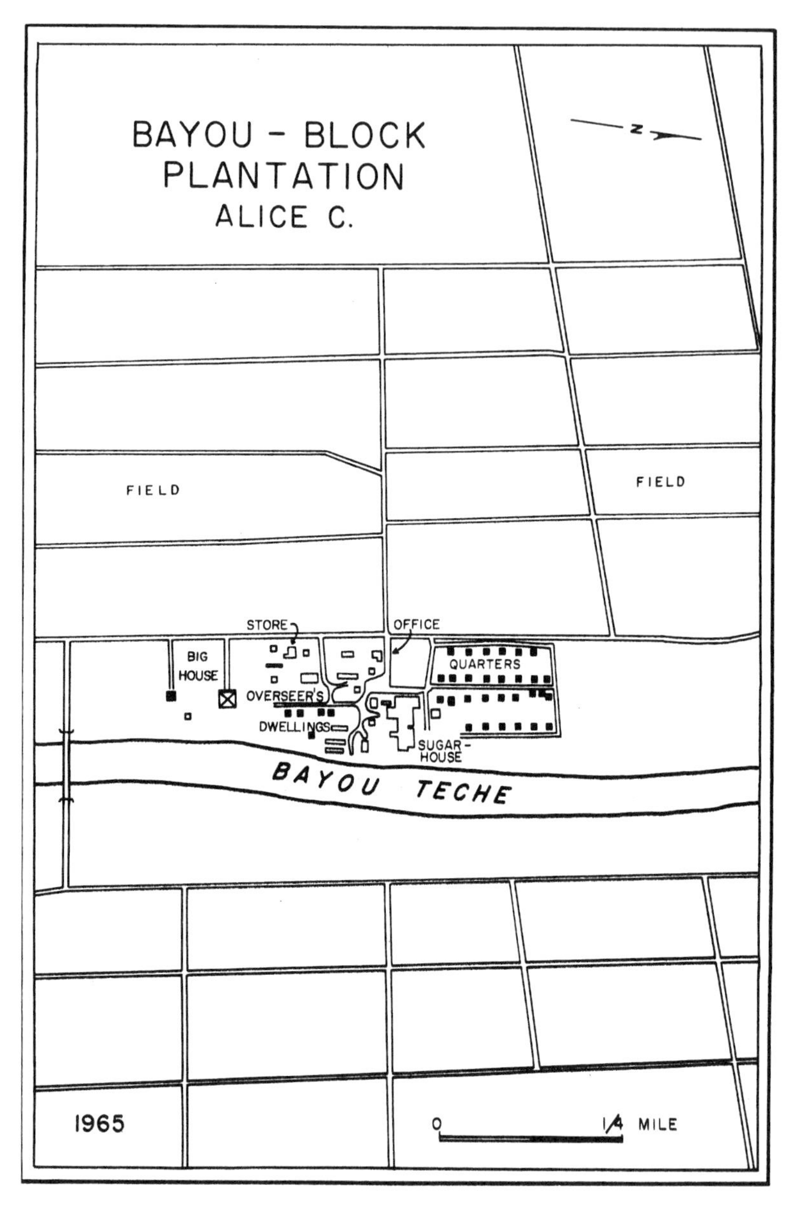

FIG. 2.18. *A good example of a bayou-block plantation was the Alice C. Plantation in Saint Mary Parish.*

in rectangular patterns, planters centrally located their buildings in a block pattern directly on the levee crests to focus building sites and processing functions nearest the bayou transportation route.

Historical map evidence indicates that block plantations coincide with areas of Anglo-planter domination. If one compares map 1.8, showing French and Anglo plantations in 1844, with map 2.1 of plantation settlement patterns, a visual correlation is clear. Anglo-American plantations represented plantation occupance patterns in 1844 and were concentrated near the extremities of southern Louisiana's rivers and bayous, where block-shaped settlement patterns emerged.

Block plantation patterns trace to Anglo-American source areas in the Tidewater and Lowland South. In the Virginia Tidewater area, tobacco plantations initially built by English settlers and their descendants between 1620 and 1795 had quarters arranged in block settlement patterns (Gray [1932] 1958, 317–22). Unlike plantations elsewhere in the South, each of the Virginia plantations had several quarters units consisting of four to fifteen houses located at various points on the properties (Morton 1941, 109). Gamble's Hundred, an estate on the James River in 1736, had six separate compact quarters for slaves (Dowdey 1939). Farther down the Atlantic coast, in the lower Tidewater region of the Carolinas and Georgia, English Tidewater planters produced rice, indigo, sea-island cotton, and a small amount of sugarcane on plantations characterized by blocky, compact settlements. Hampton Plantation, located forty miles northeast of Charleston, South Carolina, had a square settlement pattern consisting of a mansion, outbuildings, and quarter houses, all arranged in a tight, compact block (Rutledge 1941). In Tidewater Georgia in the Altamaha River area between Savannah and Brunswick and in the South Carolina low country, numerous block settlements appeared in the late eighteenth and early nineteenth centuries.[22]

Antebellum plantations in the interior Lowland South and in parts of the Upland South were characterized by English and Scotch-Irish settlers, cotton cultivation, and block settlement patterns. In 1860 the antebellum Anglo cotton plantation Barrow in Georgia had a tight, square block pattern of quarters (Barrow 1881, 832–33). Descriptions of Isaac Franklin's plantations in Tennessee and Louisiana provide connective evidence between the Upland South and Louisiana. Fairvue, Franklin's Tennessee cotton plantation near Nashville, consisted of a two-story brick main house with its yard and setting plus separate kitchen and smokehouse in one part. In another portion of the settlement was "a slave quarter . . . of fifteen or twenty substantial double brick houses laid off on the plan of a town, with the overseer's house in the center" (Stephenson 1938, 96–98). His six Louisiana plantations in West Feliciana Parish were by Franklin's orders constructed identically to the Nashville holdings.

Agglomeration was an almost universal characteristic of antebellum plantation buildings in each of the southern plantation areas. However, after the Civil War, plantations in the Tidewater and Lowland South experienced a dispersal of buildings, especially quarter houses. For example, on the Barrow Plantation in Georgia the former block pattern vanished when quarter houses were scattered among the cotton fields. Throughout much of the South, dwellings were dispersed because sharecroppers and tenant laborers worked separate parcels in the forty-acres-and-a-mule complex so common to the postbellum cotton plantation.[23] However, Louisiana's sugar plantation quarters remained agglomerated because sugarcane required gang labor throughout its cultivation and processing. The agglomerated patterns in French linear and Anglo block plantations that began in the eighteenth and nineteenth centuries continued well into the 1970s, serving as identification keys to present and past landscapes and to the culture groups who built them.

The Quarters

The quarters is a villagelike settlement established to house the workers on a plantation. The name comes from the time when the village was a slave quarters, but throughout the history of plantations in Louisiana, the settlement has always been "the quarters." Street signs labeled Katy Quarters Road and Belleview Quarters Road are among those that still mark the way to a few remaining quarters. John Michael Vlach's 1993 book, *Back of the Big House,* focuses on the quarters and brings attention to a part of the southern plantation that was often ignored as if placed downwind of the Big House in the swirling mists of historic inquiry. In this chapter, we have seen the diagnostic value of settlement geometry in linear and block patterns based on the arrangement of quarter houses. However, quarter house types are not always culturally diagnostic. If house types can be culturally diagnostic, how can quarter houses not be diagnostic? What appears to be an anomaly is in truth a paradox. Quarter houses are not universally culturally diagnostic because they were built on plantations to be simple, low-cost, reasonably efficient shelters for slaves initially and for wage-earning workers later. The planter's cultural input did not always extend beyond the Big House to the conscious architectural design for individual quarter houses. Furthermore, the planter's family might not have wanted to have slaves living in miniature replicas of the Big House. On nineteenth-century plantations, some quarter houses were prefabricated by firms like W. W. Carre's in New Orleans and shipped to plantation sites. Such houses were beyond the plantation's direct range of cultural diagnostic influence (Rehder 1971, 185–86). If quarter houses were to be built cheaply, any low-cost design could have been chosen. This is not to say that cultural traditions were entirely absent. Some French plantations, indeed, had small Creole houses for quarters, and some Anglo plantations had attached-porch quarter houses with Lowland South appearances, but the connections were too infrequent to call them culturally diagnostic for initial settler and settlement.

Antebellum slave housing and postbellum dwellings for plantation workers were rarely considered well made or even adequate for human occupation. Quarter houses could be simple, rude huts or something that was just a shade better than dreadful. Throughout the South, nineteenth-century quarter houses were likely to be crude, single-room cabins with dirt floors and few furnishings. On many plantations overcrowding was common. For example, on Charles Ball's plantation in South Carolina, 260 slaves lived in thirty-eight cabins; a slave named Louis Hughes reported that, on a plantation near Memphis, 160 slaves shared eighteen cabins (Hughes [1897] 1969). Some slaves lived in long, low sheds packed with 20 to 30 people. Unhealthy conditions and poor quality housing were so bad that some planters tore down quarters every few years and rebuilt on another nearby site. Conversely, on the plantation where Henry Watson was a slave, 100 slaves lived in twenty-seven cabins, and on the South Carolina plantation where Sam Aleckson lived, the slave cabins were neat, and some even had flower gardens.[24]

In Louisiana, the quarters have been the designated on-site settlement for plantation labor since the very origin of plantations in the region. In 1861, Howard Russell, an English traveler, described a Louisiana sugar plantation in Ascension Parish: "The sugar-house is the capitol of the negro quarters, and to each of them is attached an enclosure, in which there is a double row of single storied wooden cottages divided into two or four rooms. An avenue of trees runs down the centre of the negro street, and behind each hut are rude poultry hutches" (Russell 1863, 104). In the postbellum period, the quarters remained in situ as a village of dwellings for a res-

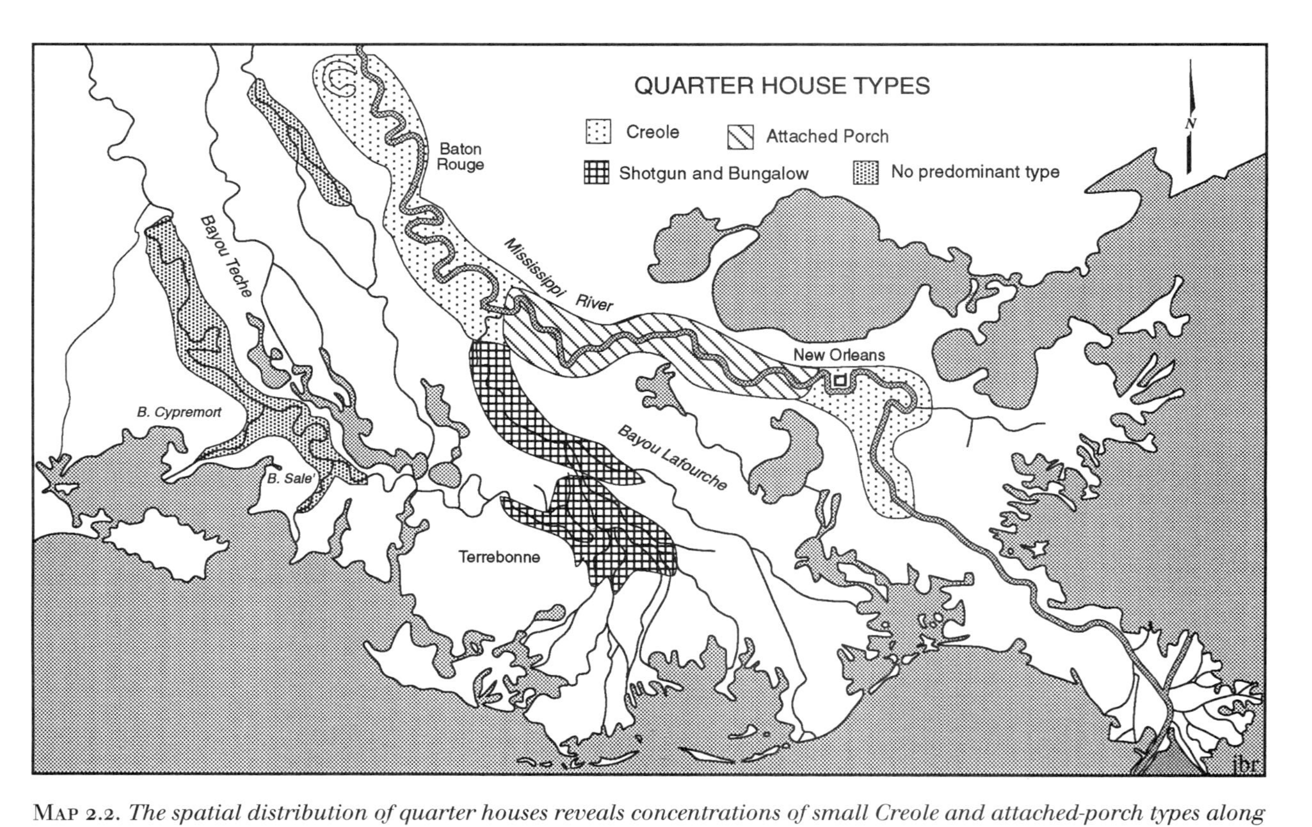

MAP 2.2. *The spatial distribution of quarter houses reveals concentrations of small Creole and attached-porch types along the Mississippi River and shotgun and bungalow types in the southern areas of the delta.*

ident labor force and later the place of residence for retirees. The status of workers may have changed from slave to wage earner, but the spatial arrangement of the quarters persisted as a relic settlement.

In the 1960s, I discovered that the number of quarter houses on Louisiana's sugar plantations ranged from four to sixty-seven and averaged about twenty-three per plantation. Since then, quarter houses have become a serious and perceived-to-be-dangerous liability for plantation owners and, as such, quarter houses have suffered wholesale destruction throughout the region. According to my landscape definition, a sugarcane enterprise required the requisite four-house minimum to qualify as a plantation. Two or three houses could form a line for a linear plantation but could not characterize a block plantation, which required four houses to form a geometric block pattern. The actual number of quarter houses on a plantation was not important as long as there were four or more. The quarters settlement still remains an essential visible landscape signature because it separates small plantations from small nonplantation sugarcane farms (Rehder 1971, 162).

Four types of plantation quarter houses were well represented in the delta region in the 1960s. The *small Creole quarter house* with a built-in porch was widespread throughout the sugar region. The *attached-porch quarter house* was a square house with a shed roof overhang attached to the roof in front of dwelling. This house was found in a small, focused area in the parishes of Saint James and Saint John-the-Baptist along both banks of the Mississippi River. Long, narrow *shotgun* houses appeared as quarter houses on plantations chiefly along Bayous Lafourche and Teche and in Terrebonne Parish. *Bungalow* quarter houses, wider houses believed to be derived from the shotgun type, were found with the same spatial distribution as the shotgun. It was not unusual, however, for several different house types to appear on the same plantation.

The Small Creole Quarter House

One of the most widespread quarter houses was the small Creole type with a built-in porch. The key signature was the roof overhang integrated into the front porch. The relatively square house had sideward-facing gables, double front doors, and a single central chimney. Outside dimensions ranged from 15 by 27 feet to 30 by 32 feet. Brick piers about two feet high supported the one-story frame dwelling.

FIG. 2.19. *A fortuitous anomaly is having all four quarter house types on one plantation. From* left *to* right: *attached-porch quarter house, bungalow, small Creole quarter house, and shotgun house on the White Castle Plantation. Iberville Parish.* Photograph, 1969

Identifying the house at first was a problem. Originally, I called it a "built-in-porch quarter house" out of respect for Fred Kniffen's use of "Creole" for folk French house types.[25] When I found the house in abundance on both French and Anglo plantations in the 1960s, the commercial nature of the house transcended cultural diagnosticity in the initial occupance phases for specific plantations. For many years, however, I had felt that the structure was indeed a Creole house type, albeit a restricted and somewhat penurious one. The house displayed the necessary traits of a gallery in the built-in porch, a central chimney, multiple front doors, and the plan of the basic Creole that had originated from earlier Creole house types. In the Gulf Coast area, the structure should be identified as a small Creole house type even though it functions as a quarter house type on French as well as Anglo plantations.

The small Creole quarter house was built originally in one-family and two-family versions. The two-family houses had two front doors that led to

FIG. 2.20. *Small Creole quarter houses at Poplar Grove Plantation date to the middle to late nineteenth century. In 1964 the quarters here included fifty occupied houses, but by 1993 only seven dwellings had tenants. The plantation closed in 1995. West Baton Rouge Parish.* Photograph, 1993

separate living quarters and a central chimney with two flues to serve the two families. The small, single-family house had a single front door, one living area, and a single-flue chimney located at one of the gable end walls (see fig. 5.6). The smaller version of the house was quite rare in the 1960s, and only two plantations had a majority of single-unit quarter houses. The two-family version rarely had two families living in it because fewer laborers were required and a surplus of dwellings was typical on plantations in the 1960s.

The small Creole quarter house with double front doors, two rooms, and a double-flue central chimney was the dominant type, and most houses like this were located on plantations along the Mississippi River from a point twenty miles north of Baton Rouge to about fifteen miles below New Orleans. In plan, the quarter house had two living areas measuring fifteen by sixteen feet located directly behind the two front doors. In the nineteenth century, each family ate, slept, and lived in a single room. An attached kitchen at the rear once served both families.[26] Eventually, the internal living space was modified to accommodate one family living in a two-unit house. The two main rooms were connected by an interior door to provide a living room on one side and a bedroom on the other. The rear kitchen built into the back porch kept its original form but served the one family in the house (Rehder 1971, 164–71).

The Attached-Porch Quarter House

Quarter houses with attached porches were scattered among plantations along the Mississippi River, but a concentration of them appeared in Saint James and Saint John-the-Baptist Parishes. In plan, the house had two living areas and a back kitchen; except for the porch, the house resembled the small Creole type. The attached porch is the important difference. Here, the porch was not integrated into the roof and appeared to be added almost as an afterthought. It was clearly much easier to build than some of the other types, and ease of construction and low cost traditionally were driving factors in quarter house construction. The relatively square house had dimensions averaging thirty by thirty feet with sideward-facing gables, usually a central chimney, and double front doors leading to two interior rooms. The attached-porch quarter house style is not as old as the small Creole type and usually had balloon framing, lightweight two-by-four studs in the walls, and commercial weatherboard. However, a few were built in the late nineteenth and early twentieth centuries with board-and-batten construction.

The Shotgun as a Quarter House

Shotgun houses as quarter houses on plantations were distributed along the southern and western bayous of the sugar region. Concentrations of them were found along Bayou Lafourche, Bayou Teche, and the many bayou distributaries in Terrebonne Parish. The shotgun house can be easily recognized, with its front-facing gable and extremely long, narrow shape. With gables set at the front and rear, the house was situated perpendicular to the road or bayou. Dimensions for shotgun quarter houses average fifteen by forty-five feet, but folk versions could reach thirty feet wide. In plan, the house was one room wide and three or more rooms long and naturally had a long series of interconnecting rooms in a single file. The single front door was usually perfectly aligned with all interior doors and with the back door.

FIG. 2.21. *Most attached-porch quarter houses were built in the twentieth century; however, this single-unit house with board-and-batten siding dates to about 1880 at Cedar Grove Plantation. Iberville Parish.* Photograph, 1969

In the folk wisdom of the area, it is said that the shotgun house was so named because if someone fired a real shotgun through the front door, the pellets would pass through the house and out the back door without causing any damage (Writer's Program of the WPA 1941, 158). Another reason for the shotgun name was the long, thin, gunlike appearance of this most unnatural, narrow house type.

The integral parts of the dwelling deserve attention. The shotgun had spatial advantages by having three or more rooms for living spaces rather than two rooms as in the other quarter houses. Proceeding from the front porch to the back porch through a shotgun, one would encounter a front living room, two or more bedrooms, and a kitchen. Among nonplantation folk versions of the shotgun, the kitchen at an earlier time would have been a structure detached from the main house. When used as a quarter house and especially as an urban dwelling, the narrow shotgun was advantageous for packing more houses into a given space.

The shotgun house has not always been a plantation house type. Early

FIG. 2.22. *The shotgun made a good quarter house because of its ease in construction and materials. Note different patterns of yellow brick tar-paper siding in contrast to the original weatherboarded front. Elmfield Plantation. Assumption Parish.* Photograph, 1967

versions of the house began as small urban houses in New Orleans and as an urban predecessor to the larger camelback house.[27] In rural, nonplantation settings small shotguns served as "camps," shelter/shacks found in bayou trapper-fisher settlements. The shotgun did not begin to appear on plantations in large numbers until the early twentieth century (Knipmeyer 1956, 82). In time, the house moved into farming areas and rapidly spread among plantations as its popularity was attributed to the ease and low cost of construction.

The shotgun house is a well-known house type, with origins tracing through African American connections to the Caribbean and to West Africa.[28] It is widely distributed throughout Louisiana as a folk house, and its distribution as a quarter house and even an urban house type takes it far up the Mississippi River and its tributaries to Arkansas, Tennessee, up the Ohio River to Louisville, Kentucky, and well into parts of southern Ohio. There are even shotguns in working-class neighborhoods dating from the

FIG. 2.23. *A shotgun house of the fisher-trapper culture serves as a nineteenth-century precursor type to plantation shotgun quarter houses. Note the cypress shake roof. Assumption Parish.* Photograph, 1967

turn of the twentieth century in Knoxville, Tennessee. How do we make the distinction between its function as a folk house and as a quarter house? As a folk house, the shotgun reflects the cultural intentions and identity of its owner-builders. But as a quarter house, it functions as a cheaply built structure that normally does not correlate directly with a diagnostic cultural identity. That is to say, as a folk house, the shotgun can be identified with French Louisiana's New Orleans urban identity and, particularly in the rural areas, with small farmers, trappers, and fishermen. As a quarter house, the shotgun had no such direct ethnic identity and could be found on French as well as Anglo-American plantations.

The Bungalow as a Quarter House

The bungalow house, also known in some forms as a double shotgun, on sugar plantations followed much the same geographic distribution as the shotgun house in the sugar region, principally along Bayous Lafourche and

Teche and along the smaller bayous in Saint Mary and Terrebonne Parishes. The bungalow house plan was two rooms wide and three or more rooms deep, with front- and rear-facing gables. Orientation to the road or bayou was the same as for shotguns, perpendicular to the transportation route. The front porch served as an important signature trait of the bungalow because on some houses part of the porch was enclosed and used as a bedroom; on others the porch roof simply formed an overhang for a wide, open porch. The bungalow had either one or two front doors, depending on whether it had a bedroom built into the front porch. Bungalows with a full front porch and a doubled floor plan were also called *double shotguns* (Oszuscik 1992a, 159–66). The bungalows on Louisiana's plantations were very different from and should not be confused with the California bungalow designs from the early twentieth century (Lancaster 1986, 79–106).

Plantation bungalows in Louisiana are believed to have derived from shotgun houses and, like them, are closely associated with urban houses in New Orleans and with rural small-farming and trapper-fisher areas. If the bungalow is indeed derived from the shotgun, we can detect similarities between the two house types. The bungalow has the same length as a shotgun house and has rooms aligned along the long axis. The bungalow has a front-facing gable and a sloping-roof porch, as seen on some shotguns. However, the bungalow is two full rooms wide and appears in plan like the urban double shotgun (Oszuscik 1992a, 159–66). It should not be called a "double-barreled shotgun" or, in the modern mobile-home vernacular, a "doublewide." Bungalows arrived on both French and Anglo plantations much like the shotgun houses seem to have done. Ease of construction and low cost were reasons for the rapid adoption of bungalows. The house type was used for quarter houses, but its size was especially attractive for housing overseers and sugar factory personnel on many plantations. In at least one instance, a large Louisiana bungalow, a double shotgun, was the house of choice as the plantation mansion at Cedar Grove Plantation (see chap. 8).

Other Dwellings for Laborers

On any landscape miscellaneous anomalies in dwellings do not match the typical categories. They come in different sizes and do not have any type affiliation with the four quarter house types already described. One such building is a multifamily structure that has the appearance of an apartment

Fig. 2.24. *A bungalow quarter house on Mulberry Plantation in Terrebonne Parish dates from the early twentieth century. Mulberry Plantation once had twenty-five bungalows like this one, but none remain.* Photograph, 1967

building or motel. Most are quite old, dating to the last century, and are built so that three or more apartments are under one roof. They have more than two front doors and more than one chimney, and some are two-story structures. In the 1960s, Smithfield Plantation had two or three multifamily houses that had the appearance of an old motel. Poplar Grove Plantation in West Baton Rouge Parish had a nineteenth-century brick apartment building that could house eight families. Over the years, the plantation declined severely but still retained most of its quarter houses; seven were still occupied in 1993. In September 1995, Poplar Grove and all of its contents except the land were auctioned off, and the fate of this plantation marks the end of an era (Frink 1995, 11–16). In the 1960s, perhaps a dozen plantations had bunkhouses and barracks for their bachelor and seasonal workers. Informants at that time told me that they had seen more bunkhouses on larger plantations in the 1930s. A good example was the one at Ashland, which looked much like a misplaced bunkhouse from a Texas ranch (see fig. 6.8).

Construction Techniques and Materials Used in Quarter Houses

Throughout the history of plantations and all stages of early development, construction methods and materials used on houses in the quarters were aimed exclusively at economy and not appearance or comfort. Quarter houses rarely were well constructed because they were placed at the lowest stratum in plantation housing hierarchy. Earliest construction techniques employed *pieux en terre,* stakes in the ground (Cruzat 1925, 664). Long, rough cypress slabs or posts were driven vertically into the ground to a depth of two feet or more to create a palisade-like wall. Interstices were chinked with mud and Spanish moss, rags, or any material that could be found. A ridge pole at the apex of the roof extended along the long axis of the hut. Joists were pegged to the tops of the posts. Roof materials were either palmetto thatch or bark shingles (Robin [1807] 1966, 122–23). Colombage, the early half-timbering construction technique, was widely used on permanent houses throughout French Louisiana, but I know of only one instance where the technique was used on quarter houses (Vlach 1993, 162, 182).

Another construction method used all-wood or lightweight balloon framing. The interior frame was completely constructed of small-dimension sawn lumber. Weatherboarding was made from sawn planks nailed horizontally or vertically as board-and-batten construction. Earlier roofs that had been covered with shakes were replaced with tin roofs in the twentieth century. The balloon frame practice probably did not enter Louisiana until after the Civil War, when planters and carpenters could obtain sawmill lumber and begin to adopt all-wood methods of construction (Knipmeyer 1956, 114). As sawmills began to appear in Louisiana, the accessibility of cheap sawn lumber meant that low-cost quarter houses would proliferate throughout the plantation landscape. Despite changes in house types, construction techniques, and materials, the tight settlement pattern in the quarters remained constant.

Contemporary siding on quarter houses includes horizontal planked weatherboard, vertical board and batten, tar paper, asbestos shingles, and some combination of these. Horizontal weatherboard has been widely used on older, small Creole house types and bungalows. Board-and-batten construction has vertical six-inch boards butted next to each other with two-inch boards nailed vertically over the interstices. The practice dates from the late nineteenth century and was a common exterior wall covering used on all kinds of plantation buildings, such as barns, sheds, blacksmith shops, stores, and many quarter houses (Bouchereau and Bouchereau 1883, xii). It is still found on some of the older nonfarm shotgun houses in the region. Tar paper, earlier in green and later in the yellow brick pattern, was ubiquitous among quarter houses in the 1960s. Some houses had a palimpsest of sidings that began with board and batten and later was covered over with wide green tar paper, then yellow brick tar paper, and then perhaps with asbestos. The asbestos siding was an innovative fireproof material that arrived in the 1950s and was used on new houses and especially to upgrade overseer's and sugar factory worker's houses. Bungalows were higher in the hierarchy of quarter houses and were the ones most likely to be regularly painted or to have asbestos siding.

Roof coverings have varied over the past two centuries. Late-eighteenth- and early-nineteenth-century quarter houses had palmetto-thatched roofs (Cruzat 1925, 664). During much of the nineteenth century, wooden cypress shakes or shingles were the predominant roofing material for quarter houses and nearly all other plantation buildings. By the 1960s few cypress shake roofs remained, and all other extant buildings had been reroofed with other materials. The tin roof, made from galvanized corrugated sheet steel, entered the region in the 1920s and continues to be the ubiquitous roof type. Common tar shingles were not as widely used on plantation buildings, especially in the quarters (Rehder 1971, 185–87).

Brick construction for quarter houses was extremely rare. In my thirty-five years of plantation investigation, I have found only two plantations that had about 25 percent of the quarter houses in brick. Those two plantations still survive, and it is comforting to know that representative brick quarter houses are preserved at Blythwood Plantation and at Evan Hall Plantation. Poplar Grove Plantation once had several brick structures, one of which was a multifamily, two-story building near the mill site. Brick sugar factories were built in the nineteenth century, and a few mansions were also in brick. Foundation piers for nearly all plantation structures were brick in the nineteenth and twentieth centuries. However, it is fair to say that brick construction was certainly uncommon throughout the plantation region for whole-house construction, and it was extremely rare in the quarters.

Cinder-block construction became the most modern method and material in the sugar region. Beginning in the 1950s and 1960s, the use of cinder blocks for new house construction arrived on plantations, for overseer's

FIG. 2.25. *This rare, early-nineteenth-century small Creole quarter house had brick construction. Recently, the house was renovated into a permanent residence. Blythwood Plantation, Bayou Goula, Iberville Parish.* Photograph, 1967

and mill workers and for replacement houses in the quarters. In the mid-1960s it was doubtful that dilapidated quarter houses would ever be replaced at all, as most plantations had surplus dwellings. But Evan Hall, Cinclare, Minnie, Madewood, Westfield, Oaklawn, and other plantations added cinder-block houses to their quarters because plantation managers decided to build new houses as an incentive to attract and keep better skilled laborers on the plantation. At the time, I made the cryptic statement, "As abandonment of older quarter dwellings increases, the need for new construction will be increased" (Rehder 1971, 189). I believed that plantation settlement would evolve into plantations with fewer but nicer quarter houses, but that the quarters settlement would endure. I was wrong. Over the next three decades, I would witness the wholesale destruction of plantation morphologies, especially in the quarters. Sadly, even on some plantations with incentive cinder-block houses, like Minnie Plantation, housing conditions were no longer important. On-site workers were no longer needed or welcome to live there, and the cinder-block quarter houses had become obsolete, no matter how well they had been built.

Quarter House Yards

The quarter house yard, as described here, applies to all quarter house types in the sugar region. The careful observer of the landscape does not overlook such things as trash piles and outhouses, for it is from such ignoble stuff that entire cultures are reconstructed in archaeology. Proceeding from front to back in a cross section of a house site, one encountered a dirt road, ditch, path, front yard, quarter house, back yard, and back ditch. Most yards were not fenced in front, but many were fenced in back. Fence materials were scrap lumber, sheet metal, chicken wire, or smooth woven wire, and sometimes any or all in combination were used in the same yard. Grass rarely completely covered the front or back yard because it was trampled by foot, tractor, or automobile traffic. Some families swept the yard to keep it free of grasses and weeds. This very old Southern trait is an African American tradition that traces to West Africa.[29] In the absence of lawn-mowing equipment, yards were kept as open sand or clay surfaces free of vegetation, vermin, and lurking snakes (Rehder 1992, 102). Yard trees were frequently chinaberry and weeping willow, mostly located in the back yard. The chinaberry tree was thought to be a qualified culture trait of French Louisiana, but it alone cannot identify French plantations (Kniffen 1963, 291–99).

Gardens for laborers date to the earliest plantation times. Small gardens were encouraged for economic reasons, and in the antebellum period gardens helped alleviate the expense of feeding slaves.[30] On plantations in the Caribbean, traditional gardens were called provisioning grounds on English plantations and *jardins à nègres* on French plantations and were vital for feeding the entire plantation population. In the 1960s nearly every quarter house on Louisiana's plantations had a vegetable garden or at least a space in the back yard for one. The gardens were planted twice a year for a winter garden and a summer garden. Winter gardens included such cold weather vegetables as cabbage, collards, mustard greens, and turnips. Rarely were beets or carrots grown in the winter garden. In summer, the same plot yielded tomatoes, okra, onions, garlic, melons, beans, and squash. Sweet corn was never grown in quarter house gardens because it "came from the field." This trait dated from the time when corn was cultivated in the field for livestock feed and as food for slaves (Rehder 1971, 192). The tradition of field corn on the plantation was still in place in the 1960s, indicating a relic trait not far removed from the time when humans and animals were both fueled by corn.

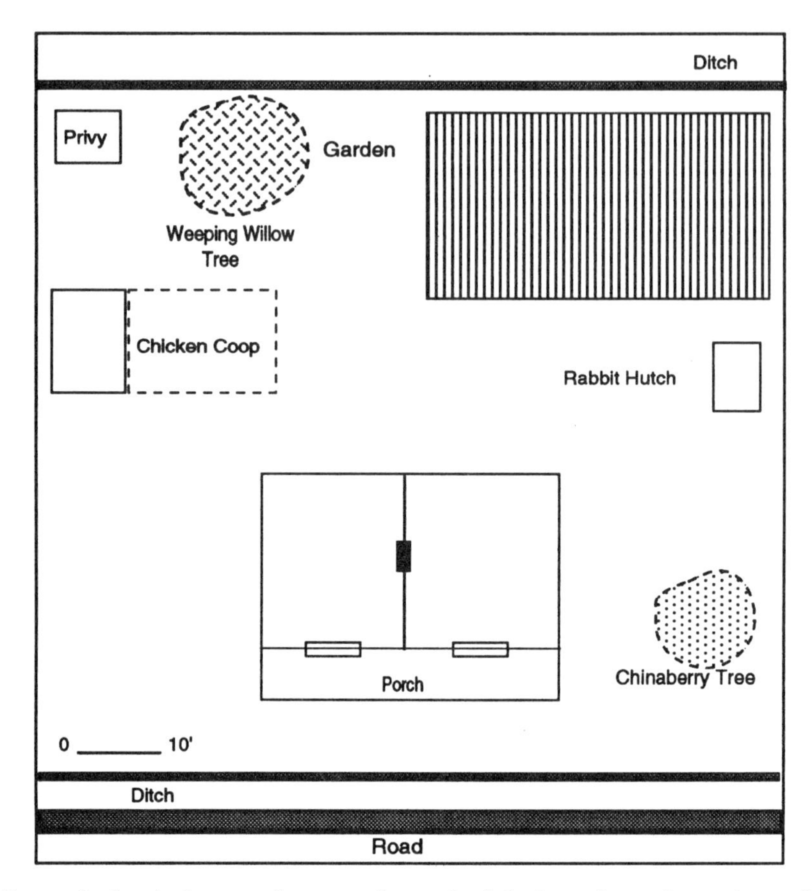

FIG. 2.26. *A typical quarter house yard contained the house located near the road, numerous chinaberry and weeping willow trees for shade, a chicken coop, a vegetable garden, perhaps a rabbit hutch, and always a privy near the back ditch.*

The quarter house back yard usually had several small outbuildings. Most plantations had outhouses or privies behind each quarter house, situated at the far corner of each yard. A ditch marking the back line of the quarters accommodated seepage. Other small outbuildings were chicken coops and rabbit hutches made with thin lumber frames and covered with chicken-wire fence. Crude storage buildings built with scrap lumber and spare tin roof sheet metal completed the back-yard ensemble (Rehder 1971, 193).

Overseers' Dwellings

Nineteenth-century Southern plantations with more than thirty slaves required a supervisor or overseer, who was furnished exclusive housing at a specific location related to the quarters (Vlach 1993, 135–41). Overseer's houses were larger, better maintained, and specifically located near but not always directly in the quarters. On some plantations, they were located nearer the sugar factory; on others the location was closer to but not on the mansion site. In the 1960s, a typical Louisiana sugar plantation overseer's dwelling was still so located and occupied by an overseer, or "white boss." The dwelling was usually not much different in type from other quarter houses, but its location, size, and upkeep often set it apart from other dwellings in the quarters. Overseer's houses on plantations along the Mississippi River were either small Creole or attached-porch types. Elsewhere in the sugar region, overseer's houses were mostly bungalows and small Creole types. Each plantation had at least one designated overseer's house, and some had five or more. Included in this group of dwelling types were the houses for sugar factory personnel, such as engineers, boilermen, and

FIG. 2.27. *The back of the quarters can be as interesting as the front at the late-nineteenth-century Laurel Valley Plantation. Lafourche Parish.* Photograph, 1995

FIG. 2.28. *The overseer's house was usually larger and better maintained than houses in the quarters. Smithfield Plantation. West Baton Rouge Parish.* Photograph, 1966

chemists, and others, like the storekeeper, who had a status ranked above field hands. Their dwellings were often identical to the overseers' houses, and unless one knew the occupants, it was difficult to determine the exact occupation of the inhabitants. A separate location in relation to the sugar factory, store, or mansion might offer a clue, but the location was not always predictable.

Yard settings for overseer's and factory workers were better kept, were more often fenced, and could easily distinguish the house site from those of the field hands in the quarters. Predominant yard trees were chinaberry and weeping willow, with the occasional crepe myrtle and fruit tree added. Grassy lawns and flower gardens enclosed by fences marked the front yard. In back, well-kept vegetable gardens with nearly the same garden varieties as found in the quarters were common. Garages and sheds, rather than outdoor privies, were more apt to be in the back yard. Indoor plumbing was not exclusive to the mansion, as it was frequently found among overseer's and factory worker's houses, but it was extremely rare among quarter houses.

The Quarters in a Social Context

My purpose in studying the quarters from the standpoint of morphology was that it told an interesting story about the landscape. But there was more to the story than could be seen. The quarters also had a contemporary social story that had not been heard recently, particularly from an academic perspective.[31] Sweltering in the heat of July and August in 1964, I made three traverses, covering about 750 miles in search of cultural landscapes in the plantation quarters. I knew little about plantations then or about the people who lived in them. I objectively searched the landscape along each traverse, expecting that the work would lead me to a reasonable sample of plantations in the region. I created a questionnaire seeking such information as the plantation name, the presence of a sugar factory, the quarters, house types, numbers of dwellings, and so on. Then something beyond my scope of inquiry interested me: "How many quarter houses are occupied by white families and how many have 'colored' families living in them on the plantation?" This was an innocuous, probably quite unnecessary question, but one that came up in the context of "while I'm out here I might as well check on the population." Later I thought that perhaps a social hierarchy would be reflected in housing types and conditions, a hierarchy that could be interpreted in the context of cultural attitudes about plantation life. Now the data that lay in my old field notes come to life in table 2.1, showing each plantation in the traverse and the number of dwellings according to the racial composition of the occupants. In 1964, fifty-six plantations in the survey had 1,269 quarter houses, which averaged 22.6 houses per plantation. Of these, 231 had white families living in them and 1,038 quarter houses were the homes of African Americans.

The plantations remaining on the landscape in the 1990s have fewer quarter houses, fewer workers of both races, and many fewer if any whites living on plantation premises. Between 1964 and 1997, a pattern emerged in the occupancy of extant quarter houses. A stratigraphy evolved from the 1964 pattern that had whites occupying the mansion, the manager's house, the storekeeper's house, the overseers' houses, and most if not all of the sugar factory workers' houses. By the 1970s and 1980s, all whites, except for some still living in the mansions and managers' houses, had left the plantations. Former white overseers' houses, almost all of the sugar factory workers' houses, and even the storekeepers' houses were now occupied by blacks. It was a form of social payment to the better skilled black workers to

TABLE 2.1

Racial Composition of Resident Laborer Households from a Sample of Fifty-nine Louisiana Sugarcane Plantations in 1964

	No. Resident Laborer Households	
Plantation	*White*	*African-American*
Mississippi River		
Smithfield	12 (36%)	21 (64%)
Orange Grove	4 (10%)	37 (90%)
Barrowza	4 (18%)	18 (82%)
Westover	8 (20%)	31 (80%)
Allendale	1 (3%)	27 (97%)
Catherine #1	9 (36%)	16 (64%)
Poplar Grove	15 (30%)	35 (70%)
Cinclare	12 (26%)	34 (74%)
St. Delphine	1 (10%)	9 (90%)
Myrtle Grove	3 (10%)	28 (90%)
Evergreen	3 (15%)	17 (85%)
Dunboyne	3 (50%)	3 (50%)
Upper Einer	1 (25%)	3 (75%)
Tally Ho	5 (31%)	11 (69%)
Catherine #2	6 (20%)	24 (80%)
Cedar Grove	2 (6%)	34 (94%)
Texas	2 (7%)	28 (93%)
Cora	13 (48%)	14 (52%)
New Hope	4 (21%)	15 (79%)
Evan Hall	5 (14%)	33 (86%)
Riverside	1 (10%)	9 (90%)
Salsburg	1 (7%)	14 (93%)
Lauderdale	5 (42%)	7 (48%)
Minnie	1 (10%)	9 (90%)
St. James	2 (7%)	27 (93%)
Pikes Peak	3 (21%)	11 (79%)
Laurel Ridge	2 (8%)	22 (92%)
St. Joseph	3 (15%)	18 (85%)
Armant	11 (34%)	21 (66%)
Whitney	5 (34%)	10 (66%)

TABLE 2.1 *(continued)*

Plantation	*No. Resident Laborer Households*	
	White	*African-American*
Columbia	3 (13%)	20 (87%)
Gold Mine	2 (10%)	20 (90%)
Bayou Lafourche		
Belle Terre	2 (12%)	16 (88%)
Belle Alliance	1 (7%)	13 (93%)
Sweet Home	2 (17%)	10 (83%)
Elmfield	1 (10%)	9 (90%)
Madewood	2 (13%)	13 (77%)
Rosedale	2 (11%)	16 (89%)
Woodlawn	3 (12%)	23 (88%)
Laural Grove	4 (12%)	28 (88%)
Dixie	3 (25%)	8 (75%)
L.T.	1 (17%)	5 (83%)
Greenwood/Elmer	2 (7%)	28 (93%)
White	1 (10%)	9 (90%)
Leighton	8 (15%)	25 (85%)
Georgia	6 (23%)	20 (77%)
Abby	2 (12%)	15 (88%)
Coulon	3 (15%)	17 (85%)
Rienzi	4 (19%)	17 (81%)
Laurel Valley	4 (9%)	42 (91%)
Bayou Teche		
St. John	6 (30%)	14 (70%)
Albania	2 (14%)	12 (86%)
Alice C.	3 (9%)	32 (91%)
Kraemer	2 (15%)	11 (85%)
Justine	5 (45%)	6 (55%)
Shadyside	2 (8%)	23 (92%)

keep them on the plantation. By 1997, nearly everyone, black or white, would be leaving the plantation quarters because maintaining living quarters was no longer a priority for plantation management. When plantation owners and their insurance companies realized the liability of maintaining a quarters, they wanted out of the housing business. It was too dangerous.

The better workers were gone already, and the responsibility of housing retirees and welfare families in the quarters was not in the best interest of the plantation. The era of traditional patriarchical concern, care, and oversight for plantation laborers in the quarters was over.

Sugar plantations have been sites for human habitation in an ecumene that reflects culture and form in southern Louisiana. Mansions, settlement patterns, and dwellings in the quarters expressed a unique range of settlement elements over time and space. Culture-bearing builders of sugar plantations inscribed into the Louisiana landscape residual, culturally diagnostic landscape traits. Categorically, French Creole mansions represented initial occupance patterns of eighteenth- and nineteenth-century French planters. Tidewater and architect-designed mansions were contributed by nineteenth-century Anglo-Americans. Plantation settlement patterns as internally arranged buildings became diagnostic landscape traits from their geometry and historic cultural associations. Mansion types and agglomerated settlement patterns together served as the most reliable diagnostic traits for sugarcane plantations in Louisiana. French plantations marked by Creole mansions and linear settlements were the result of French Creole planters' efforts to initiate plantations along the Mississippi River. Initial occupance, as revealed in the examples of Armant, Cedar Grove, and Whitney Plantations, testify to initial French ownership and directives for construction. Anglo plantations, identified by Anglo mansions and block-shaped settlements, were established on the western, southern, and northern extremities of the sugar region. The examples at Ashland, Oaklawn, and Madewood Plantations provide sufficient visual and documentary evidence to support their Anglo identity. Reflecting the cultures of their initial owner-builders, Louisiana's sugar plantations also pointed to extraregional source areas as evidence of their origins. House types and settlement patterns from the French West Indies and the Lowland and Tidewater South accompanied migrating planters and served as keys to cultural diffusion.

CHAPTER 3

The Morphology of the Functional Plantation Landscape

The Sugar Factory

For centuries, the sugar factory has been called the sugarhouse, sugar mill, boiling house, *central, ingenio,* and sugar factory. The structure is one of the most striking cultural features on the sugar plantation landscape. This immense building measures 300 feet long by 150 feet wide by 75 feet high and clearly is the tallest building on the plantation. From a distance, the sugar factory, with its shiny galvanized steel siding and roofing and tall black smokestacks, gives the appearance of a imposing industrial factory-in-the-field. Since the 1970s, sugar factory owners have replaced tall smokestacks with much shorter but highly efficient precipitator stacks to meet air quality standards required by the Environmental Protection Agency. Even so, sugar factories still loom out of the mist on billiard table flat terrain.

Grinding season is the frantic harvest time between October and December. It is always the busiest time of the year for plantations, especially at

FIG. 3.1. *A sugar factory in full operation at grinding season takes on a dynamic, noisy life of its own. Compare this photograph of the Saint James Sugar Cooperative in 1996 with figure 3.2 of the Saint James factory in 1853 at the same plantation site. Saint James Sugar Cooperative. Saint James Parish.* Photograph, 1996

the sugar factory. Sugarcane stalks are cut in the fields, loaded into cane carts and trucks, and transported to the sugar factory, where a continuous stream of canes are washed, crushed, and fed through banks of heavy grinding rollers to extract cane juice in the sugar mill. The cane juice passes through several processing steps of clarifying and boiling to reach the desired primary product, crystalline brown raw sugar. By-products of molasses syrup and bagasse, the residual cane stalk pulp, are diverted to storage areas. The factory operates continuously–twenty-four hours a day, seven days a week–for the short, three-month grinding season. During the other nine months of the year, workers repair broken machinery, clean boilers, refurbish the twelve to twenty-four rollers in the mill, and in general prepare the sugar factory for the next grinding season.

Forty-four sugar factories and 202 plantations were on Louisiana's plantation landscape in 1969 (Rehder 1971, 196). The ratio then was approximately one factory for every four plantations. In 1997, the ratio was much the same, with nineteen sugar factories and about 80 plantations. Ten surviving factories are cooperatively owned; the other factories are privately owned. On the positive side, plantations without sugar factories do not bear the heavy expense of maintaining costly sugar factory equipment. In the nineteenth century, every plantation had a sugar mill and boiling house, collectively called a sugarhouse, so at that time there was a one-to-one correlation between plantations and sugar factories. Technologically more efficient but fewer sugar factories began to appear around the turn of the twentieth century. Known in the Caribbean as *centrals,* these centrally located sugar factories in Louisiana later collectively served a wider number of sugar farmers in the region.

Numbers of sugar factories and production of raw sugar followed a divergent paradigm over time. Factory numbers illustrated an evolving trend. Seventy-five sugar factories in 1801 grew to 691 by 1830 and reached the maximum figure of 1,536 operations by 1849. One year later, Louisiana had 1,495 sugar factories, of which 907 had steam-powered mills and 588 had horse-driven mills (Champomier 1850, 43). The Civil War dealt a heavy blow to the sugar industry; by 1865 only 188 factories were clinging to life. By 1869, four years after the Civil War, 817 sugar factories survived in Louisiana (Bouchereau and Bouchereau 1896, 81). The number of factories in 1880 had reached a postbellum high of 1,144, but by 1882 the number had declined to 910 (Bouchereau and Bouchereau 1882, 54). The declining trend in the number of factories began as the *central* concept gained acceptance, so that by 1898 there were 347 sugar factories. By 1910 there were 214 factories.

The declining number of sugar factories was not a sign of declining productivity; on the contrary, it was a clear indicator of higher milling efficiency. Daily capacity for mills in the 1880s was 300 tons, but by 1900 milling capacities had improved to 700 tons and even to 1,500 tons per day (Sitterson 1953, 262–63). The declining number of mills, however, was exacerbated by the mosaic disease that struck the sugar region in the 1920s and the economic chaos that came with the Great Depression in the 1930s. In 1922, there were some 112 mills of various sizes and capabilities in the region, but in 1926 only 54 were functioning because of the mosaic disease. From these disasters, recovery was slow, and with it emerged a smaller number of surviving sugar factories. In 1930 the operations were back to 70 sugar factories.[1] The number of sugar factories in Louisiana was 43 in 1970, 24 in 1980, and then 20 from 1990 to 1996. In 1997, only 19 sugar factories

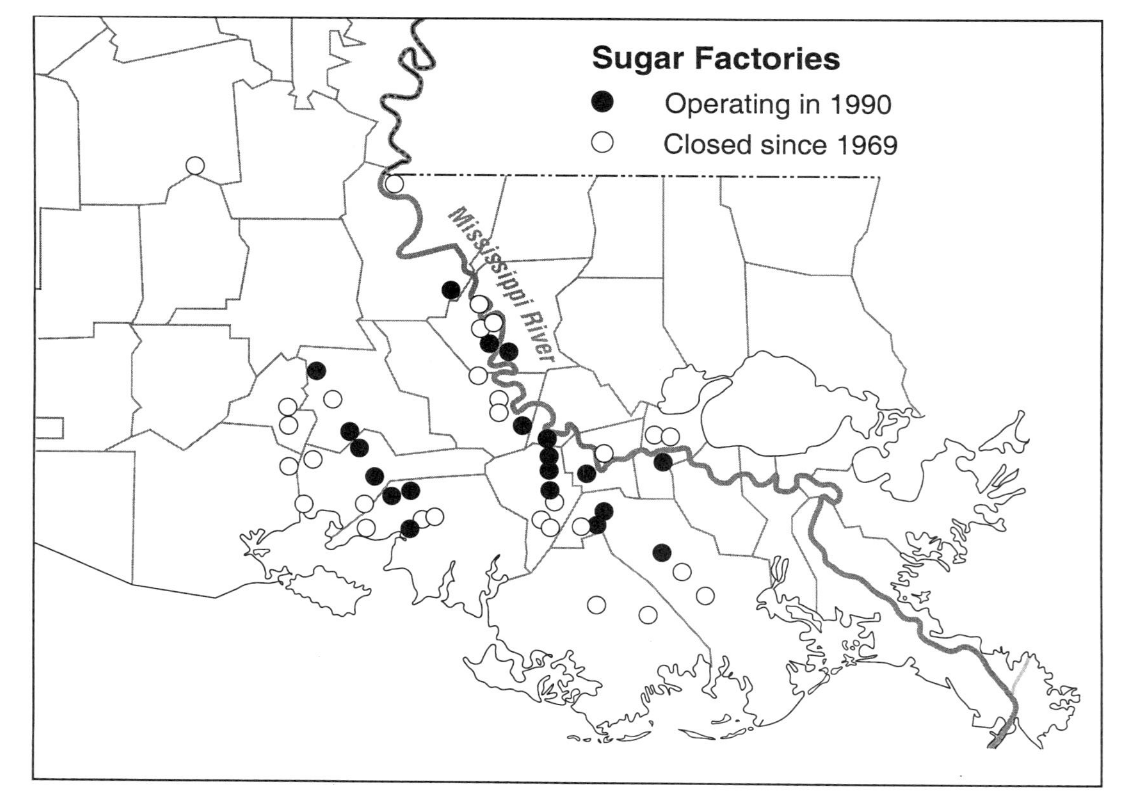

Map 3.1. *The spatial pattern of Louisiana's sugar factories depicts declining distributions since 1969.*

TABLE 3.1

Louisiana Sugar Factories and Raw Sugar Production, 1801–1995

Year	*No. of Sugar Factories*	*Production (tons)*[a]
1801	75	2,500
1810	91	4,833
1824	193	–
1827	308	–
1828	691 (120)[b]	50,590
1830	691	37,500
1833	–	37,500
1840	–	60,000
1841	668	62,500
1844	762 (408)[b]	114,999
1845	1,240	159,849
1849	1,536	134,921
1850	1,495	115,484
1853	–	247,578
1856	–	36,988
1859	1,308	127,019
1861	1,291 (1,027)[b]	264,159
1865	188	10,401
1867	769	42,128
1869	817	49,707
1870	1,100	72,372
1880	1,144 (871)[b]	136,512
1889	746	143,745
1894	449 (446)[b]	355,382
1900	300	302,778
1909	214	301,763
1910	214	294,905
1922	112	295,095
1926	54	47,000
1929	–	199,609
1930	70	183,693
1933	68	205,000
1937	92	188,000
1940	74	235,095

(continued)

TABLE 3.1 *(continued)*

Year	*No. of Sugar Factories*	*Production (tons)*[a]
1942	–	397,000
1944	–	369,000
1946	–	331,000
1948	–	393,000
1950	–	451,000
1952	–	451,000
1954	–	478,000
1956	–	429,000
1958	47	443,000
1959	47	440,000
1960	46	470,000
1961	46	650,000
1962	46	472,000
1963	46	759,000
1964	48	573,000
1965	47	550,000
1966	46	562,000
1967	46	740,000
1968	46	669,000
1969	44	537,000
1970	43	602,000
1971	43	571,000
1972	43	660,000
1973	39	558,000
1974	37	594,000
1975	36	640,000
1976	35	650,000
1977	33	668,000
1978	28	550,000
1979	25	500,000
1980	24	491,000
1981	23	712,000
1982	21	675,000
1983	21	603,000
1984	21	452,000

TABLE 3.1 *(continued)*

Year	*No. of Sugar Factories*	*Production (tons)*[a]
1985	21	532,000
1986	21	671,000
1987	21	731,000
1988	21	797,000
1989	20	844,000
1990	20	438,000
1991	20	762,000
1992	20	868,000
1993	20	890,000
1994	20	1,018,000
1995	20	1,075,000

Sources: Champomier 1844–50; Bouchereau and Bouchereau 1882–96; Buzzanell 1993; Farr 1960; Gilmore 1933–40; Gray [1932] 1958; Heitmann 1987; Sitterson 1953; U.S. Department of Commerce, Census of Agriculture; USDA Economic Research Service, Sugar and Sweetener Reports.

[a] For comparative reasons, all production has been converted from hogsheads @ 1,000 pounds to tons @ 2,000 pounds.

[b] Steam-powered sugar mills

were still functioning in Louisiana.[2] In Louisiana's sugar industry today, 10 sugar factories are cooperatively owned and operated by groups of sugar farmers. The other 9 are private and for the most part are corporately owned. Louisiana's sugar factories are modern, highly efficient, high-capacity, environmentally regulated enterprises that effectively process Louisiana's sugar production, the highest in the state's history.

The Nineteenth-Century Sugar Factory

The nineteenth-century sugar factory was a large, brick structure that measured about 100 to 160 feet in length by 60 or more feet in width. The building had one or two appendages placed at right angles to the long axis of the main structure, thus creating an L- or T-shaped structure. The appendages served as purgeries and for raw sugar storage (Silliman 1833, 30). Functional areas included the sugar mill, cane juice clarifiers, boiling apparatus (originally as open kettles and later as vacuum pans), crystallizing troughs, and purging facilities. The mill was located at the far end of the structure nearest the cane fields. This arrangement conveniently allowed

FIG. 3.2. *A brick sugar factory in 1853 at the Saint James Plantation was representative of a "sugar house in full blast." Compare this 1853 factory with figure 3.1, the contemporary Saint James Sugar Cooperative on the same plantation site.* Reproduced from *Harper's New Monthly Magazine* 1853, 7:761

cane to be sent directly from field to mill. Ironically, sugar factories traditionally still follow this same spatial arrangement, even though larger quantities of cane from other plantations are trucked through the front of the plantation settlement and taken around to the back of the sugar factory to the mill yard.

Earliest sugar mills were the familiar three-roller type, consisting of a wooden frame and three vertical cylinders driven by animal power. Cylinders in the late 1700s were solid oak and measured about three feet long; later ones were covered with metal sheets. By 1820, solid metal cylinders, better known as rollers, began to be used. Grooves cut parallel to the axle enabled the roller to take a firm grip on the canes as they passed through the mill. Cogwheels were fixed at one end of the rollers, with the middle roller receiving the motive power. By the 1820s, rollers were being set horizontally in the mill framework, arranged into a triangular formation with one roller set atop two rollers below, and were much larger, measuring five feet long or more. Power applied to the top roller was transferred to the lower ones via cogs.

Motive power for all eighteenth-century mills and for many nineteenth-century ones in Louisiana was supplied by animals. Normally, oxen were the animals of choice (hence the name *cattle mills*), but horses and mules became more widely used in the nineteenth century. The wind- and water-powered mills that were so important to the Caribbean sugar industry for so long were entirely absent from Louisiana.[5] Power from cattle was supplied to the center roller on vertical three-roller mills. Two to four long poles called sweeps were attached at one end to the center roller. From the other end of the sweep, the harnessed pair of oxen turned the mill. On horizontal roller mills, a vertical axle was geared to the top roller, which in

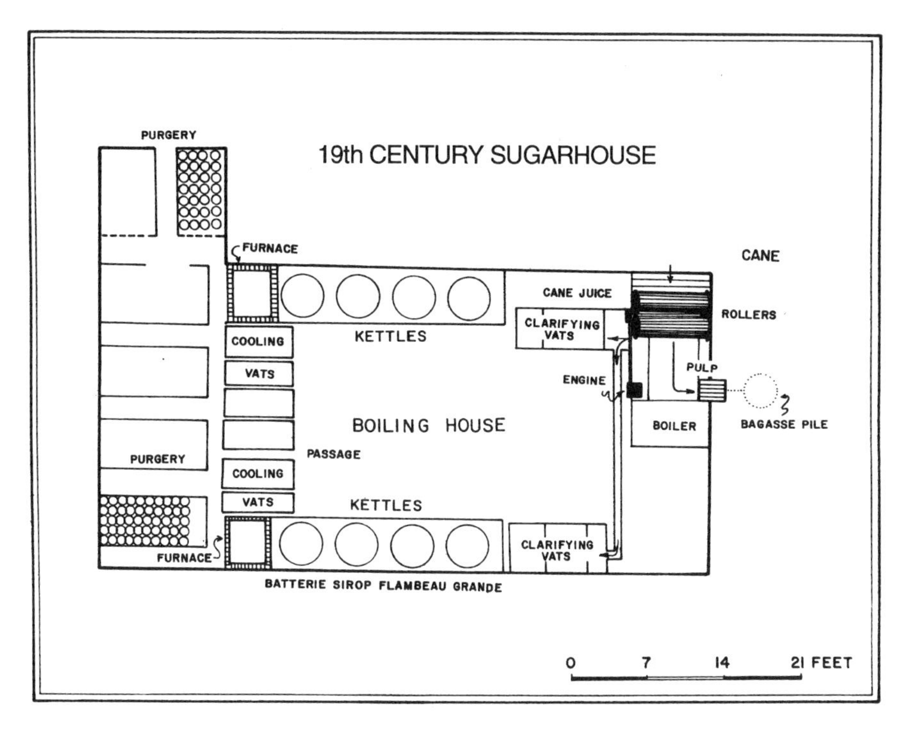

FIG. 3.3. *The floor plan of a nineteenth-century sugar factory demonstrates the flow of sugar processing from the sugar mill* (right) *to the two rows of open kettles in the boiling house* (center) *and to the cooling vats and purgery* (left). Source: Silliman 1833

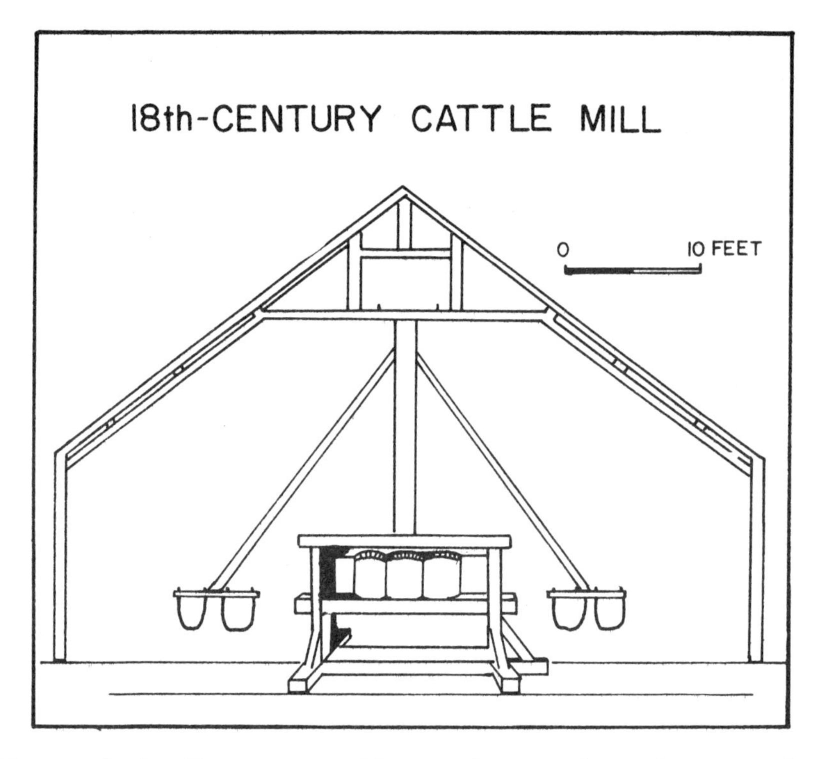

FIG. 3.4. *Cattle mills were powered by oxen, horses, mules, and sometimes humans. The mills were used on sugar plantations in Louisiana between 1742 and 1905. Cattle mills have been used widely on plantations in the Caribbean since the seventeenth century.*

turn was geared to the two rollers below. One end of the sweep was attached to the vertical axle, and the other end to the animals (Silliman 1833, 31). Until the steam-powered mill arrived in Louisiana after 1822, cattle mills dominated the grinding function, and they continued to be utilized on small cane farms well after the Civil War. By the 1880s, some 273 sugar operations were still grinding cane with animal power; by 1900 the number had dropped to 5, and after 1905 there were no more (Browne 1938).

In Louisiana's delta sugar country, the animal-powered mill no longer exists operationally. However, in other parts of the South, especially in Mississippi, Alabama, Arkansas, the Carolinas, Georgia, and Tennessee, mule-powered mills still operate as syrup mills. Such extraregional mills extract the juices of sugarcane and especially sorghum cane, juices that are made only into syrups. These mule-powered syrup mills in the Lowland and Upland South belong to separate culture regions and landscapes that are quite distant spatially, temporally, and culturally from the sugar plantation's form and function on the southern Louisiana landscape. That is not to say that syrup mills were ever absent from Louisiana. On the contrary, Louisiana produced 4,125,083 gallons of syrup in 1909, which was second only to Georgia's 5,533,520 gallons among the twelve syrup-producing states in the United States (Yoder 1919, 2). In 1933, Louisiana's sugarcane region had ten syrup mills, but as a wartime effort in 1941 twenty-three mills were exclusively producing syrup (Gilmore 1933; 1941). In the late 1960s, Romano's and Steen's were two syrup mills still in operation. The C. S. Steen Syrup Mill in Abbeville, Vermillion Parish, remains the best known commercial syrup operation in southern Louisiana.

Steam-powered mills, with their more efficient power, gear systems, and pressure controls, began to enter Louisiana in 1822. By 1828, there were 705 sugar mills, of which 67 were powered by steam. By 1860, 992 of the 1,308 sugar mills in the region were steam powered.[4] Since steam mills were more effective with horizontal rollers, the number of mills with vertical rollers declined. Besides their efficiency, another reason for the rapid acceptance of the steam mill was the introduction of ribbon cane in 1825. The hard rind of this new cane required the more powerful steam-driven mills for effective crushing (DeBow 1856, 275). Steam mills were expensive to purchase and maintain, so small planters continued to use animal-powered mills until well after the Civil War period.

Between 1905 and 1950, all sugar factories of any consequence in Louisiana were powered by steam. Horizontal rollers were set into groups of three, with each mill having four to eight sets, called *tandems,* aligned in a row. Twelve to twenty-four rollers, with each roller weighing about one ton, truly pulverized, if not puréed, the canes passing through the mill. Contemporary sugar factories now have gas turbines for power, computer-controlled automation for some of the sugar-making processes, and additional equipment controls to meet clean air and water regulations.

Processing fundamentals in the manufacture of raw sugar during the 1800s differed little from those in modern sugar making. Sugarcane stalks were ground at the mill; juice was extracted, clarified, boiled, and crystallized; and raw sugar was separated from molasses. Essentially these same fundamentals have been used for more than a millennium throughout the sugar-making world. Even as technological changes in equipment and processing techniques improved, procedures in Louisiana still followed a special requirement to rush to process the cane before first frost. No other sugarcane region in the world had to deal as much technically or emotionally with this recurring climatic problem.

Nineteenth-Century Sugar Making

One of the best descriptions of antebellum sugar making came from an early sugar manual published in 1833 by Benjamin Silliman. The process began with hand-cut cane arriving from the fields in large, mule-drawn, two-wheeled carts. At the mill yard, canes were hand carried from a massive pile to be fed into the turning rollers of the sugar mill. After the grinding stage, bagasse cane pulp trash passed down a trough to a waiting cane cart to be hauled away to a trash pile. Bagasse sometimes was used as a supplemental cattle feed and as a temper for levee construction; otherwise, it was discarded in early-nineteenth-century Louisiana. Bagasse later came to be used as supplemental fuel in sugar factories, and in the twentieth century it was made into pressed wallboard and ceiling tiles. The all-important cane juice drained off into large rectangular cypress vats, where stalk debris and trash were removed from the juice. From here the juice passed on to the kettles.

Kettles and furnaces occupied the main section, the *boiling room,* of the sugarhouse. Along one or both sides of the long room, cast-iron kettles ranging in size from thirty-six to seventy-two inches in diameter were set into a foundation of fire brick and masonry. An arched flue beneath the foundation directed heat to the bottoms of the kettles. Just as in the Caribbean, kettles were assigned names from the largest to smallest: *grande, propre, flambeau, sirop, batterie.* The grande was located nearest the mill to receive the first cane juice and to receive the lowest temperatures from the furnace. The purpose of the lower temperature was to heat the cane juice gradually to initiate the process of clarification and boiling without scorching the mixture. The batterie, or smallest kettle, was located at the opposite end of the boiling room, where the furnace heat was hottest. Heat thus passed through the arch flue under the bottoms of kettles to provide a gradual heat pattern from the coolest grande to the hottest batterie, with other kettle sizes and variations in heat between. At the end of the furnace, smoke passed up through a large, square, sixty-foot-high brick chimney. Sugar factory chimneys reflected a technological evolution from the initial relatively

FIG. 3.5. *Three-roller sugar mills were steam powered by the mid–nineteenth century.* Reproduced from *Harper's New Monthly Magazine* 1853, 7:764

low, square chimneys to taller, octagon-shaped ones to tall, round, brick smokestacks in the late nineteenth century to very tall twentieth-century steel ones to, most recently, once-again short, steel stack precipitators.

The boiling process began with the grande kettle filled with clean cane juice. Six to twenty-four cubic inches of slacked lime were stirred into the juice to release impurities. As juice temperatures increased, a greenish scum formed on the top, which was ladled off by one or two workers with copper skimmers. Eight to ten minutes later, the contents of the grande were ladled into the propre to be boiled and then ladled into the flambeau. After more boiling and skimming, the mixture went to the sirop kettle. In these middle kettles, the juice continued to boil off its water content and more green scum was skimmed off. In each successive kettle, the cane juice became thicker (Silliman 1833, 30–40). In the last kettle the syrup eventually reached the consistency for granulation. At this most critical time, the batch was ready for *striking,* to be drawn off and placed into cooling vats.

FIG. 3.6. *The open kettle method of sugar making was the traditional technology of the seventeenth through nineteenth centuries in the Americas.* Reproduced from *Harpers's New Monthly Magazine* 1853, 7:765

The skill of the sugar maker was tested at two points in the sugar-making process, first in the liming at the early stage and second, but most important, at the batterie kettle in making the precise decision to strike the batch.

Into six or more cooling vats went a thick, grainy-textured syrup. Each vat, measuring seven feet long by five feet wide by one foot deep, received a three-inch layer of the grainy syrup. After successive layers had cooled and granulated, a crystallized mixture of sugar grains and syrup called *massecuite* was then removed from the vats and placed into wooden barrels, or hogsheads.

The final step in the sugar-making process was purging the molasses from the raw sugar crystals in the massecuite in the hogsheads. Purging took place in a separate room, the *purgery*, which was an appendage to the main building of the sugar factory. Open cypress cisterns some twenty feet square covered the purgery floor. Wide planks placed over the cisterns at eighteen-inch intervals supported the draining hogsheads. Each hogshead had small holes drilled into the bottom of the barrel so that the molasses would ever so slowly drain out, leaving the raw sugar crystals inside. Twenty to forty gallons of molasses would drain from each hogshead. Undrilled hogsheads would be filled with brown raw sugar, capped, sealed, and made ready for shipment to complete the process at the sugar factory (Silliman 1833, 30–40). Each hogshead held about one thousand pounds of raw sugar when the process was completed; thus, the hogshead was the standard unit of measure of production throughout the nineteenth century.

The open kettle method, as this process was called, continued into the twentieth century even though the process was crude and inefficient when compared to the better technologies that were to come later. Poor-quality sugars resulted when temperatures could not be kept even, and if the juice cooked too long, a batch would be ruined. Two basic faults of the method were high temperatures and problems in clarifying the syrup (Rillieux 1848, 285–88). Between 1830 and 1860 several inventions advanced the manufacture of sugar. In 1830, the vacuum pan entered the sugar-making process. Everything was the same as in the open kettle method until the final boiling stage when, instead of using the batterie kettle, the syrup was granulated in a vacuum. Sugar crystals were made larger through controlled granulation in the pan, and the result was an improved raw sugar product that brought a higher selling price. In 1843, the Rillieux apparatus from Louisiana was patented. Norbert Rillieux, a free man of color, invented the device, which had two vacuum pans connected so that steam given off by the first pan provided heat for boiling in the second pan; vapor from the second pan collected in a condenser. The advantages of the Rillieux apparatus were favorable lower temperatures that prevented scorching, more reliable high-quality sugar, and fuel efficiency (289–90).

Impurities in the juice and syrup was another problem facing sugar makers. Clarification processes ranged from skimming to liming to filtration and other methods in the attempt to rid cane juice of trash, chemical impurities or scum, and water. The latter was easily removed by boiling. Until about 1850, liming was used to release the impurities in juice in the grande kettles. Boneblack filters came into widespread use after 1850. In this process, cane juice passed through strainers in the initial clarifying vats and then was filtered through iron cylinders filled with boneblack, a charcoaled bone material. The charcoal filtration system was used until about 1900 (Sitterson 1953, 150–52).

An ever-increasing shortage of fuel wood was still another problem. Every steam-driven system revolved around wood burning, to fuel boilers operating steam mills, to fuel the boiling apparatus at the sugar factory, and to fuel steamboats plying navigable streams throughout the region. The bagasse burner was developed to meet the fuel problem. Between 1830 and 1854, planters experimented with attempts to burn bagasse in their steam operations. Some planters in the 1840s erected silos in which to store and dry the cane pulp before burning it; others wanted to burn bagasse immediately as it came wet out of the mill. In 1854, a furnace was invented to ignite wet bagasse, but it still required a significant amount of wood. After the Civil War, bagasse burners became more efficient, largely because mills became more efficient at extracting cane juice and producing drier bagasse (Sitterson 1953, 152–53). In the 1960s, natural gas was the main fuel, and some of the region's bagasse was dried, baled, and shipped to the Celotex factory near New Orleans, where acoustic ceiling tiles and wallboard were made from the pulp. Contemporary sugar factory furnaces are fueled with natural gas and are abundantly supplemented by bagasse burning.

With technological advances attacking these and other problems, each step in the process became an improvement. To conclude our discussion on nineteenth-century sugar making in factories known as sugarhouses, perhaps the words of T. B. Thorpe written in 1853 best convey the sense of quiet conviction signaling the end of the grinding season. "But to the sugar house, the crop has just been gathered; and by a thousand wings of commerce, it has been scattered over the world; the engines of the sugar house, therefore, are lifeless; its kettles are cold, its store-rooms are empty, and the key that opens to its interior hangs up in the master's house, where it will remain until the harvesting and manufacturing of the new crop" (754).

From Field to Factory: Modern Sugarcane Processing

In the contemporary sugar world, the stages involved in bringing a crop from planting to harvest to processing follow many of the same fundamental steps as in the past but now with improved technology in the field and in the sugar factory. In the agricultural field, the stages in sugarcane production are planting, cultivation, harvesting, cane handling, and transport. Sugar factory processing stages include cane handling, cane washing, cane preparation, milling, evaporation, sugar boiling, centrifugal separation of raw sugar from molasses, storage, and product shipment. Interspersed throughout the process are changes in the ways things are done as well as changes in mechanical technology. For example, at the sugar factory the newer components in sugar making include cane sampling and testing, computer-guided instrumentation, improvements to boilers, and pollution controls (Center for Louisiana Studies 1980).

Planting is one of the primary stages in crop agriculture. However, with sugarcane only a portion of the annual crop is newly planted. Sugarcane stalks ratoon by regenerating from the root stock and stubble. Once planted, cane can be expected to ratoon over the next three or four years into usable annual crops that will diminish with each successive ratoon. In the initial planting of a crop of sugarcane known as *plant cane,* stalks are cut into segments and planted in the tops of rows. For centuries, this was all done by hand. Then mechanical planters were developed, beginning with special carts in which workers hand-fed canes into the ground. The process evolved to methods that have mechanical chain rakes and drum planters that open the row, drop the stalk in the ground, and cover the stalk with soil in one continuous operation (Allain 1980, 124).

Cultivation is the caring process by which fertilizer is applied to and weeds are removed from a crop. For centuries, various techniques in hoeing and plowing were used to cultivate a crop; in the twentieth century herbicides have been added to cultivation. Hoeing was an exceedingly labor-intensive method and was always done by hand, usually by the hands of slaves. Nineteenth-century plowing was exclusively done with animals, earlier with oxen and horses and later with mules. Since the 1950s, tractors have done all the plowing. But today's larger, more powerful two- and four-wheel-drive tractors plow, shape, and fertilize three six-foot-wide rows at a time. Sugarcane, because of its height and thick foliage, can be cultivated by machine only over a limited period of months, after which the crop is too tall for farm equipment to pass through the field. As sugarcane matures, the canopy helps shade out weeds. Fertilizing sugarcane fields has been an ever-evolving process, ranging from the past uses of livestock manure and bagasse to modern liquid chemical fertilizers, herbicides, and insecticides applied by tractors with boom sprayers. Aerial applications of chemicals to fields are done using aircraft crop-dusting techniques.

In the field, cane harvest and cane handling functions run in tandem. Sugarcane stalks are cut a few inches above the ground, laid across the furrows, and left for the foliage to dry for a few days and to be burned. Then the dry canes are loaded into huge metal cane carts or trucks to be transported to the sugar factory. Cutting cane has never been an easy task, and throughout the sugar world much of it is still done by human hands wielding cane knives, machetes, cutlasses, or bill knives. Louisiana is blessed with flat, sturdy terrain that allows cane cutting with mechanical harvesters. In Florida, where all the most modern methods of sugarcane production are found, cane is still cut by hand because the soft muck soils there bog down heavy mechanical harvesters. The use of cheap, imported Caribbean laborers to harvest Florida's cane is considered by the sugar corporations there to be not only the best way to cut cane but also the most cost efficient (Wilkinson 1989, 4–6). The amount of hand-cut cane there is enormous; Florida's 428,000 acres are largely cut by hand, whereas Louisiana's 364,000 acres are all harvested by machines.[5]

Cane loading in the field and transport to the mill are important steps in the process. In the past, heavy sugarcane stalks were loaded into cane carts by hand, but now mechanical hydraulic grabs reach out from the front of tractors, pick up cane stalks and place them into either cane carts or long-haul tractor-trailer trucks. Until the 1970s, tall (thirty to seventy feet high) metal cane derricks were regular features on the landscape. Cane was loaded into cane carts in the field and then taken to cane derricks located elsewhere in the field or at the site of an old sugar factory. At the derrick, the canes were loaded into trucks that then hauled the stalks to the sugar factory. On a few plantations in the late nineteenth and early twentieth centuries, transport from field to factory was via small steam locomotive-powered railroads, as at Armant, Laurel Valley, and Enterprise Plantations (Wade 1995, 180–81, 267).

Processing Stages at the Sugar Factory

Tons of cane stalks arrive at the mill yard at one end of the sugar factory, where the processing stages begin with cane handling, cane washing, and cane preparation. Inside the factory, the stages are milling, evaporation, sugar boiling, and centrifugals. In nearby warehouses and molasses tanks, storage and product shipment preparation take place. In earlier times, cane-handling cane derricks, cranes, or booms in the mill yard unloaded canes from cane carts and trucks and then loaded the canes onto a washing table. Newer techniques now involve a hydraulic unloader that tips the cane cart and dumps the cane over a retaining wall onto a concrete pad. The canes are stored there until being loaded onto the washing table by front-end loaders with grabs.

In the cane washing stage, canes were once loaded onto a conveyor unit

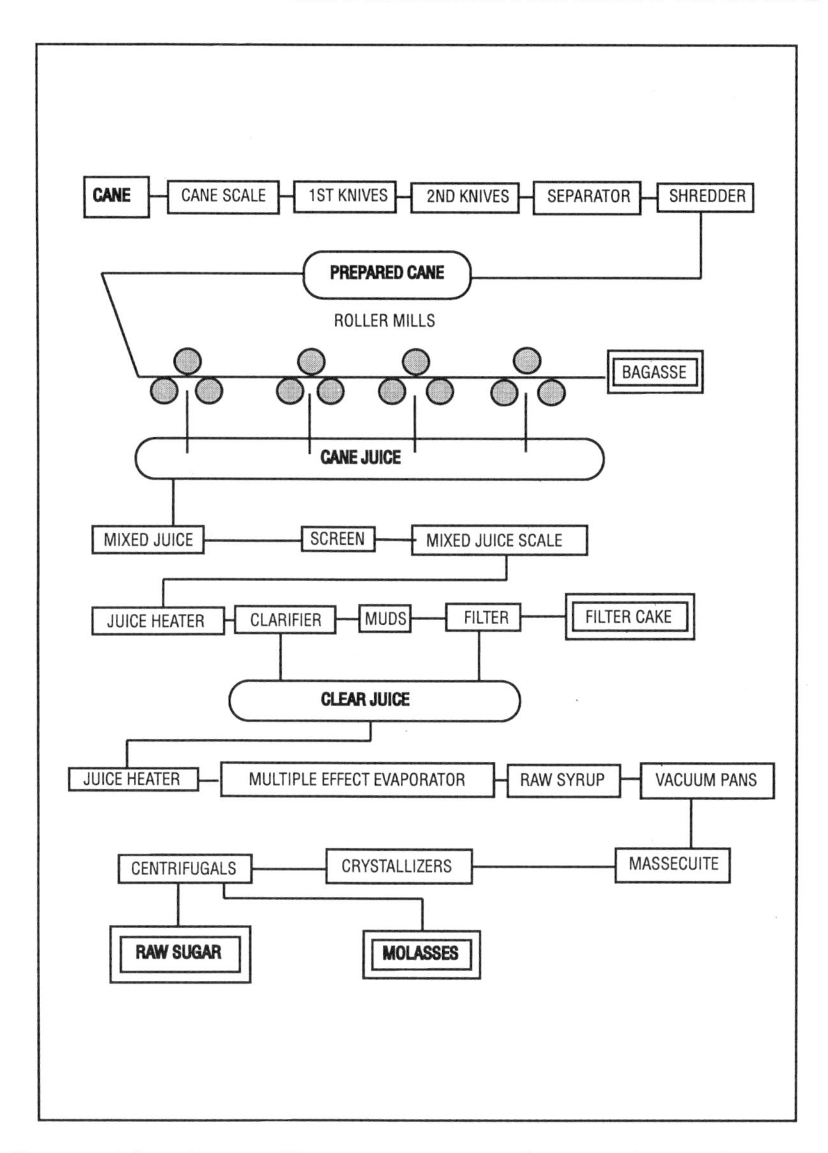

FIG. 3.7. *A flow diagram illustrates sugar-processing stages in a modern sugar factory.* Source: Barnes 1964, 349

called a cane-feeding table; in later years feeder tables were converted to washing tables. Since the 1960s, better designed cane washing tables set at a steep forty-five-degree angle receive canes, where high-pressure water is forced over the stalks to remove soil particles, soot, and trash leaf matter. Canes are reported to be 60 percent cleaner using this method. High-speed

FIG. 3.8. *The Westfield Plantation factory represented 1960s technology in the mill yard and overhead crane system. Assumption Parish.* Photograph, 1967

conveyors move the canes up to the cane preparation area, where machines with rows of knives cut and shred the stalks.

The next stage is milling, where cane pieces are pulverized and pressed through a milling tandem, a series of four to eight sets of horizontal three-roller mills. A milling tandem may have as many as twelve to twenty-four rollers or crushers. Milling is much the same as it was in the past century, except that the grooved rollers are much heavier now and there are more banks of rollers. Bagasse passes on through the mill rollers to another destination. Some bagasse, called *cush-cush,* falls onto a screen below the rollers, from which it is conveyed back up to the start of the milling process to be remilled. A few sugar factories have screw presses that extract additional juice from cush-cush and bagasse; this produces a dryer bagasse that burns better (Birkett 1980, 103).

Bagasse has had many different uses. In the past century, it was placed on cane carts and either fed to livestock or burned in the sugar factory furnace. Until the 1970s, much of the bagasse was sent to a drying room, where it was baled and then stored outside under tin roofs to await shipment to the Celotex Corporation in Merrero near New Orleans. At the Celo-

FIG. 3.9. *A contemporary mill yard has cane sampling equipment (*left*) and more efficient cane loading and cane washing equipment (*right*). Saint James Sugar Cooperative. Saint James Parish.* Photograph, 1996

tex plant, bagasse continues to be made into acoustical ceiling tiles, wallboard, and insulating materials. Since the energy crisis in the 1970s, however, more bagasse remains at the sugar factory to be burned in a bagasse burner along with natural gas, providing steam energy to the factory.

After the milling stage, cane juice is strained and pumped into clarifying tanks where waste steam from the powerhouse heats the juice. Lime or phosphoric acid precipitates impurities, and a vacuum filter rotating inside the tank removes the scum, called *muds.* This part of modern sugar making corresponds to the grande kettle in the old open kettle method.

The cane juice then goes through the next series of processing stages: evaporation, sugar boiling, and centrifugals. The evaporation process corresponds to the middle kettles in the old open kettle method of sugar making. Cane juice is pumped into several multiple-effect evaporators, large vessels sixteen feet high with six- to eight-foot diameters; most sugar factories have four or more evaporators. Multiple-effect evaporators were invented and patented by Norbert Rillieux in Louisiana in 1843 and are called multiple effect because the process takes place in a series of interconnect-

FIG. 3.10. *A sugar mill has four to eight tandems with three rollers per tandem. Each roller weighs one ton. Westfield Plantation. Assumption Parish.* Photograph, 1967

ing steam-heated tanks. Juice is pumped through tubes inside the evaporator, and steam is added to the tank above the tubes. The heated juice then percolates through the tubes to evaporate water in quantity equal to about 70 percent of the gross cane tonnage. The output from the multiple-effect evaporator is cane syrup.

Sugar boiling takes place in vacuum pans where cane syrup is heated to remove additional moisture; it is here that sugar crystals and molasses form and the strike is made. In modern sugar making, a three-boiling system produces A, B, and C strikes. Each consecutive strike is based on the molasses content and the concentration of sugar crystals in the batch. From the vacuum pans, the mass of grainy crystals and sugary molasses, now called *massecuite,* is transported to cooling vessels, or crystallizers, where further crystallization takes place.

The separation of molasses from the sugar crystals takes place in machines called *centrifugals,* which operate in principle much like the spin cycle on a clothes washing machine. In the last century, purging the molasses from sugar crystals took place over a period of weeks in barrels in a separate room called the purgery. Before the 1960s, sugar factories had only batch centrifugals, in which a batch of massecuite was placed into the machine, spun to remove the molasses, and stopped to remove the remaining several hundred pounds of raw sugar crystals. This stopping and starting was costly in time and energy. Within the past thirty years, continuous centrifugals with cone-shaped baskets inside have replaced the older technology. Massecuite is continuously fed into the machine; as the massecuite slides up the cone by centrifugal force, molasses is spun through a perforated screen, and the raw crystals pass over the lip of the cone (Birkett 1980, 108). From this point the separated products can be transferred to separate storage facilities.

Raw sugar is either bagged or stored in bulk in warehouses, and molasses is piped into large, round storage tanks adjacent to the sugar factory. From here transport of the raw brown sugar to distant refineries can be made via truck, by railroad, or even by water on barges. Molasses is usually transported in railroad tank cars, but at times it has been transported in barges.

Refining raw sugar into white sugar was not performed on nineteenth-century plantations and was performed only rarely on plantations in the twentieth century. This special job was done in large refineries located in New Orleans, Savannah, Baltimore, Philadelphia, New York, or Boston, far from the rural sugar plantation landscape. In the 1960s, Louisiana had five sugar refineries: American Sugar Refining Company at Chalmette near New Orleans, Colonial Refining Company in Gramercy, Godchaux Sugar Refining Company in Reserve, South Coast Corporation's Georgia Refinery at Mathews on Bayou Lafourche, and Southdown's refinery at Houma in Terrebonne Parish. Only the latter two were located on plantations far from New Orleans. Three refineries remain in Louisiana, and only the one called Colonial at Gramercy has not changed appreciably (Buzzanell 1993, 28). Domino Sugar Company has taken over the operations at Chalmette, and Supreme Sugar Company has placed its refinery at Leighton Plantation at Labadieville in Assumption Parish near Thibodaux.

Five trends characterize recent technological improvements in sugarcane processing: (1) improvements in sugar factory equipment, (2) automated instrumentation and computer processing, (3) the trend toward energy-efficient methods and equipment in sugar processing, (4) the installation of pollution controls for improved air and water quality, and (5) the trend to larger sugar factories in an effort to reach a balance in the economy of scale (Birkett 1980, 99–114). In this chapter I have touched on the first four trends. However, the fifth trend, to fewer but larger sugar factories, has had a significant effect on the landscape, the closing and disappearance of smaller, less efficient factories (see chaps. 4 and 11). The milling capacity of a sugar mill refers to the number of tons of sugarcane stalks that can be processed in one day and serves as a credible index of the economy of scale. Between 1962 and 1982, milling capacities doubled, enabling larger, more efficient factories to survive. Smaller mills that did not meet environmental regulations and those that did not upgrade to reach higher capacities simply succumbed. Table 3.2 illustrates the trends in milling capacities for the twenty functioning sugar factories in Louisiana in 1992. Ten were privately owned and operated, and ten were organized into cooperatives owned and operated by collective groups of sugar growers. Whereas Louisiana's sugar factories at the turn of the twentieth century had milling capacities of 700 to 1,500 tons, today's fewer but larger factories process between 5,000 and 14,000 tons each per day.

Barns, Outbuildings, and Other Landscape Features

All sugar plantations have a collection of barns, sheds, warehouses, and other structures that together form an outbuilding complex. Plantations with a sugar factory have more outbuildings clustered around the factory. Even in the absence of a sugar factory, sugar plantations have more outbuildings than perhaps any other plantation type. A typical nineteenth-century sugar plantation had one or more large mule barns, a blacksmith shop, stables for horses, various implement sheds, feed and corn storage sheds, poultry houses, pigeonniers, pigpens, and many other outbuildings of various shapes, sizes, and uses. T. B. Thorpe in 1853 described some of the outbuildings: "Here are often seen stalls for fifty and sometimes a hundred mules and horses arranged with order and an eye for convenience. The vast roof that covers these necessary appendages to a plantation, together with the granary, sheds, and a score or more of useful, but scarcely to be recollected structures, form, of themselves, a striking picture of prodigal abundance and suggest the immense outlay of capital necesssary to carry on a large sugar plantation with success" (754).

Contemporary plantations require fewer functional outbuildings, and obsolete buildings are neither maintained nor preserved. Most mule barns,

TABLE 3.2

Louisiana Sugar Factories, 1962–1992

		Grinding Capacity in Tons per Day			
Factory	*Parish*	*1962*	*1972*	*1982*	*1992*
Private Factories					
1. Alma	Pointe Coupee	2,400	2,400	5,500	7,000
2. Cinclair	W. Baton Rouge	2,000	4,200	3,100	5,500
3. Columbia	St. John-the-Baptist	1,600	1,900	1,800	2,200
4. Cora–Texas	Iberville	2,000	3,000	6,450	11,000
5. Enterprise	Iberia	3,000	4,250	7,000	14,500
6. Leighton	Lafourche	4,200	6,300	8,000	9,000
7. Lula	Assumption	2,800	3,800	4,500	6,800
8. Raceland	Lafourche	5,000	5,000	7,500	9,500
9. Sterling	St. Mary	5,500	6,000	7,500	9,500
10. Westfield	Assumption	3,300	4,200	4,550	6,600
Others[a]		43,300	41,750	5,000	0
Subtotal capacities		75,100	82,800	60,900	81,600
Average capacity/factory		2,347	3,312	5,536	8,160
Cooperatives					
11. Breaux Bridge	St. Martin (1938)[b]	1,500	2,400	2,900	4,800
12. Cajun	Iberia (1963)	–	6,000	5,500	8,000
13. Caldwell	Lafourche (1946)	3,700	4,800	6,000	6,800
14. Evan Hall	Ascension (1936)	4,500	5,000	5,600	7,500
15. Glenwood	Assumption (1932)	4,000	4,200	4,900	5,600
16. Iberia	Iberia (1937)	2,800	4,250	4,250	6,500
17. Jeanerette	Iberia (1972)	1,750	1,750	4,500	6,500
18. St. James	St. James (1947)	2,800	5,000	5,500	7,000
19. St. Martin	St. Martin (1973)	2,400	3,500	4,350	5,200
20. St. Mary	St. Mary (1947)	2,200	3,200	4,250	5,500
Others[a]	–	7,380	9,900	7,800	0
Subtotal capacities		33,030	50,000	55,550	63,400
Average capacity/co-op		2,541	3,571	4,629	6,340
Total capacity		108,130	132,800	116,450	145,000
Total average/mill		2,402	3,405	5,063	7,250

Sources: Buzzanell 1993, 28; Durbin 1980, 31; Gilmore Sugar Manuals.

[a] Others are any private or cooperative factories operating in 1962, 1972, or 1982 that closed by 1992.

[b] Inauguration date is the year that each cooperative was formed. All Louisiana sugar factories were private before 1932.

FIG. 3.11. *An outbuilding complex can have several different functional buildings (from* left *to* right*): a small Creole quarter house, a cane harvester shed, a mechanics shop and warehouse, and a former mule barn converted into an equipment shed. Allendale Plantation. West Baton Rouge Parish.* Photograph, 1964; source: Rehder 1978, 138. Used with permission from the Louisiana State University School of Geoscience

blacksmith shops, stables, and other structures in the outbuilding complex have disappeared from the landscape. Beginning in the 1930s, metal buildings began to appear, especially to replace sugar factories and associated structures. Until the late 1960s, some plantations still had mule barns that had been converted to tractor sheds, a remaining blacksmith shop, and a few other relic buildings that had been converted to other uses. But by 1970 most wooden buildings had been torn down and replaced by fewer, galvanized sheet metal–covered specialty structures.

A complete sugar plantation has two functional areas of outbuildings. At the sugar factory site are molasses tanks, a water tank, large warehouses for raw sugar storage, a mechanics shop, the mill yard, a mill office, the weigh station, and other metal storage buildings in addition to the actual sugar factory building. In the second area is an agricultural complex of outbuildings usually located at or near the sugar factory. On some plantations the agricultural complex may be located some slight distance from the sugar factory so that traffic to one functional area will not interfere with that of the other. On a very few large plantations, multiple agricultural outbuilding complexes are separated to serve widespread areas of cultivated land. On others, complexes of agricultural buildings are sited at the old outbuilding and factory sites of preexisting plantations. Regardless of location, an agricultural outbuilding complex in the 1990s has one or more large, open-sided tractor sheds with sheet metal roofs, sheds for mechanical harvesters, a mechanics shop, a chemical/fertilizer storage tank area, and a few other small storage sheds, as needed. Most significant on the "new" landscape is a designated spot or parking area for leased farm equipment left out in the open to all the rust-causing elements of nature. On some plantations now the only shelter for farm equipment is a shade tree.

Until about 1940, every plantation had one or more large barns to shelter the work animals. Mule barns were large wooden buildings 100 to 150 feet long by 40 to 60 or more feet wide. Floor plans of mule barns would have shown numerous animal stalls on the ground floor and an open loft above. The barns had a central passage with stalls on either side and were, in effect, enormous transverse crib barns. Mule barns were built with either box stalls or hitch stalls. Box stalls could be built in either single- or multiple-stall plans, but for large plantations multiple stalls accommodated larger numbers of animals. Some barns also had single box stalls reserved for saddle horses and mean mules. The box stall was important because, with kicking mules, you always wanted to meet the mule head first. Hitch stalls were used to tie animals to lines, posts, or rings inside a section in the barn (Scoates 1937, 14). Entrances were on the gable ends, but some barns had other openings on the sides. For example, on Ashland Plantation south of Houma, one of the smaller barns had open sides that provided cooler, shaded shelter and easy access to the animals. Later, such open-sided barns were converted into tractor sheds.

Barn construction was of heavy wooden timber frames with either horizontal or vertical board-and-batten siding. The saddle roof, originally covered with cypress shingles or shakes, was now covered with galvanized steel sheets. Eighteenth-century plantation outbuildings had riven slabs or pit-sawn wood planks for siding and were roofed with palmetto or bark roofing materials.[6] A few plantations in 1969 and fewer yet in 1996 still had remnant mule barns that no longer sheltered mules. The remaining old buildings now store fertilizers, tools, and other farm equipment.

FIG. 3.12. *The barn and mule lot at Palo Alto Plantation in Ascension Parish was typical of nineteenth-century plantations.* Reproduced from *Louisiana Planter and Sugar Manufacturer* 1892, 9:105

On the Whitney plantation is perhaps the oldest surviving barn in Louisiana, dating from about 1790. It is a small structure about fifty-three feet long by thirty-four feet wide with a Norman truss-hipped roof, cypress internal framing that is mortised and pegged, and wide, horizontal cypress plank siding (Rykels 1991). The original cypress shake roof had been covered later with a tin roof. In plan the building has a large central room and four storerooms around it. Doors and windows are on all four sides of the building, but no door is large enough to deliver draft animals, wagons, or large implements. The brick foundation raises the main floor of the structure to four feet above the ground level and suggests that the building was not a barn for livestock but rather may have been used for general but compartmentalized storage. In 1969 the building was being used as a mechanics shop and for storing farm chemicals, feed, and other supplies. When I reexamined the barn in 1991, it was empty.

The Cajun barn type was common to small Cajun sugarcane and livestock farms, and only a few may have made their way to plantations. The Cajun barn was quite small, measuring not much more than thirty feet on

FIG. 3.13. *A view of, perhaps, Louisiana's oldest surviving barn, built about 1790 at the Whitney Plantation. Saint John-the-Baptist Parish.* Photograph, 1991

a side. Diagnostic traits of the barn were a recessed entrance, a central corn crib, and animal stalls or hitching stalls on either side of the long corn crib (Comeaux 1989).

Blacksmith and mechanic shops have always been an integral part of a plantation. The number of tractors, harvesters, plows, and the myriad of implements on sugar plantations mandate such buildings and their important functions, which continue on some plantation landscapes. However, with the growing trend of leasing farm equipment, some plantations neither shelter nor repair equipment beyond such general maintenance as oil changes and lubrication, and on some of these properties a shade tree suffices for shade tree mechanics.

Tractor sheds began to arrive on the landscape in the late 1940s as mules were being phased out and more farm machinery and equipment began to appear. Tractor sheds are simple, seventy-foot-long and narrow shelters with sectioned drive-through open stalls (see figs. 3.11 and 10.2). They look like unfinished buildings with a roof, support posts, and no walls, and the ones I observed in the 1960s are no different from those seen today. Roof materials are universally galvanized steel, and the wooden support posts on older sheds have been replaced with steel ones.

All other sheds and small outbuildings are rectangular, one-story structures with openings on the gable ends. Their functions are many–storing fertilizers, chemicals, machine parts, implements, and many other materials needed at the plantation's agricultural outbuilding complex. In the past century all sheds of this nondescript kind were wooden with horizontal siding. They were rebuilt with vertical board-and-batten siding in the post–Civil War period, but by the 1940s, galvanized sheet metal became the material of choice for siding all outbuildings ranging from the largest sugar factories to the smallest storage sheds in the outbuilding complex.

Stores and Churches

The plantation store and church are two socially important buildings that appear on some plantations. The building designated as the store may have more value placed on its function than on its form because it has different functional roles as the center of small but constant economic activity and as a focus for social gathering. The plantation store is the first place a visitor or vendor goes when arriving on a plantation because the store is a flagship public place. As an outsider, I used the store as an opportunity to buy field food–crackers, cheese, beans, candy, and soft drinks–and to be introduced to the local happenings on this and nearby plantations. In some small way my purchases were, in the words of anthropologists, "pay to informants." It was something I would have done naturally anyway, and the store provided a convenient introduction to its plantation. The store served the needs of the plantation's workers, providing provisions of canned goods, bread, flour, milk, and, strangely enough, sugar. The store carried work clothes, black rubber boots, patent medicines, cloth and sewing materials, and hand implements like hoes, rakes, and long-handled cane knives, among other dry goods. Every plantation store had a special cage or counter covered with lots of penny candy for sale to children living on the plantation.

Stores were run as company stores and provided credit to workers. Prices were always high because of the captive market. Many a plantation laborer "owed his soul to the company store," but he had the option to receive cash payment for wages earned. In 1964, I witnessed something I did not believe still existed, the issuance of scrip paper as wages to workers. Scrip was a guarantee that workers would spend their wages only at the company store and absolutely nowhere else. I was shocked that on some plantations the management still stooped to such unfair practices.

The store was a place for social gathering where children played, old men sat and spat, people met to talk, mail could be picked up, and all store-bought items were acquired. On small plantations, the store was also the plantation business office. Here, workers received weekly or bimonthly wages and paid debts to the storekeeper. Outside vendors and suppliers called at the store for either the store's or the plantation's business. Rarely was the store without people in or around it.

The origin of the plantation store dates to the late nineteenth century, a time of reconstruction in the plantation South with a changing economic and social climate. Earlier, under the slave system, food, clothing, and most necessities were provided by the master. Plantation stores were unnecessary because goods were available from passing steamboats and traveling peddlers (Sitterson 1953, 109–10). In the postbellum era of the 1800s, however, plantation stores were established to provide products to sell to wage-earning laborers.

Store architecture varied throughout the sugar region. On some plantations, an existing building or storage shed was the initial store, but on other plantations a special building was built to become the store (Pulliam and Newton 1973). Store buildings were similar in form and plan. The store was a long, rectangular structure with the entrance at the gable end. A raised, wide porch yawned across the front and often had someone sitting, standing, or playing in the shade of its wide, overhanging roof. Some stores were large structures measuring fifty or more feet wide by as much as one hundred feet long. The buildings were of all-wooden construction, weatherboarded in horizontal clapboards or with vertical board-and-batten siding, and for some unknown reason many stores were painted white with dark green trim. The roof was universally galvanized steel.

Inside, a large room that took up almost all of the square footage had tall shelves lining the walls, covered with goods of all descriptions. Along one side was a counter where the storekeeper sat, collected money, and registered items bought on credit. Nearby was a large case or cage with fifty dozen kinds of penny candy. In the back of the store was an office and a sizable storage room. Some stores had a meat case in the back where cold cuts, pork, sandwich meats, cheese, and any other items needing refrigeration were kept. Stores had pickled pig's feet, pickled eggs, pickled sausages, or just sour pickles in big, clear glass jars on top of the meat counter. Cold drink boxes somewhere inside the store had a variety of soft drinks–Barq's root beer, peach soda, strawberry soda, grape Nehi, Orange Crush, chocolate soda, and the big three colas, Royal Crown Cola, Coca Cola, and Pepsi Cola. I was especially partial to plantation stores for field food.

Over the past thirty or more years, plantation stores have been disappearing at a rapid rate, mainly because the captive market is gone. Few people live on plantations, and most people have cars and better roads on which to drive to town, where supermarkets and discount stores get the local trade. As the way of life on plantations changed, stores simply could not control or keep up with the external changes that we call progress.

Churches began to appear on Louisiana's sugar plantations in the late 1800s and especially in the 1900s. During the antebellum period, most planters prohibited religious activity among slaves; it has only been since Reconstruction that churches have been formed and buildings erected as

FIG. 3.14. *Every store had a wide, shaded porch, and most had wooden shutters on doors and windows. The late-nineteenth-century Pikes Peak Plantation store served several nearby plantations in Saint James Parish but had been closed for several years before it was razed in 1995.* Photograph, 1967

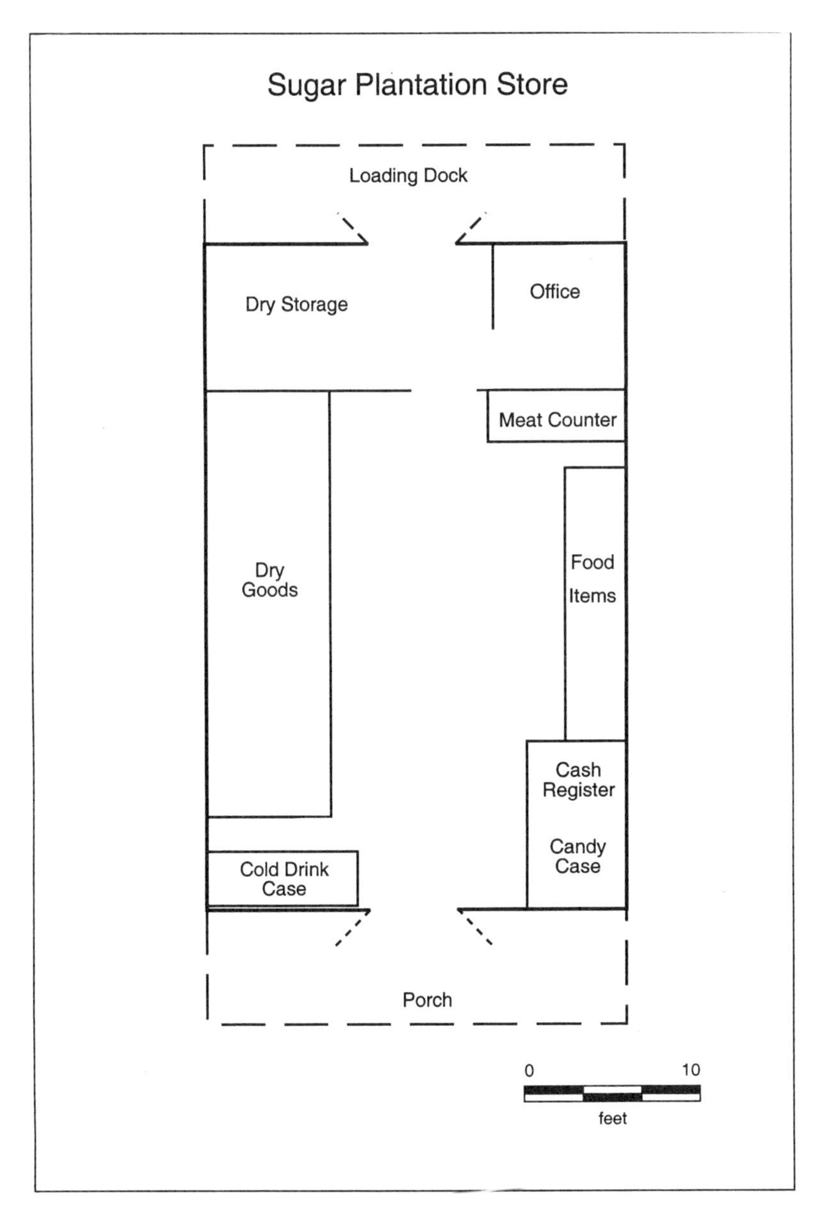

FIG. 3.15. *In plan, a sugar plantation store was a simple structure with dry goods lining shelves on the long axis walls, meat counter and storage in back, candy counter and cash register, and a cold drink box.*

places of worship (Genovese 1976, 187). By the 1960s, plantation churches served African American congregations, but few churches were ever located within the plantation settlement. The church was built some distance away from the settlement on land obtained from the plantation management or off the property entirely in nonplantation, free villages. The buildings have always been small, sided with weatherboard, and painted white with green or blue trim. Some churches have painted window panes to add a stained glass appearance; blue panes are said to keep the devil out.

These small but functional structures were a special symbol of freedom in the period of Reconstruction, and their traditional strength lay in a resident plantation population. In recent years, as plantations have experienced planned obsolescence, the overall destruction of nonessential buildings has taken stores and a few churches just as surely as quarter houses and mule barns. The detachment of laborers, their families, and retirees from plantations precipitated the removal of residential, commercial, and social functions and their concomitant forms from the plantation landscape.

Fences and Fields

A fence is an inherent part of most agricultural landscapes. But is the function of the fence to keep animals out of cropland or is it to pen animals into an enclosure? Fences may serve to keep animals out or to keep animals in, to create a man-made separation between properties, and especially in Louisiana to mark the landscape with a symbol saying "This is mine, but you are welcome to open the gate and visit here for a while."

Earliest plantation fences were made of stakes driven into the ground and were called *pieux debout,* or stake fences (see fig. 6.5). The pieux debout was a palisade of cypress slats used for penning livestock and for enclosing small garden plots. It was also used in urban palisades. In 1727, the modest city of New Orleans was described as being surrounded by a large ditch and fenced with sharp stakes wedged together (Gayarre 1851, 387). In rural areas nearly all fences of this early type were unsupported, but for some a horizontal stake was affixed to the top of the fence for additional support.[7]

Another fence type that had widespread popularity on plantations was the *pieux traverse,* a horizontal fence that was used concurrently with the pieux debout stake fence. The pieux traverse was similar to post-and-rail fences of the Middle Atlantic states in that posts had holes cut into them

FIG. 3.16. *Plantation churches were relatively small, simple buildings like this one at Belleview Plantation on Bayou Teche. Saint Mary Parish.* Photograph, 1993

with horizontal rails fitted into the holes. This was a solid, sturdy fence with posts driven deep into the ground and horizontal rails affixed to extend the linear coverage. Four or five rails produced a fence about five feet high.[8] This field fence enclosed pastures and limited agricultural cropland on plantations from the mid-1700s until the 1880s, when barbed wire entered the region (Pierce 1851, 606).

Of the Anglo-American fence types, the snake fence, a portable split-rail fence, would have been found in areas north of Baton Rouge near the Anglo-French margins of the region. Also known as worm or zig-zag fences, snake fences did not appear in southern French Louisiana on the sugar plantation landscapes of the delta region where the pieux traverse was in widespread use on most all plantation types. The old fences are now gone, and only one or two rare examples remain, mostly for ornamental purposes.

Since about 1880, barbed wire has been the fence of choice for some stock pens and especially for fencing cattle in pastures (McCallum and McCallum 1965, 33). Sugarcane fields are simply too large to fence in, and livestock is so scarce that fences have a limited utility. Prized livestock (horses now and mules in the past) would have been corraled with either wooden fences or smooth wire fences to prevent stock from injuring themselves on barbed wire.

Sugarcane fields follow a well-defined and established pattern. Fields here are long, narrow strips that follow the almost imperceptible gradient of the natural levee from the frontlands at the levee crest to the backswamp. All arable land, any land with sufficient drainage and fertility, has now or once had the potential of being part of a cane field. Superimposed on the surface is a network of ditches and roads in a rectangular, gridded pattern in which fields are divided by size into plots and sections.

The smallest, most basic field unit is the cane plot, a field bordered on two sides by lateral ditches that are parallel to the furrows. A plot measures from one to five acres and contains eighteen to twenty-five rows of cane. A grouping of fifteen to twenty plots is called a section, or cut. This larger unit is bounded by headlands or turnrows that are twenty-four feet wide to enable farm tractors to turn around. Most headlands have become roads. Throughout the delta, plantations with field plots and sections incorporating five hundred to eight thousand acres in cultivation create an impressive pattern on the landscape.

Rows and furrows are naturally important to sugar cultivation, but row widths govern farm machinery technology. Since row widths are set at a standard six-foot distance, all tractors, harvesters, cane carts, and other farm machinery that must go into the fields are built to accomodate the six-foot standard. Row widths have not always been standardized; in the nineteenth century, widths were variable. Before 1830, cane rows were set to narrow distances, ranging from two to four feet wide. The narrow rows fit well with the technology of the time–an abundance of hand labor but limited numbers of mules and plows. The narrow rows also worked well for single-mule plows and one-mule cane carts. Furthermore, the Creole and Otaheite cane varieties were thought to be better suited to narrow row cultivation. After 1832, row widths were widened to six to eight feet to accommodate the new ribbon and purple cane varieties. W. W. Pugh, a prominent sugar planter, explained that the new cane varieties required wider rows to allow air and sunlight to reach the ripening cane (Pugh 1889, 62). More importantly, the change from single-mule plows to two-mule plows warranted wider rows. This combination of new cane varieties and a change in technology prompted the six- and eight-foot-wide rows, which became the norm until the 1880s. In 1901, W. C. Stubbs of the Sugar Experiment Station

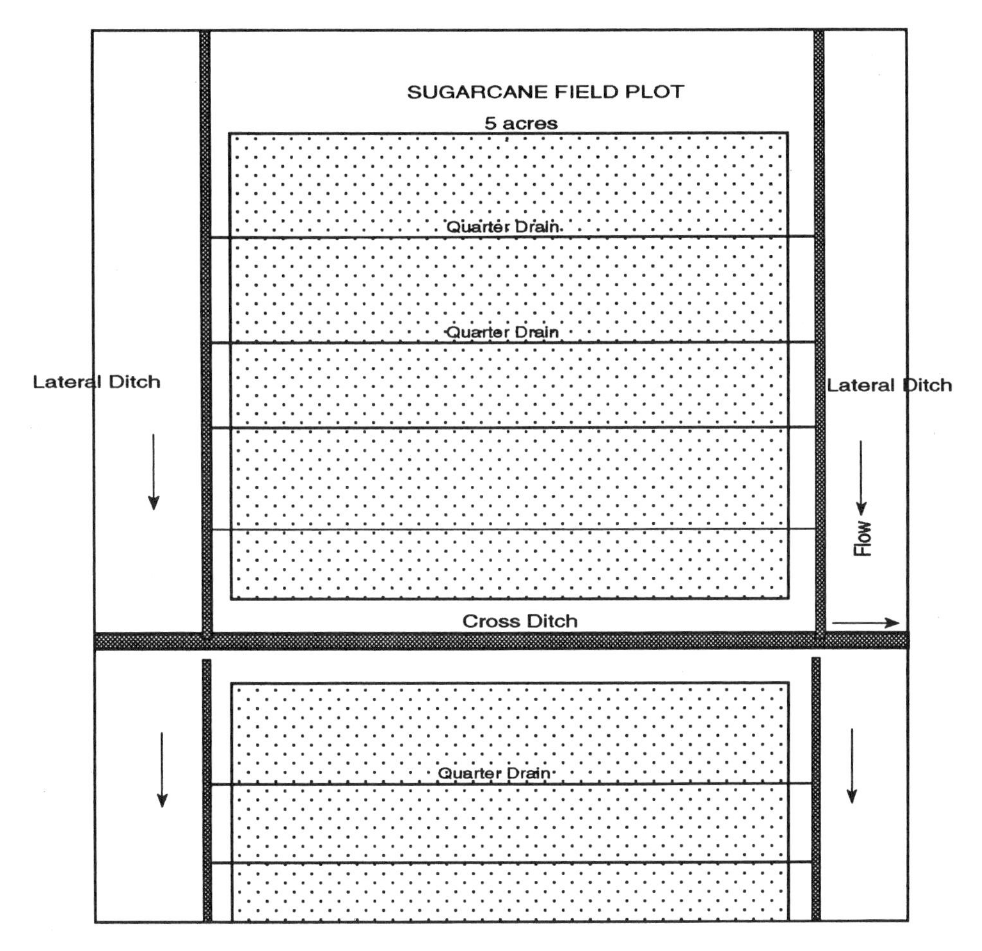

FIG. 3.17. *Sugarcane fields are divided into one- to five-acre plots and laced with a gridwork of ditches.*

reported that, even though cane could grow and mature in narrow rows, the practice required more seed or plant cane, and planting twice as much cane for three-foot rows instead of six-foot rows was not cost efficient. Stubbs suggested a five-foot row, but planters did not like working their two-mule plows in the five-foot rows so they expanded the width to six feet (Stubbs 1901, 112). After 1900, the six-foot row became the standard width, and equipment manufacturers have since standardized cane carts, wagons, tractors, harvesters, and other implements and farm equipment to meet that six-foot width.

The height of soil rows is a conspicuous characteristic in cultivated sugarcane fields. Row heights are set to eighteen to twenty-five inches above the base of the trough in the furrow to provide proper drainage and aeration of cane roots in Louisiana's relatively wet alluvial soils. Fields nearest the backswamp require the tallest ridged field rows because of backswamp flooding and moisture-retentive Alligator clay soils.

A natural part of cane cultivation in Louisiana is the moisture and drainage required of sugarcane. Sugarcane's moisture requirements are quite high, about sixty or more inches of rainfall per year; paradoxically, the plant does not survive in a constantly wet soil environment. Root rot and root diseases accompany wet soil conditions, so an elaborate system of ditches and canals is needed to clear the fields of excess water. Natural drainage is hampered by the low slope gradient between the natural levee crest and the backswamp. For the widest natural levees in the delta, the gradients are about six to nine inches of elevation change per mile, a gradient ratio of 0.5:5,280 feet to 0.8:5,280 feet (Russell 1936, 47). Imagine a surface that is so flat that a point one mile away is only six inches lower than the starting point and that those six to nine inches must be spread over the mile distance. Better yet, extend your fingers as wide as possible while holding your hand in a vertical position. Now imagine walking a distance of one mile while slowly closing your fingers until your thumb and little finger touch. Water here almost has trouble running downhill!

Each small field plot has three or four shallow ditches called *quarter drains* that run at right angles to the furrows. Water from the quarter drains runs into lateral ditches located on either side of the field plot. Lateral ditches run parallel to the furrows and measure three to four feet wide and two to three feet deep. Water from the lateral ditches runs slowly to larger ditches called *cross ditches* because lateral ditches seldom carry water effectively for more than 2,400 to 3,000 feet. The large cross ditches are placed at right angles to the lateral ditches and carry water to large twenty-foot-wide canals that extend into the backswamp. Field maintenance is always important, especially in the wet spring, when soils are exposed to erosion by raindrop, and in summer, when hurricane season approaches.

Except for a fewer number of cross ditches on the landscape today, contemporary ditches and canals are similar to the widths, depths, and numbers of ditches in the nineteenth century (Delavigne 1848, 139–41). South of Baton Rouge, the number of ditches per acre increases because the delta's slope gradients are less and soils have more clay content. In the southern parishes of Lafourche, Terrebonne, and Saint Mary, lateral ditches are

FIG. 3.18. *This cane field has a lateral ditch at* center. *The ratoon crop stage is in April, and the view is toward the levee.* Photograph, 1991

placed about 50 feet apart. Conversely, in the vicinity of Baton Rouge and northward, lateral ditches are 150 to 200 feet apart.

Ditches and canals alone were never enough to drain Louisiana's sugarcane fields. Mechanical devices were put to use in draining plantation lands, especially during heavy rains and under flood conditions. In the nineteenth century, draining wheels, which looked like huge grist mill water wheels, operated on the backlands of larger plantations. These great wheels, fifteen to twenty-five feet in diameter and powered by steam engines, threw water over small man-made levees in the backswamp.[9] On contemporary plantations, gasoline- or diesel-fueled pumps move water over the same small backswamp levees.

Agricultural Implements

Agricultural implements are considered part of the material culture that inscribes cultural evidence into a physical landscape. Farm implements and farm machinery on Louisiana's sugar plantations included hoes, plows, tractors, cane knives, mechanized harvesters, cane loaders, cane carts, cane trucks, and loading derricks. Implements reflect technological changes in motive power over time from hand implements to animal-powered equipment and then to gasoline- or diesel-powered machines. Hand implements and animal-pulled plows were the primary farm equipment before the 1830s. The earliest plows–crude, heavy, wooden tools drawn by a single ox or horse–were used for land preparation and plowing. Brush drags, also pulled by animals, did the harrowing. All other cultivation was done by slaves wielding hand hoes, spades, and shovels (U.S. Department of Commerce 1913, 9). The hand hoe was extensively used on antebellum plantations, but the use of hand hoes rapidly diminished after the Civil War because the captive hand labor in slavery was no longer present. Two types of hoes were commonly used in cane cultivation. A short-handled, narrow-bladed hoe was for grubbing stubble (ratoon) cane, and a long-handled hoe with a five- to eight-inch-wide blade was used for weeding (Pugh 1888, 143–67).

No implement was put to greater use than the steel plow (DeBow 1848, 131–32). The steel plow was lighter, stronger, sharper, and more maneuverable than any previous plow. Field hands could prepare land, open furrows for planting seed cane, and do a much better job cultivating land. The plow could also be used to clean quarter drains and shape up headlands after harvest. In the Midwest it was "the plow that broke the Plains," but in Louisiana it was the plow that opened backland areas of stiff clays near the backswamp. Horsepower was increased to two to four draft animals, usually large, wide-footed mules. Two people were required to operate the plow; a strong man worked the plow, and another person, usually a young boy, drove the mules. The gang plows, multiple hitches, and use of more than four draft animals so common on nineteenth-century Great Plains grain farms were not found on Louisiana's sugar plantations because cane fields had prohibitively higher ridges and wider rows (Maier 1952, 8).

Field implements specifically designed for sugarcane were not developed until after the Civil War, when labor became scarce. The first of these was the slide shaver, a device that cut cane stubble at a uniform length and angle (Bouchereau and Bouchereau 1875, xii). The implement had two runners fifteen to eighteen inches apart and, fixed between the runners, a sharp blade that cut or shaved the old, stubbled cane stalks. At the rear of the device was a "V-drag" that loosened the soil to expose cane nodes for later germination (Maier 1952, 8).

Other innovations were large turn plows for land preparation, left-hand turning plows that covered plant cane, iron-toothed harrows, and double-moldboard plows. The double-moldboard plow came into widespread service because it opened a furrow in a single operation, unlike earlier plows that required two trips over the field to accomplish the same result. The rotary hoe was introduced in the late nineteenth century to replace hand labor in building up rows into higher ridges. This mule-drawn implement had radiating paddles in a fan-shaped arrangement that scooped up soil from both sides of the ridge and in so doing worked both sides of the row (Maier 1952, 8). In later years, disk harrows and disk cultivators came into accepted use and, except for their number and motive power, the disks then differed little from those currently in use.

Mechanized farm equipment for cultivation was slow to enter the sugar region. Initial attempts to use steam tractors failed because the heavy equipment mired in deep muck soils. Moreover, steam tractors were good for only one job, breaking land. Attempts to use caterpillar tractors in the 1930s and 1940s also failed because the equipment was heavy, it failed to have adequate clearance over tall ridges and tall canes, and the caterpillar's tracks lacked the maneuverability of wheeled tractors.

Contemporary tractors are high-clearance vehicles designed to fit a six-foot row and fitted with hydraulic devices to lift gang plows clear off the ground. It is important for the tractor to be able to turn around on existing narrow headlands. And it is particularly important for the equipment to be roadworthy so that it can go from field to field and from plantation to plantation with tilling equipment still attached to the tractor but clear of the road. Tractors come in two- and four-wheel drive models that, when properly outfitted, can cultivate two, three, or even four rows on a single pass through the field. Many have enclosed, air-conditioned cabs with cellular phones, AM/FM stereo radio cassette systems, and comfortable seats. Such tractors may be efficient labor-saving devices, but they are expensive, at nearly $100,000 per unit, so it is not surprising that many planters choose to lease high-cost tractors and mechanical harvesters.

Sugarcane throughout the sugar world was cut exclusively by hand until about 1935. Paradoxically, cane is still cut by hand in many places, including Florida, much of the Caribbean Basin and South America, Africa, India, and Southeast Asia. After the invention of the mechanical cane harvester in 1935 in Louisiana, the need for hand-cut cane diminished; however, cane knives continued to be used to clean up fields, catch stalks

Fig. 3.19. *Cutting cane with a long-handled cane knife here was just to "clean up the field" by cutting those stalks missed by the mechanical harvesters on Poplar Grove Plantation. West Baton Rouge Parish.* Photograph, 1966

missed by the harvester, and prepare plant cane. Cane knives come in many different types. The ones in Louisiana have long wooden handles and long, wide blades with a hook at one end of the wide, blunt blade. The Louisiana blade measures eighteen inches long and has a single cutting edge, and the hook on the dull side of the blade is used for picking up cane stalks and for stripping leaves. In the Caribbean, cane knives called cutlasses have long, pointed machete-like blades. An ancient cane knife used in the Caribbean is the bill blade, a heavy, round metal blade shaped much like a frying pan and weighing about as much. Another Caribbean knife appears much like those in Louisiana except that it is shorter in both blade and handle and lacks the hook. When Jamaican cane laborers were brought into Louisiana to harvest storm-damaged cane by hand and were issued the traditional long Louisiana knife, the Jamaicans immediately cut the handle and filed off the hook to convert the Louisiana cane knife into a proper traditional "Jamaican cane knife."

The mechanical cane harvester was invented in Louisiana in 1935. Several reasons account for its late arrival in the evolution of sugar technology, perhaps the most important being that labor remained plentiful and cheap in Louisiana. The cane mosaic disease in the 1920s had reduced the crop to such an extent that there had been a labor surplus. Moreover, the Great De-

FIG. 3.20. *The long-handled cane knife has been the traditional cane knife in Louisiana's sugar country. Note the hook on the blade (at* left*) and my five-inch pocket knife for scale. Poplar Grove Plantation. West Baton Rouge Parish.* Photograph, 1966

pression was creating a large labor surplus throughout the sugar region as plantations and cane farms were foreclosed. Gasoline-powered technology was just coming into its own for vehicles of all kinds and tractors in particular. The time was simply not quite right for mechanized harvesting to leap to the forefront during the troubled early 1930s. By 1935 high-wheeled tractors were entering the region and a wave of technological developments were taking shape, with mechanical hoeing machines, stubble machines, cane loaders, and other mechanization. Early harvester models built on a tractor chassis were ugly, crude contraptions that cut cane but not in a pretty way–tearing, ripping, and snatching the stalks and flipping them to the ground. The harvester would require years of modification to make it as efficient as the harvesters in use today. But the 1935 models worked well enough that, by 1939, sixteen harvesters were in operation. By the 1940s, twenty-five machines were cutting cane in Louisiana. By the end of World War II, some four hundred mechanical harvesters were chopping their way across Louisiana's cane fields. Sixty-three percent of the 1946 crop was harvested by mechanical means.[10]

The process of cutting cane by machine is a noisy yet fascinating operation. On the contemporary landscape, two types of harvesters are used: (1) the older single- and double-row harvesters and (2) the newest state-of-the-art machine, a cane combine harvester built by Cameco Industries. The older one-row machine aligns to the row much like a corn harvester; as it moves forward, a V-shaped guide gathers the cane stalks between sets of conveyor chains and moves them along to a circular blade, which cuts the tops of the stalks. Another circular blade simultaneously severs the stalks at the ground. Freed at last, the canes move through more sets of chains that convey them to the back of the harvester. The stalks take a right-hand turn and are laid back onto the field across the ridges at right angles to the row. After this short but important journey, the canes are allowed to dry in situ and then are set fire to burn off the now-dried leaves. The placement of the stalks across the rows allows a mechanical loader to gather canes in a grab or clawlike device and to place the now dry, leafless, but soot-covered cane stalks into waiting cane carts or cane-hauling trucks that will take the canes to the sugar factory.

The Cameco cane combine harvester originated in 1965 in Australia but is manufactured now in a new 294,000-square-foot factory in the heart of cane country in Thibodaux, Louisiana. The Cameco harvester is powered by a 250-horsepower Caterpillar engine, and drives can come either with

FIG. 3.21. *Mechanical cane harvesters in the 1960s were single-row machines.* Photograph, 1969

rubber-tired wheels or with caterpillar tracks (Cameco Industries 1996). A newcomer to the landscape, the Cameco differs from other harvesters by its ability to cut the cane at top and bottom, strip the leaves and blow them back down to the ground, chop the stalks into smaller, ten-inch pieces, and load directly into a much larger cane cart. The unique feature of stripping the leaves beforehand means that standing cane can be harvested immediately without having to wait for the leaves to dry and later be burned. An added benefit is the lack of cane-field smoke that pollutes the fall harvest atmosphere. Also important is that the chopped cane output packs tighter in cane carts and tractor-trailer haulers, enabling the cane operator to speed the cane to the sugar factory.

Cane loading operations and devices evolved as time and labor-saving elements of grinding season became more important. Before mechanization, hand-cut cane was hand loaded into cane carts. This simple but arduous, time-consuming task was especially burdensome because of the short harvest season faced by Louisiana every year. At no time has the Louisiana sugar industry enjoyed the luxury of harvesting at any time during the year

Fig. 3.22. *Contemporary cane harvesting uses very different equipment in the Cameco cane combine.* Photograph, 1996

on the flexible timetable of Caribbean sugar planters and planters elsewhere in the tropics. Louisiana's planters are compelled to wait as long as possible to allow canes to attain the highest level of sucrose; then they harvest the crop as quickly as possible before first frost. It is the old story of "hurry up and wait" but in reverse. Wait as long as you possibly can and then hurry like hell to cut that crop before frost. Such pressure for years has made planters lose sleep but develop better and faster means for cutting and moving the cane to the sugar factory.

The first mechanized cane loader arrived in Louisiana in the late nineteenth century. This mule-powered implement was simple, with a boom and hoist mounted on a wagon bed. A scissorslike grab attached to the end of the boom was connected by cable to a mule in the field. Three people worked the loader: one man placed the grab over the cane, one man worked the boom, much like a crane operator, and a third person guided the mule to pull the cable.[11] Even as motive power changed to steam and gasoline, the basic equipment and principles of operation remained the same. It was not until after tractors gained greater favor in the 1940s and

1950s that a motorized loader built around a tractor chassis entered the operation. A front-end loader with a hydraulic grab reaches out and gathers cane stalks that have been laid crossways to the rows. Once the claw is full of cane, the operator turns the machine to one side to load a tandem of two cane carts that have been pulled into the field by another tractor.

Cane carts have been a part of sugar cultivation for centuries, and until recently their material culture has changed little in form. These two-wheeled carts designed specifically to haul cane from field to sugar factory began with wooden wheels, then changed to steel-rimmed, spoked wheels, and eventually evolved into carts with inflated rubber-tire wheels. The unusually tall wheels with six-foot diameters provided the high clearance needed for the cart to pass over the high ridges in the field and to roll on and out of miry cart paths and roads to the mill. On nineteenth-century cane carts, the bed of the cart measured about four feet by eight feet; when loaded high on the sides, it carried as much as a ton of cane (Kellar 1936, 166). Changes in cart size came as changes in row widths were instituted in the region. Before 1825, rows were narrow to accommodate Creole and Bourbon (Otaheite) cane varieties, single-draft plows were used, and cane carts were small and narrow. When wider rows came into existence, cart axles were made longer to accommodate the six-foot standard. Cart beds

FIG. 3.23. *An early cane cart and cane loading equipment in 1906 represented the best technology for the time.* Reproduced from *Louisiana Planter and Sugar Manufacturer* 1906, 36:165

FIG. 3.24. *Contemporary cane carts and cane loading operations are faster and on a much larger scale.* Photograph, 1996

became larger, set to a standard of five by twelve feet, a standard that held in the region until the 1970s. Recent innovations in cane cart design reflect larger dimensions; they are eight feet wide, twenty-five or more feet long, and twelve or more feet high. These so-called tandem chain-net carts, which are now the size of a small moving van, have grown to four large wheels. Even larger are cane trucks, which are the size of an eighteen wheeler's trailer; when loaded with cane, they have the appearance of a see-through box of pretzels. These sugarcane supertankers dominate the long-distance hauling of cane to one-way distances as much as seventy to one hundred miles from field to factory.

Cane derricks, once a major part of the sugar plantation landscape, were used to load cane from cane carts onto larger trucks. The derrick was built of two twenty-five- to thirty-foot-long steel booms. One boom was the support post held vertically in place by guy wires or by two smaller, fixed support posts. The second boom, a working lifting beam, had one end attached to the base of the vertical support boom and the other end rigged with cables, block and tackle, and chains so that a mass of cane stalks could be lifted from the cane cart and transloaded into a large cane truck.

Cane derricks date from the nineteenth century, when they were used principally in the mill yard at each sugar factory. Cane would arrive directly from the field in cane carts, and the derrick, then made of heavy wooden beams, would unload the cart at the mill yard. In the 1920s and 1930s, with the closing of so many sugar factories, cane derricks were pressed into wider service. On plantations with closed sugar factories, the derrick was retained so that cane could still come to the mill site as it had always done and be loaded into large trucks to be transhipped to a working sugar factory several miles away, beyond the range of cane carts. On other plantations, the derrick was relocated to a site nearer the center of the plantation or closer to the highway, where canes in cane carts were loaded into large trucks. Large plantations had several cane derricks strategically located at convenient sites to efficiently load truck trailers that would be hauling cane to sugar factories many miles away.

Plantation Roads and Railroads

With all the movement of cane over short and long distances, roads and railroads have played a major role in the function and forms of transport over the landscape. Every plantation has a two-lane main road that connects the plantation to the outside world's parish- or state-maintained public road. On plantations along the Mississippi River, the main road bisects the linear plantation settlement. Along Bayou Lafourche, a plantation's main road begins at a T intersection at the bayou road and leads back to the block plantation settlement. This main road, sometimes called a quarters road, is the focus of the settlement, providing access into and from the plantation, but it is also the main stem to which field roads connect. Each plantation has a gridwork of internal roads interconnecting fields with the main road and the plantation headquarters, a central place that may or may not have a sugar factory, an outbuilding complex, an owner or manager's residence, or a quarters. Single-lane field roads carry tractor traffic, for the most part, and the occasional overseer's truck. Most plantation roads are unsurfaced, and only the main roads are ever paved. The internal road patterns on plantations have not changed appreciably since the beginnings of plantation sugar in Louisiana. Early roads were smaller and narrower than the slightly wider and better maintained roads on current plantations.[12]

The role of internal railroads on Louisiana sugar plantations was never large nor really effective but not from a lack of trying on the part of a few planters. On-site plantation railroads simply did not work. Why were internal railroad systems attempted on Louisiana plantations, and why did they fail? Planters in the 1870s began to experiment with crude railroads using portable wooden tracks and horse-drawn carts in Saint Mary and Terrebonne Parishes (Sitterson 1953, 264). By the turn of the twentieth century, steam engines and fixed steel rails on narrow gauge lines were appearing on the landscape. In 1900, 64 of the 723 sugar producers in the state had railroads (Rightor 1900, 687–726).

The economy of scale, or the size of operation, governed the efficiency of and need for a railroad. Cane carts proved to be the most economical for hauling cane short distances and were used on most plantations. The purchasing and maintenance costs for locomotives and rolling stock were simply too high for most plantations, and railroad tracks took up space that other vehicles could not use. Furthermore, the soft muck soils in the backlands would not support tracks carrying heavy locomotives and rolling stock. The few plantations that had some success with internal railroads were those linked by rail over large areas, where cane could be hauled to a central mill. The railroad phenomenon that flowered around the turn of the century eventually began to die in the 1930s from economic problems in the Great Depression and from a change in technology, the arrival of the gas-powered tractor. In the 1960s, three plantations still had narrow gauge railroads that continued to operate on the landscape. But for much of the region, tractors hauled cane economically up to a distance of five miles from field to mill; for distances beyond five miles, trucks took over the transportation. External transportation via main line railroads naturally continues to be important for transporting bulk cargos of raw sugar in freight cars and molasses in tank cars.

Landings and Levees

For external transportation in the nineteenth century, planters relied almost exclusively on water transport on the Mississippi River and bayous throughout the sugar region.[13] Plantations fronting navigable streams had landings. The plantation landing was more of a geographic location than a form. A landing was that point on the landscape where the internal focus of a plantation turned and reached out to meet the outside world brought to the scene by boat. "Steamboat comin roun da bend" was a call that excited everyone on a plantation. Whenever a steamboat approached the landing or

FIG. 3.25. *Waterways have been crucial for transporting bulky molasses and raw sugar, as from the former Louisa Plantation on Bayou Cypremort. Saint Mary Parish.* Photograph, 1967

was moored at the site, a higher level of excitement and energy pervaded the plantation scene. Planters embarked as they made their way to New Orleans or beyond. Others were returning from New Orleans or from trips abroad, some with families and some with slaves.

For most plantations, a landing was merely a cleared opening on the stream bank where wagons could be driven directly to the river at a site where a steamboat might moor (Barry 1997, 40). Landings were important places where primary commerce took place. Plantation raw sugar and molasses products were loaded on river boats to be physically exchanged for external supplies, machinery, and slaves arriving at the plantation. Fuelwood sat in huge stacks at the landing to be loaded aboard the steamboat to fuel the boilers, which provided steam to drive the massive pistons. Some landings were more than mere mud. A few evolved into commercial stores, such as Caire's Landing in Saint John-the-Baptist Parish. On streams with less flow, landings took the form of wooden wharves and small docks. Planters flew flags and banners at the landing to identify the plantation name. Current landings exist in name only where river charts still have the

FIG. 3.26. *Nineteenth-century landings sometimes marked the presence of stores as well as plantations, such as this one at Caire's Landing in Saint John-the-Baptist Parish.* Photograph, 1993

names of plantation landings as points of reference. Except for ferry boat landings and barge terminals, the landscape presence of plantation landings has disappeared.

The Mississippi River has an infamous history of flooding the Lower Mississippi Valley. The very existence of the delta and its fingers of natural levees is due to the natural patterns of alluvial flooding and deposition. Just as Herodotus explained "Egypt is the gift of the Nile," one can counter that "southern Louisiana is the gift of the Mississippi." However rich that gift might be, the river exacts a price through flooding. The fight against floods in the Mississippi floodplain has been a long, arduous, and expensive battle. The present levee system confines the Mississippi River continuously between a pair of man-made levees from Venice, Louisiana, to Cairo, Illinois. This massive levee system was constructed and is maintained by the U.S. Army Corps of Engineers. For much of its controversial history, the army's levee system has prevented major flooding in the Lower Mississippi Valley (Barry 1997, 190–91).

The man-made, or artificial, levees are extrordinarily imposing well-engineered features on the cultural landscape. Proceeding from the chocolate-colored Mississippi, you will encounter the following features: the steep stream bank, the batture, the actual man-made levee with a road on top, a fence, a ditch, and perhaps a paved river road on the landward side. The batture is a strip of wooded, flood-prone land from which earth is excavated to construct the levee. The batture may be only a few yards or up to a mile or more wide, but it is pocked with large, deep borrow pits from which the sandy, silty earth is dug to produce the levee structure. The batture is traditionally a no-man's land, but at times it has been settled by squatters.

Levee dimensions exceed the expected, with every mile of levee containing at least 421,000 cubic yards of earthen materials. The sides of a levee are built with a three-to-one slope, so that a levee 30 feet high has a base that is 188 feet wide. The crown, or top, of the levee has an 8-foot-wide road paved with rangia clam shells in Louisiana or with gravel elsewhere. A special vegetative cover is given to the levee. In the batture willows are encouraged to thrive. On the levee itself is planted bermuda grass, a short grass that binds well with levee materials, requires less mowing and upkeep, and provides a surface good for inspection for potential breaks by the levee police (Du Terrage 1920, 632).

The need for man-made levees was recognized early in the European settlement of the region. Between 1718 and 1900, various levee regulations required property owners to be responsible for building and maintaining their own levees (DeBow 1851, 530–34). This reliance on the individual was unsuccessful because landowners constructed dikes of variable heights and quality, and some neglected the job altogether. It would not be until after 1900 that a system of permanent uniform levees would be built and then only after years of appeals made to state and federal authorities for financial aid and the formation of the Mississippi River Commission. Elsewhere in the region, bayou plantations were able to maintain their own small levees for the most part because flooding was not so severe. Planters maintained their own backland levees near the backswamp throughout the delta region. The artificial levee was and will continue to be a necessity in a land that is so close to water at the margins of terra firma on the Gulf of Mexico exposure and so close to being flooded on the other by its maker, the Mississippi River.

CHAPTER 4

A Prescription for Landscape Decline

American agriculture seems to thrive on change. Vigorous, structurally dynamic, and becoming too impersonal, the ways of farming are part of a rural revolution in the United States in which changes in the ownership and operation of family farms lead to fewer farms farmed by fewer farm owners (Hart 1991b). Progressive changes are perceived to arrive with corporate agribusinesses operating large landholdings with greater efficiency through improved technology and management.[1] In this chapter I explain a sequential pattern of land occupation in Louisiana's sugarcane industry in the light of landscape models and transformations. When John Fraser Hart spoke in 1982 of the death of the southern plantation, I disagreed with him because the plantation concept had endured so well on sugar plantations in southern Louisiana. But since the introduction of a California agribusiness model and the emergence of what I call the CALA model, the concept of *plantation* as a functional and cultural form in Louisiana admittedly has been rapidly declining and is probably approaching death. For much of the delta area, structural changes in the plantation landscape were fueled by change agents in the 1970s, and some changes were made by entrepreneurs from California and Arkansas.

When I synthesized 202 Louisiana sugar plantations and analyzed 6 of them as case studies in 1969, I was convinced that I had a complete story and that the plantations were safely in situ (Rehder 1971). Returning in 1989 to reflect on my earlier research, I expected the plantation landscape to be well preserved, locked in time and place. After all, it had changed very little morphologically since the 1930s and 1940s, and spatially it was the same as the mid-nineteenth-century plantation landscape. But dramatic structural changes had occurred in apparently negative ways. More than half of the plantations were missing, and the number of sugar mills had declined from forty-four to twenty. Two case-study plantations had disappeared, three were in ruins, and only one was still intact. Archaeologists were already digging sites that I had studied as plantations just twenty years before.[2] What was going on in the Louisiana sugar business? Was it really declining, or was it in some strange way improving? If the traditional plantation model that had been the focus of my original research was dying, what was taking its place?

Models of Land Occupance

Louisiana's 690 sugarcane-producing units are cane farms, plantations, agricultural operations, cooperative cane growers, and sugar corporations. They can be classified into models of land occupance based on historical parameters of ownership and scale of operation,[3] both of which serve as indicators for phases in the sequent occupance, or settlement succession, of the landscape.[4] Some units move through each historical sequence, while others retain the form of a traditional plantation. Briefly, the models emulate the following pattern:

1. The traditional plantation (since 1795) represents a relic from antebellum times.

2. The Louisiana plantation corporation model (since 1930) develops as a multiplantation corporation with an "old plantation" patriarchal pattern of management and settlement for individual plantations.

3. The California agribusiness corporation model (since 1930) emerges

and remains on the West Coast as an impersonal, centrally managed agribusiness factory-in-the-field. Conceptually, it enters Louisiana as a management overlay in the 1970s.

4. The CALA (**CA**lifornia-**L**ouisian**A**) model (since 1974) results from the lamination of the California corporation model to the Louisiana corporation model and creates a streamlined, almost antiseptic, but functionally efficient landscape that loses the old plantation patriarchal management and settlement qualities of the southern plantation.

5. The small plantation + cooperative sugar factory model (since 1932) is based on the traditional plantation and the symbiotic relationship between the small plantation and the cooperatively grower-owned sugar factory that ensures a degree of plantation survival in an era of uncertainty, when even the largest corporations have failed.

The Traditional Plantation

The traditional plantation model, at the height of its antebellum form in 1860, was a singularly owned, family-occupied enterprise. Within a rectangular landholding of several hundred acres, the unit consisted of a sugarhouse (or sugar factory), warehouses, sheds, barns, implements, oxen, mules, horses, and crops of sugarcane and corn (Sitterson 1953, 45–47). Three to five overseers oversaw the work done by an average of seventy slaves (Schmitz 1979, 270–85). Dwellings of variable size and with identifiable cultural identity sheltered the planter's family, his overseers, and his slaves. The latter were housed on the premises in the agglomerated village settlement called the "quarters" (Rehder 1989a, 576–77). In 1860, Louisiana had more than thirteen hundred sugar plantations, with many generally fitting this description (Champomier 1860).

In chapter 2, we saw how diagnostic landscape traits in settlement patterns and house types marked significant cultural differences between French and Anglo-American sugar plantations. Plantations with French origins had Creole mansions and linear settlement patterns. Anglo plantations had block-shaped settlement patterns with mansions having a single front door, multiple outside chimneys, some with a front-facing gable, and all with symmetrical floor plans of the English pen tradition.[5] The traditional plantation model of 1860 not only mirrored earlier models of the late eighteenth and early nineteenth centuries, but also was reflected in twentieth-century plantations. As plantations after the Civil War changed from slave to wage labor, the settlement patterns in the quarters remained the same throughout the temporal range from 1795 until the 1970s. But after 1970, the quarters settlements were aggressively removed from the landscape.

Sugar factories effectively served as technological, spatial, and temporal symbols of the sugar industry. Each nineteenth-century plantation operated its own sugar factory (at the time called a *sugarhouse*), but plantations of the twentieth century did not (Heitmann 1987, 11–12). In 1870 fewer than 300 of the 1,300 prewar sugar factories were functioning, but 1,100 were back in operation briefly by 1879, only to drop in number to 275 in 1900 and to 70 by 1930.[6] The numbers and distributions of sugar factories were noticeably reduced as technological improvements led to a centralization of factories. The decline in number of sugar factories continued, with 43 factories in 1970, 24 in 1980, and 20 until 1996, when the number dropped to 19 sugar factories.[7]

The Louisiana Plantation Corporation Model

The Louisiana plantation corporation model developed from the accumulation of formerly family-run traditional plantations, largely acquired from foreclosures and bankruptcies in the 1930s. Clearly corporate in upper management and ownership, the model represented a collection of separately operated plantations. Collective holdings in different geographic regions were identified with division names, such as Oaklawn Division or Thibodaux Division. Within divisions, plantations retained original property boundaries and identity by name. Those with sugar factories became divisional centers, but those without sugar factories were identified as agricultural operations. Despite upper management corporate authority, an "old plantation" patriarchal concern in plantation management was clear.

One of Louisiana's largest sugar corporations, Southdown, best exemplifies the model. Built from the ashes of the Depression by the acquisition of property from foreclosures and bankruptcies, Southdown began in 1930 under the name Realty Operators (Farr, Whitlock & Co. 1960, 208–9). For more than forty years, the corporation exclusively owned and operated one of Louisiana's largest sugar refineries, three sugar factories, and twenty-one plantations in three divisions.

Within the corporate structure each plantation had a personality of its own. For example, Southdown's Armant Plantation, located on the Mississippi River at Vacherie, illustrated the sequent occupance from a traditional

TABLE 4.1
Southdown Divisions and Plantations, 1974

Division and Plantation	*Land Area*
1. Houma Division	
Ardoyne Plantation	28 arpents[a] front, both banks
Concord (eastern part)	724 acres
Crescent Farm	751 acres
Greenwood Plantation	952 acres on Bayou Black
Oak Forest Plantation	2,845 acres on Bayou Black
Southdown Refinery (capacity, 1,250,000 pounds)	
Southdown Factory (capacity, 4,000 tons/day)	
Mandalay Plantation	33 arpents front, left bank
Waterproof Plantation	20 arpents front, left bank
Hollywood Plantation	1,400 acres
Southdown (western part)	1,842 acres
2. Thibodaux Division	
Abby-Highland Plantation	1,309 acres
Breaux Plantation	507 acres
Edna Plantation	30 × 40 arpents
Greenwood Factory (capacity, 3,200 tons)	
Greenwood Plantation	2,075 acres
Nora Plantation	9 × 40 arpents
Orange Grove Plantation	1,283 acres
St. Rose Plantation	14.5 × 80 and 7 × 40 arpents
3. Vacherie Division	
Armant Factory (capacity, 3,800 tons)	
Armant and St. James	54 × 80 arpents or 4,320 acres

Sources: Terrebonne Parish 1974, book 602, p. 611–47; 1977, book 697, p. 540; Gilmore 1975.
[a] arpent = 192 feet

plantation into a divisional center in the Louisiana plantation corporation model and later exhibited the consequences of its transformation into the CALA model (see chap. 5 for the complete narrative on Armant). Except for cane fields, Armant Plantation no longer exists. But in 1969, at the zenith of its landscape presence, Armant contained 2,876 acres of cultivated land within a landholding of 4,320 acres. The French Creole mansion, built in 1795 by Jean-Baptiste Armant, was deteriorating and had to be razed. At the heart of an operational outbuilding complex was a modern sugar factory with a daily grinding capacity of 3,800 tons and accompanying storage tanks, barns, warehouses, and sheds. With tall, black smokestacks towering over the corrugated metal buildings, the enormous sugar factory complex may have resembled a silver steamboat sailing a sea of sugarcane, but realistically and functionally it was clearly a factory-in-the-field. On site, Armant also operated a bagasse factory for drying cane stalk pulp for the Celotex Corporation to make ceiling tiles and fiberboard. The settlement for the primary quarters contained twenty-one quarter houses. These and other clusters of quarters housed fifty-seven resident workers on the plantation (Rehder 1971, 237–76).

When Armant Plantation was established in 1795, with the initial owner, Jean-Baptiste Armant, the shape of the property was the familiar long rectangle from the French arpent land survey and it was oriented perpendicular to the Mississippi River. The linear settlement pattern and French Creole mansion traced directly to Jean-Baptiste Armant. Through successive years, the plantation passed from father to sons in the Armant family until ownership eventually went to John Burnside in 1860, Oliver Beirne in 1883, and then William Porcher Miles in 1889.[8] The Miles family maintained ownership and operations until 1934, when Southdown acquired Armant Plantation on assumption for $310,000 (St. James Parish 1934, book 62, p. 452). Except for modernizing the sugarhouse and outbuildings, the Southdown corporation made few changes to the structure, settlement pattern, integrity, and identity of the plantation. It was one of twenty-one plantation units under the umbrella ownership and management of Southdown.

Southdown, the parent corporation, diversified extensively between 1967 and 1989, acquiring oil interests, cement companies, a brewery in Texas, a candy company, a soft drink company, additional land parcels, and California vineyards and wineries (Moody's 1990, 4239). In the midst of these departures from sugar, Southdown executives decided to get out of the sugar business entirely, and by 1980 the corporation had sold all of its sugar-related holdings. The current Southdown corporation exclusively operates a large cement business with eight quarries and eighteen distributors of cement and concrete products.[9]

Meanwhile, Armant and many of its sister plantations began to disappear morphologically from the landscape. Major changes occurred on former Southdown plantations after the sugar portion of Southdown was sold to a group of California entrepreneurs. By 1981, nearly all buildings related to Armant Plantation had been razed. All former resident laborers and their quarter houses were gone. The Armant sugar factory site of 129 acres was sold to a Costa Rican company, which leased the site to a small chemical firm (St. James Parish 1981, book 235, p. 370). The only remaining evidence that the property had ever been a sugarcane plantation were two metal warehouses at the sugar factory site and the continued cultivation of cane fields, now under separate management. Seeking an explanation for these dramatic changes, I looked elsewhere in American agriculture to another plantation-like farming structure, an agribusiness model in California. It was a model that in part had diffused to California in the 1930s from the Deep South cotton culture (Gregor 1962, 1982). Remarkably, the model appeared to have diffused from California to Louisiana in the 1970s.

The California Agribusiness Corporation Model

The California model, an agribusiness landscape model for more than sixty years, illustrates a pattern of agricultural operation and rural occupancy in California that diffused to Louisiana in the 1970s. The California model represents a one-unit, centrally managed enterprise containing large landholdings, operated by teams of management experts, with labor furnished by nonresident workers.[10] Two California companies, separated in time, are exemplary for comparison to the Louisiana corporations. In the 1930s, the Earl Fruit Company in Kern County, California, owned and operated twenty-seven farms, leased eleven others, and contracted for fruit from independent growers (Kirby 1987, 7). The company kept 350 dwellings for use by its three thousand seasonal workers (McWilliams 1942, 17–20). It owned eleven packing houses and an interest in a packing box company and was a subsidiary of a company that owned warehouses and distributorships in major market cities. It also owned a third interest in a major winery (Kirby 1987, 7). Thus, from tree to table, the Earl Fruit Company controlled a large share of the business and nearly all of the processes in between.

A more recent California example is Tejon Agricultural Partners (TAP), also in Kern County in the San Joaquin Valley of California. TAP is a 21,000-acre operation with agricultural specialists, managers, owned and leased land, leased equipment, and landless workers (Kramer 1977, 183–256). It is a farmerless farm because the owners do not farm it and the farmers do not own it. Much like a factory in an industrialized city, neither group resides on site. Even as TAP was being developed, tenants were forced off the property (210).

TAP began in 1971 as a reaction to an opportunity to tap irrigation water from California Water Project canals that were supposed to come through TAP land. Corporate executives from the parent company, Tejon Ranch, created a financing and management arrangement so that TAP could be incorporated as a limited partnership owned by limited partners and a general partner (Stanley 1979, 14–18). The limited partners were thirteen hundred mostly nonfarming investors who invested $16 million (with a minimum of $5,000 each), primarily for tax advantages. The general partner was Tejon Agricultural Corporation, a corporate subsidiary, which provided twenty-one thousand acres of land worth $13.7 million and $3 million for operations but controlled all operations of the partnership. The property was then encumbered and mortgaged to the John Hancock Life Insurance Company as collateral for a $27 million loan (Kramer 1977, 202–8). Once established, TAP became an agricultural factory-in-the-field (McWilliams 1939), owned by nonfarm investors and managed at the corporate level from a parent company with headquarters in Los Angeles.

Tejon Ranch Corporation, the parent company and one of California's oldest, continues as a 300,000-acre conglomerate that includes a seed company, oil reserves, a cement plant, cattle, and feedlot operations (Moody's 1990, 6301). Although in its diversity the parent company resembles the earlier Southdown, TAP is not at all analogous to Louisiana's traditional plantations. TAP is a California fabrication much more like the CALA model that surfaced in Louisiana in the 1970s.

The CALA (California-Louisiana) Model

The CALA model emerges from the combined characteristics of California agribusiness corporation models and the Louisiana corporation plantation model. It develops from the lamination of managerial skills and philosophy from California to the old plantation landscape structure of corporate sugar in Louisiana. Both California and Louisiana models maintain immense landholdings, large capital outlays for production costs, lease agreements, corporate structure, and an image that large size usually means success. Gregor argued in 1962 that large farms in California rightfully should be called *plantations* because of their patterns of operation

(Gregor 1962, 1–14). However, I argue that the California agribusiness model (Gregor's *plantation*) carries with it a different way of life, a different attitude in management, and a different historical development from that of the Louisiana sugar plantation corporation. By combining the two corporate models, a California-Louisiana hybrid, the CALA model, emerges.

Our original, home-grown Louisiana corporation plantation model had been characterized by (1) origins from acquisitions of foreclosed plantations in the Depression of the 1930s; (2) traditional identification of plantation units; (3) patriarchal management at three levels–central, divisional, and plantation; and (4) local product identity with initial exclusive interests in sugarcane and later interests in Louisiana petroleum. Key characteristics of the California model that reached Louisiana were (1) creative financing through insurance lenders; (2) centralized corporate management throughout the system; (3) multiple internal corporations; and (4) landscape streamlining by consolidating agricultural units and equipment and by removing resident laborers and their houses from corporate land.

Transformation

The best example of the CALA model developed from the acquisition of Southdown in 1974–77 by California agricultural entrepreneurs. In the transformation of the Louisiana corporation model at Southdown into the CALA model in the 1970s, large plantation tracts under a corporate umbrella became available at a reasonable price because the original Southdown was anxious to divest itself of the sugar business and to diversify. A bonus was the large infrastructure of a refinery, three working sugar factories, and agricultural equipment that came with the sale. Financing came through partnerships with such large life insurance companies as Prudential and Northwestern Mutual. This practice arrived directly from California.

Under these arrangements, the California group conveniently purchased the entire complex. Management then looked for ways to simplify and streamline the system and operations. Refineries and outdated sugar factories were closed. Sugar factories from competing plantations were purchased separately and then closed. Tenants were moved off the land.[11] Unnecessary buildings were leveled because vacant houses were taxable and every acre of land was needed for cane cultivation. Most important, the liability of having agricultural workers and their families living on the premises was a risk the corporation did not want to take. The fewer things and people to be managed, the more efficient the operation. But removal of quarter houses meant the destruction of one of the key landscape signatures of plantation morphology. Without quarters, the plantation had become a corporate landscape.

While it appears simple and streamlined, the CALA model is a myriad of complex business connections with many facets of the operation separately incorporated. The trucks become one corporation; the sugar factory is another corporation. Most agricultural equipment, additional land, and crops are leased. Operational management is in the hands of a few savvy agribusiness experts with expertise in managing people, equipment, and land.

Chronology

The initial linkage between Louisiana and California was made in 1969 when the Louisiana corporation, Southdown, bought a California company called McCarthy and Hildebrand Farms. In 1972, Southdown acquired Chillagoe Land Company from McCarthy Brothers and began to invest in the wine business by acquiring more land in California and forming Santa Clara Vintners (Moody's 1990, 4239). Meanwhile, in California, Leland McCarthy, the same McCarthy who had been involved in the earlier transactions, was a major grape producer having problems marketing his grapes to a single market–Gallo. McCarthy sold the land but continued managing the vineyards.

By 1974, there was a direct linkage from California to Louisiana among Southdown, the Prudential Life Insurance Company, and a group of entrepreneurs and managers including the McCarthy brothers. Prudential Life Insurance Company financed the purchase of Southdown for $11 million in the name of Prudential Southdown Partnership. The Southdown that they bought consisted of one sugar refinery, three sugarcane factories, and twenty-one plantations covering more than twenty thousand acres (Terrebonne Parish 1974, book 602, p. 647).

In 1977, McCarthy Joint Venture, headed by California vineyard entrepreneur Leland McCarthy and financially backed by Prudential and other investors, purchased the former Southdown package from Prudential Southdown Partnership for $11 million (Terrebonne Parish 1977, book 697, p. 523). By 1979, McCarthy Joint Venture had sold off nine of its twenty-one plantations. Seven of the nine units were sold to Thomas B. Goldsby Jr. and James G. Robbins, land speculator-entrepreneurs from West Memphis,

Arkansas, and Memphis, Tennessee (Terrebonne Parish 1979, books 744, 749, 769).

By late 1979, McCarthy and Cook as the California connection and Goldsby and Robbins as the Arkansas connection teamed up again with Prudential financing to purchase the largest of Louisiana's sugar corporations–the SouthCoast Corporation. Led by Goldsby in the name of Mid South Mortgage Company, the entrepreneurs acquired SouthCoast, a 42,000-acre complex consisting of one sugar refinery, three sugar factories in three divisions, and many plantations, for $42.5 million (Terrebonne Parish 1979, book 768, p. 45). By the early 1980s, Goldsby reportedly was having financial difficulties and left the operation, returning to Arkansas. The dream of an empire in southern Louisiana was apparently over. Meanwhile, Californians McCarthy and Cook continued to be powerful sugar entrepreneurs, repurchasing four plantation properties from Robbins in 1989–90 and effectively merging Louisiana's two largest sugar corporations–Southdown and SouthCoast–to form the current CALA model, now called *South Coast Sugars.*

In short, the California corporate agribusiness model neatly matched the Louisiana corporations. Between 1974 and 1977, Southdown was transformed into the CALA model through the management strategies of its new California owners. By 1979, through entrepreneurial actions by two land speculators from Arkansas along with the California owners of Southdown, the immense 42,000-acre SouthCoast Corporation was added. The subsequent combination and transformation of the two Louisiana corporations, Southdown and SouthCoast, represented over 80,000 acres and nearly 30 percent of the Louisiana sugar industry and became the greatly modified mega-corporation called South Coast Sugars (Gilmore 1981). The outcome was a singular corporate giant with more than 80,000 acres of Louisiana sugar land under its control.

Consequences of the CALA model were streamlined operations, radically modified landscapes, and the disappearance of traditional plantation culture. Two refineries were closed at the time of acquisition, and seven of eight sugar factories have been closed. The Oaklawn sugar factory permanently closed in 1989, leaving the large-capacity sugar factory at Raceland as the only remaining operating sugar factory in the CALA model. In 1993, financial problems reportedly forced South Coast Sugars to revert to the original lenders so that Prudential controls Louisiana's largest sugarcane corporation and its operation.

TABLE 4.2
Chronology of the Transformation of Louisiana Sugar Corporations into the CALA Model

SOUTHDOWN		SOUTHCOAST	
1930	Realty Operators is established	1935	SouthCoast is established.
1974	Prudential Southdown Partnership purchases Southdown.		
1977	McCarthy Joint Ventures acquires Southdown and becomes the initial CALA model.		
		1979	Mid South Mortgage Co. purchases SouthCoast.
1979	Mid South Mortgage Co. + McCarthy Joint Ventures combine Southdown + SouthCoast to form SOUTH COAST SUGARS, a mega-corporation CALA model.		
1989–90	South Coast Sugars repurchases Crescent, Waterproof, Hollywood, and Mandalay Plantations.		

Sources: Terrebonne Parish 1974, 1977, 1979, 1989–90; St. James Parish 1934.

The plantations in the CALA model and elsewhere in the region had developed into something that was neither a family farm nor a traditional plantation. Workers were merely employees unattached to the land. Workers did not live on it; they did not own it; they only worked it from the seats of rented tractors, cane harvesters, and trucks. It takes time to comprehend a farmerless farm. Traditional topophilia (Tuan 1974) rarely exists because the land one works belongs to the corporation. Even to corporate management, the land conjures no personal interest, no more so than what the pavement on a highway means to a motorist. Detached from the traditional values of home, homeplace, granddaddy's plantation, my place, or even your place, there is something alien about a farmerless farm.

Louisiana Small Plantation + Cooperative Sugar Factory Model

With the emergence of the CALA model, the concept of *plantation* as both a functional and a material cultural property seemed to be proceeding toward imminent death. Yet in a smaller and quite separate model, the

Louisiana small plantation + cooperative sugar factory model had evolved from original traditional plantations and was surviving. This model developed separately from the larger corporate types, and in my study of thirty years of change in Louisiana, this model changed the least. The model is small, identifies with cane farms and small plantations, and maintains a symbiotic relationship to a cooperative sugar factory. Ranging in size from two hundred to more than two thousand acres, this model may have originated as an antebellum plantation that retained its original size and identity with the family farm. It survived by becoming part of a larger infrastructure that maintains all utilitarian aspects of sugar making. The model is owned or more often leased and is operated much like a family farm. Survival depends on the ability of the operators to continue cultivating sugarcane at a commercial level by maintaining a symbiotic relationship with the larger infrastructure and the cooperative sugar factory.

The small operator cannot go it alone in the sugar business. Small feeder growers contract to sell their cane to nearby sugar factories. Sugar factories in Louisiana must have such arrangements because the capacity of each factory usually exceeds the amount that can be supplied by company land. All small growers need sugar factories, and all sugar factories need outside growers. The grower supplies a specified number of tons of cane stalks to the factory and receives 61 percent of the proceeds from the sale of the raw, unrefined, brown sugar and about 50 percent of the proceeds from molasses after processing. The factory retains 39 percent of the proceeds from the sale of the raw sugar and about 50 percent of the proceeds from molasses sales (Chapman, Heagler, and Paxton 1987, 6). In effect, the grower pays the sugar factory a "processing charge" to grind his cane, and the factory "buys" the cane as raw material.

Cooperative sugar factories, however, are owned and managed by groups of cane growers. A group may have twelve to two hundred members, with farms and plantations ranging in size from a few acres to several thousand acres (Gilmore 1981). Some units are cane farms, some are plantations, and some are multiplantation operations all planting, cultivating, and harvesting sugarcane at a commercial scale but processing the cane at a sugar factory that the growers collectively own.

A good example of the small plantation + cooperative sugar factory model is Madewood Plantation (see chap. 10 for a complete narrative on Madewood). Madewood is located on the left bank of Bayou Lafourche in Assumption Parish and has 550 acres of cultivated land, of which 200 acres are in sugarcane. The plantation is one of three in a multiplantation enterprise owned by John Thibaut. The others are Glenwood and Woodlawn Plantations. Cane grown at Madewood is processed at the Glenwood Cooperative sugar factory. The cooperative, the first one in Louisiana, was founded in 1932 as a response to the needs of local small growers and as a result of the consolidating and centralizing functions of modern sugar processing technology.

In my 1969 research, Madewood Plantation was a case study of an Anglo-American sugar plantation in French Louisiana. The property traces to Thomas Pugh, a North Carolinian, who migrated to Louisiana in 1818. Between 1823 and 1848, Pugh purchased separate land parcels for Madewood, eventually amassing a landholding of 1,000 acres of arable land with 1,700 acres of backswamp. The plantation remained in the Pugh family until 1896, when Leon Godchaux, the post–Civil War sugar magnate, purchased it. Subsequent ownership went in 1910 to Henry Delaune and Alcee F. Delaune and then in 1916 to Emile Sundberry and to Robert Baker. Baker kept Madewood until 1946, when it was sold to D. Bronier Thibaut, the father of the present owner, John Thibaut.[12]

Morphologically, Madewood contains Thomas Pugh's 1846 mansion, which architecturally traces to the English Tidewater region of the Atlantic seaboard. Built of cypress wood and brick, the impressive house has a front-facing gable, paired end chimneys, a symmetrical floor plan, and paired pavilions as wings on each side. The mansion and its six-acre site, now operated as a stop for plantation tours, has been separately owned by the Marshall family of New Orleans since 1964.[13] The quarters and outbuilding complex occupy a nodal-block settlement of buildings located one-half mile from the mansion in the direction of the backswamp, at the center of the landholding. Nineteenth-century Anglo planters in the South arranged quarter houses in a square or block-shaped settlement pattern based on a grid of streets.[14] In 1969, the quarters contained eleven dwellings, more than enough to house eight resident workers. The outbuilding complex had a drive-through tractor shed, two storage sheds, and a very old wooden blacksmith shop.

By 1989, subtle changes in the settlement had occurred. The quarters, now down to six houses, had also gained two new houses. Cinder-block bungalows were added to convince better agricultural equipment operators to stay on the plantation. The plan ultimately failed, and tenants were now only four retired residents. The blacksmith shop had been replaced by a metal harvester equipment shed. The overseer's house was still intact but unoccupied because overseers, managers, and active farm laborers now chose to live away from the plantation in town. The plantation bell, which once had summoned workers from the fields, was gone. John Thibaut decided that it should be preserved at his home in Napoleonville. Even though the internal morphology of the quarters had changed, Madewood had changed the least of my case-study plantations. The majority of traits observed in 1969 remained intact, largely because the ownership and management had been under the same man. The management skills and traditions that were passed down through three generations from D. Bronier Thibaut to his son John and thence to his son Steve accounted for the maintenance of stability in this plantation enterprise.

Land occupance models evolved from traditional plantations into corporate forms in Louisiana and separately in California and then merged and grew into the mega-corporation CALA model. Meanwhile, small plantations and cane farms symbiotically attached to cooperative sugar factories plodded along with enough inertia to achieve a degree of success and, more importantly, survival. Not all of the 690 sugarcane operations in Louisiana fit the evolutionary stages in the models. Perhaps more models exist. But much of the landscape change over the past thirty years can be attributed to the functional evolution chronicled here.

CHAPTER 5

Armant Plantation, 1796–1998

Armant Plantation is located forty miles upstream from New Orleans on the west bank of the Mississippi River in Saint James Parish, about a mile above Vacherie. Between 1974 and 1993 South Coast Sugars owned Armant as part of the CALA model discussed in chapter 4. The property is no longer a division-centered plantation and, other than cane land, Armant no longer exists. The Armant that I observed in the late 1960s was a large French linear plantation with 2,876 acres of cultivated land, a modern well-maintained sugar factory, and quarters for more than fifty-seven workers and their families. The plantation had a particularly advantageous location on the Vacherie alluvial crevasse, a tongue of alluvium that provided extended arable lands reaching from the Mississippi River to the backswamp for a distance of 4.8 miles. Armant was a double concession eighty arpents deep instead of the customary forty arpents. The plantation benefited greatly from the additional 1.75 miles of arable land that lay beyond the comparable 3-mile natural-levee width found elsewhere along this stretch of the Mississippi River.

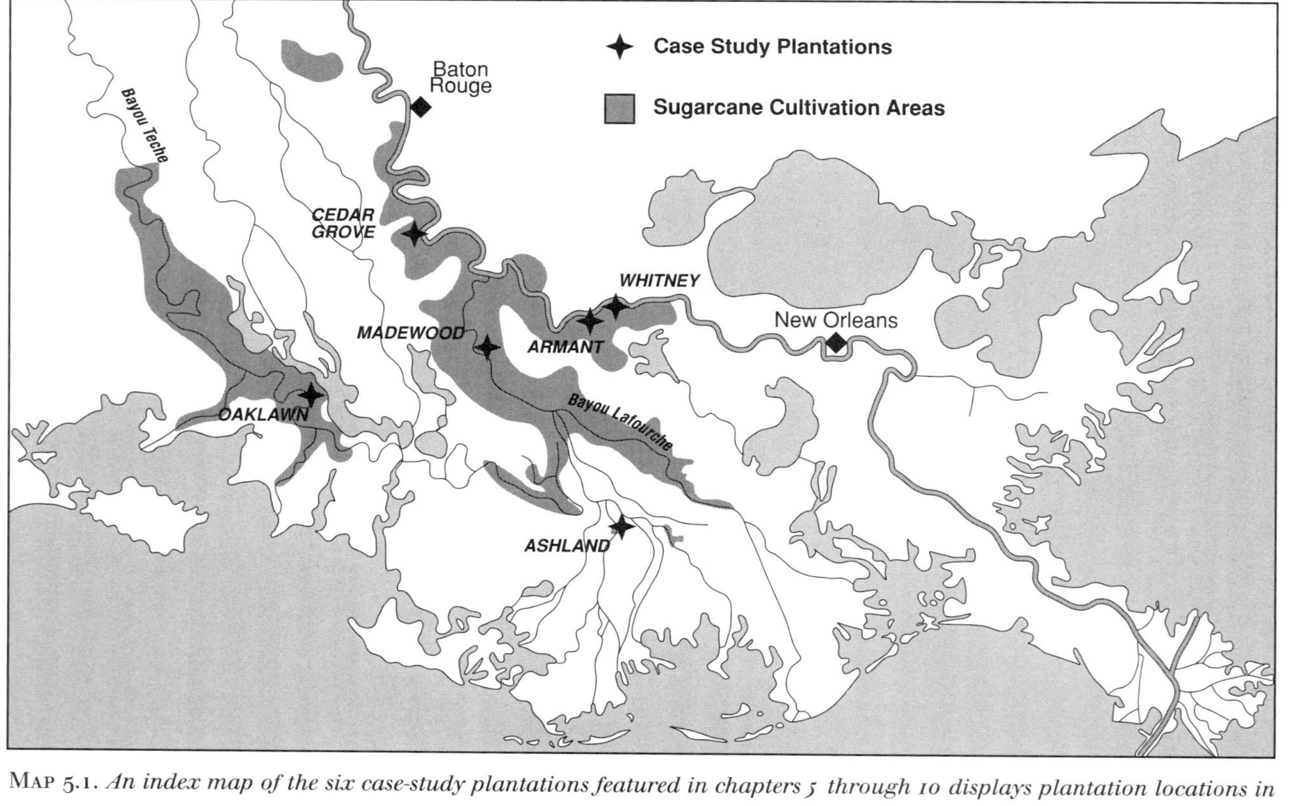

MAP 5.1. *An index map of the six case-study plantations featured in chapters 5 through 10 displays plantation locations in relation to sugarcane cultivation areas in the delta region.*

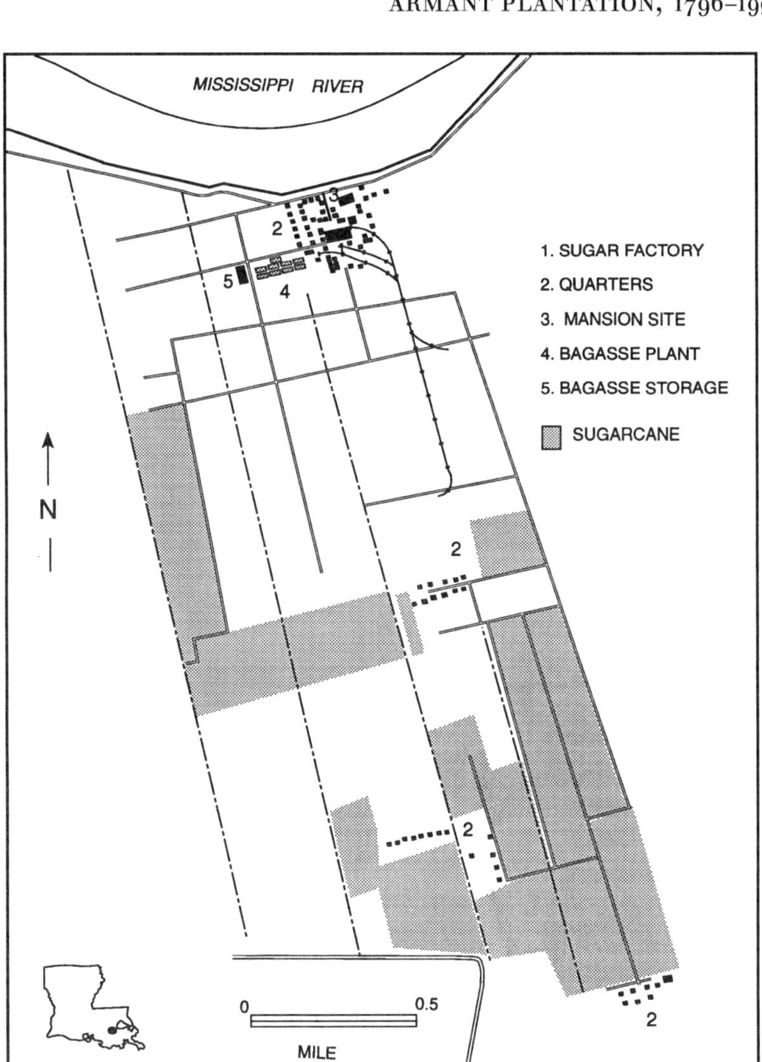

MAP 5.2. *Armant Plantation in 1969 had a complement of plantation traits: three settlements of quarters, a sugar factory, an outbuilding complex, a bagasse plant and storage area, fields, and other landscape elements.* Source: Rehder 1971, 262

Historical Succession

Armant Plantation was named for its initial owner, Jean-Baptiste Armant Sr. The date of origin for the plantation remains unknown, but earliest map evidence points to a date of 1796, when the property was already deemed a plantation. According to Daspit, Jean-Baptiste Armant Jr. built a Creole cottage for his parents in 1808 after they deeded the primary mansion and twelve thousand acres to him when he wed Rose Carmelite Cantrelle (Daspit 1996, 196). Saint James Parish courthouse records indicate that, for the several years between 1809 and 1818, both Armant Sr. and his son J.-B. Armant Jr. continued buying small parcels of land to expand the plantation. Even after Armant Jr. gained full ownership of the properties, he continued to purchase lands adjacent to the Armant Plantation between 1820 and 1830.[1] A notice of foreclosure on the plantation mortgage described Armant Plantation in 1840: "The property is an habitation established in sugar situated in Saint James Parish on the right bank of the Mississippi at 18 *lieues* [1 lieue = 2.5 miles] from New Orleans having 32 arpents *de face* [in front] by 80 in depth bordered on the sides above by heirs of Jacques Roman and below by heirs of Duparc. Together with all buildings, steam-powered sugar mill, animals, implements, and 124 slaves." In January 1847, seven years before his death, J.-B. Armant Jr. donated the plantation to his eight children.[2] The "Armant Brothers" successfully maintained the plantation for the next thirteen years.

John Burnside, one of Louisiana's prominent landowners, purchased Armant in a sheriff's sale in 1860. Burnside acquired all of the lands except a four- by four-arpent plot. The sellers reserved this mansion site, the mansion, and other building improvements for their mother, Mrs. Jean-Baptiste Armant Jr. In 1883 Burnside willed Armant Plantation to Oliver Beirne. The transaction indicated an increase in the size of the holdings by a tract of 6,156 acres in backswamp lands beyond the eighty-arpent line.[3]

In 1889 the Miles family gained the right to succession from the Beirne family. The five owners included Sally Beirne Miles, Susan Wailey Miles, Margaret Melinda Miles, Betty Beirne Miles, and William Porcher Miles (St. James Parish 1889). Management of the plantation went to William Porcher Miles, who between 1884 and 1915 technically improved operations. He modernized the sugar factory by introducing vacuum pans, centrifugals, and double-effect evaporating equipment. Miles adopted improvements in the field by installing a portable railroad to facilitate hauling harvested cane

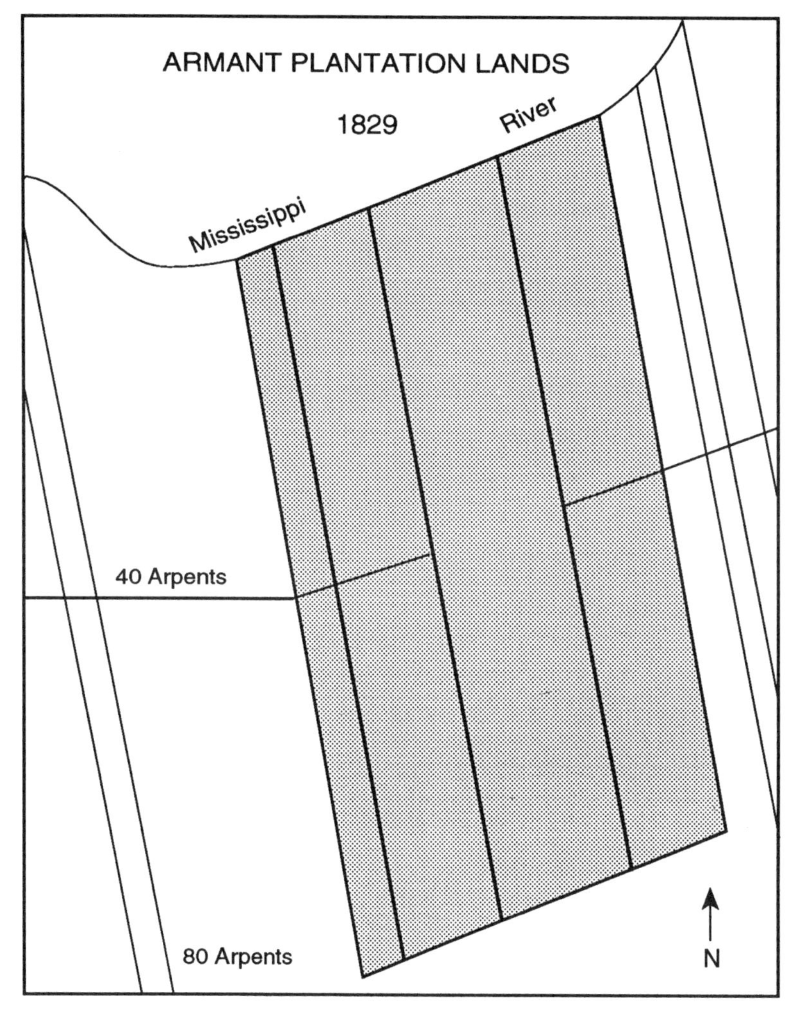

MAP 5.3. *By 1829, Jean-Baptiste Armant Jr. had developed Armant's plantation lands into a double concession that measured eighty arpents deep.* Source: Louisiana State Land Office, surveyor's township plats T12S, T13S, R17E for 1829

to the mill. In 1891 Miles experimented unsuccessfully with drain tiles in the sugar fields. He even experimented with tenant labor, despite the overwhelming use of gang labor on other Louisiana plantations.[4] Tenants were still part of the Armant Plantation in the 1960s and probably traced to the William Porcher Miles era of ownership.

In 1934 mounting debts during the Great Depression precipitated the sale of Armant for $310,000 to Realty Operators Company. All of Armant's buildings and land went with the transaction, except 5,436 acres of back-swamp lands behind the eighty-arpent line, which were to be kept by the Miles Planting and Manufacturing Company. In 1948, Realty Operators became Southdown Sugars, who were the owners of Armant until 1974.[5]

Major changes took place at Armant and other sugar plantations held by Southdown in the early 1970s. Executives at Southdown decided to get out of the sugar business entirely in 1974 by selling the corporate complex of twenty-one plantations including Armant, three sugar factories, and one sugar refinery to Prudential Southdown Partnership for $11 million. By 1979, the former Southdown plus the former SouthCoast corporations were merged into the mega-corporation that I call the CALA model discussed in chapter 4. Under an ever so slightly different name, this became South Coast Sugars assembled by entrepreneurs Goldsby, Robbins, McCarthy, and Cook. Early in the 1980s, Goldsby and Robbins left the operation to McCarthy and Cook. In 1993 the mega-corporation reverted to the Prudential insurance corporation, the primary lenders.[6]

Landscape Morphology, 1967–1969

Armant Plantation in the 1960s was a compact linear settlement complex with quarter houses arranged in a line. Linear plantation settlements in the area usually indicated a French identity, and Armant's linear settlement pattern and initial French ownership in the Armant family clearly supported this conclusion. House types built through traditional folk patterns and techniques are the better cultural indices, and the mansion at Armant was no exception. For 175 years, the Armant mansion was the cultural icon for typing the plantation. It was a raised Creole mansion featuring a bricked, raised basement that was a full-sized floor above ground. Diagnostic French architectural traits included multiple front doors, central chimneys, a hipped roof, walls of wood-framed construction in colombage half-timbered techniques for the main floor, and galleries at the front and back of the structure incorporated into the roof design. Columns were bricked below and wooden above to complete the outward appearance of the dwelling. The original structure measured sixty feet long by thirty-six feet wide, with eight-foot galleries extending beyond the thirty-six-foot measurements. In the 1880s kitchen and laundry appendages had been added to the rear of the main structure.

FIG. 5.1. *The French Creole mansion on Armant Plantation, as seen in these front north and west elevations, was built about 1795. Saint James Parish.* Photograph, 1967; source: Rehder 1978, 141. Used with permission from the Louisiana State University School of Geoscience

The roof had originally been built of shakes or cypress shingles and later had been reroofed with slate. The slate roof was probably added under William P. Miles's ownership in 1890 because at that time the wooden shingle roof on the sugarhouse was replaced with slate. The upper floor main living area was built in the French tradition of colombage half-timbered construction; cypress timbers became a support frame, and bousillage in a matrix of mud and Spanish moss filled interstices in the walls. Exterior walls were covered with weatherboarding to protect the half-timbering inside.

On a sweltering August day in 1969, I arrived at Armant to discover the mansion in the process of demolition. The brick raised basement was the only remaining part of the house. Workmen confirmed that all of the main-floor walls had been half-timbered. I used these remains and earlier photographs that I had taken to verify the floor plan and to examine the brickwork, walls, and dimensions of the main structure. In plan, the Armant

FIG. 5.2. *Armant mansion at the rear south elevation displays the kitchen appendage* (right) *and laundry and bathroom* (left) *that were added to the structure in 1880. Saint James Parish.* Photograph, 1967

mansion had front and rear galleries, a formal front parlor, three bedrooms, and another parlor that doubled as a dining room. Multiple doors on the galleries led into rooms with a floor plan quite consistent with French Creole architecture. Rooms were asymmetrical, and no two rooms were the same size. The kitchen, laundry, and bathroom were attached to the rear gallery about 1880. Following French tradition in the area, a door from each room opened onto a gallery. The galleries were much more than just shaded porches. They were an integral part of the human traffic flow; people routinely went out onto the galleries just to move from room to room in the structure.

Quarters

The quarters was a settlement of nearly identical dwellings for the workers at Armant and was located 300 yards from the mansion and sugar factory. A double row of cabins that extended 275 yards long formed the pri-

FIG. 5.3. *Armant mansion was razed in the summer of 1969.* Photograph, 1969

mary front quarters, the core of the linear settlement pattern near the river. In the primary quarters were twenty-one dwellings, sixteen of which were occupied by agricultural workers. Twelve more houses separated from the quarters were occupied by managers, overseers, and sugar factory workers. All quarter houses measured thirty-two by thirty feet and were of the small Creole type. Chimneys were set at the center of the roof line, and the construction of each structure, except for the tin roof, was of wood frame and weatherboarded siding. Many yards had spaces for gardens, but most were not cultivated. Fences of smooth wire, chicken wire, and scrap-lumber slats enclosed yards, and chinaberry and willow trees shaded houses throughout the quarters.

Such were thc settlement characteristics of the main Armant quarters, called Armant Front. Three additional quarter-house settlements with many of the same house types were located at sites 1.25, 2.25, and 2.75 miles behind the main agglomerated center of Armant. In the vernacular of southern Louisiana, they were "back quarters" in reference to their location. It had been a matter of convenience to have additional quarters in the center and back units on some plantations, especially at Armant, where 4.8

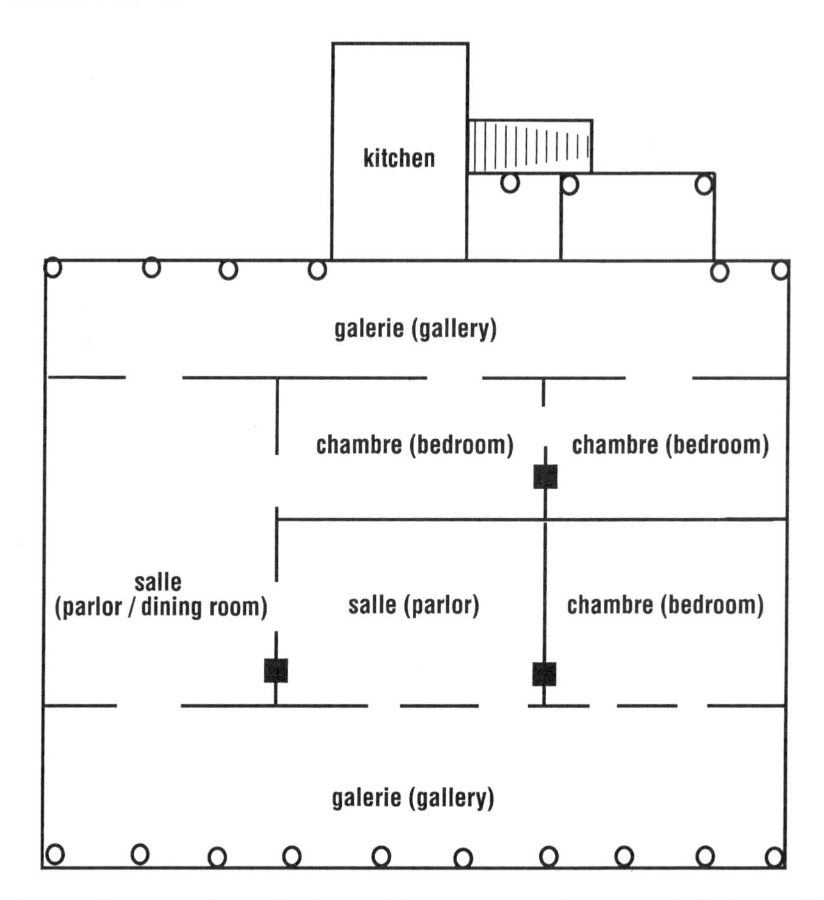

FIG. 5.4. *The floor plan of the Armant Plantation mansion was typical of a Creole mansion, with rooms opening onto the galleries.*

miles of arable land lay at length from end to end. Back quarters were rarely found materially or functionally on the landscape at other plantations, but at Armant such back quarters were regularly used by tenants and hired hands. A small green road sign labeled Back Quarters Road is the only reminder of these tenant quarters.

Other dwellings at Armant used by mill workers, overseers, and managers were built of better materials and were kept in better condition. Representative house types were eight bungalows, a new cinder-block house, two small Creole types, and a shotgun. Most of the sugar factory workers and overseers had well-kept yards, flowers, and vegetable gardens.

The Sugar Factory

An immense sugar factory and outbuilding complex was located at the center of the front of the plantation near the mansion and close to the primary quarters. The factory dominated the scene with its black smokestacks, galvanized metal surface, and enormous dimensions of 460 feet long by 260 feet wide by 70 feet high. Armant's sugarhouse was truly a modern sugar factory. Modernity in this case meant that it had been built since the 1930s, when sugar factories in Louisiana had been updated technologically.

The milling capacity, expressed in tons of cane ground per day and used as an index of technological efficiency, had risen from 900 tons in 1922 to 3,500 tons in 1968. Armant was above average compared to Louisiana's other plantations, where milling capacity averaged 2,400 tons and the number of roller mills averaged thirteen. Part of Armant's success had been attributed to its milling equipment–sixteen rollers arranged in two sets of two rollers each and four three-roller mills. Power for most of the machinery was supplied by eleven steam boilers that collectively produced 4,110

FIG. 5.5. *Armant Plantation's main front quarters had small Creole quarter houses.* Photograph, 1969

FIG. 5.6. *The last quarter house at Armant Plantation was built in the back quarters about 1890 and was razed shortly after this picture was taken in February 1993. Note the kitchen appendage.* Photograph, 1993

horsepower. The loading or feeder tables, sets of cutting knives, sixteen-roller mill, and boiling house equipment were powered or heated, or both, by steam engines and boilers. In the boiling house portion of the factory were two 8-foot-diameter liming tanks; five cane-juice heaters; three clarifiers at 20, 18, and 8 feet in diameter; two evaporators at 9 and 11 feet in diameter; five vacuum pans at 8 to 10 feet in diameter; four condensing pans; twelve crystallizers with 800 cubic feet of capacity each; and nine centrifugals with 34-inch baskets for the separation of crystallized sugar from the molasses (Dupy 1967, 26–32).

Various sources supplied sugarcane to the Armant sugar factory in the 1960s. The grinding season, averaging seventy-three days, began about mid-October and ended by the third or fourth week in December. Not all cane ground at Armant came from its own fields. Company-controlled plantations in the Vacherie Division, including Armant, Saint James, Salsburg, and New Hope, supplied 25 percent of all cane processed at the factory; another 15 percent came from tenants living and working at Armant,

FIG. 5.7. *One of the late-nineteenth-century houses that was assigned to overseers, factory personnel, the storekeeper, or persons in middle management at Armant Plantation.* Photograph, 1969

and the remaining 60 percent was supplied by fifty independent growers. A symbiotic relationship existed between independent growers and Armant's sugar factory. Without outside growers, Armant could not operate economically by grinding only company and tenant cane. Moreover, independent growers could not maintain a separate sugar factory for each plantation because the cost of machinery, equipment, and manpower would have been prohibitive.

The last Armant sugar factory was refurbished into an all-steel building in the late 1930s. The factory had always been at the same site and for much of its history had been kept up to date. As early as 1828, the factory had had a steam-driven grinding mill, and by 1843, while still owned by J.-B. Armant Jr., the structure had been refurbished with $51,000 worth of boiling equipment and Rillieux apparatus (Sitterson 1953, 149). In 1880, with John Burnside the owner, the sugarhouse had brick walls and a cypress shake roof. It remained this way until the 1890s, when William P. Miles had the sugarhouse reroofed with slate (Bouchereau and Bouchereau 1911). The

Fig. 5.8. *The Armant sugar factory was at the peak of its landscape presence in 1969.* Photograph, 1969

brick structure at Armant was replaced in the late 1930s with metal buildings. Between 1974 and 1981, the sugarhouse was closed and razed as new owners South Coast Sugars took over the plantation. The sugar factory site of 129 acres was subsequently sold to a Costa Rican entrepreneur, Enrique Uribe, who then leased the site to Toth Chemicals, a small chemical firm that has modified warehouses at the otherwise deserted mill site.[7]

Outbuildings at Armant were related to both sugar factory and agricultural functions of the plantation. A complex of separate but related buildings included storage sheds, warehouses, machine shops, and tractor, implement, and equipment sheds. No animal barns functioned as such at Armant, although there was a relic barn in a weed-covered former feedlot. The last of the plantation's mules disappeared in 1958, but in 1934 there had been sixty or more mules, all named on the plantation.[8]

Within the outbuilding complex, structures related to the sugar factory included six tanks for molasses storage and two steel warehouses 180 feet long by 160 feet wide by 50 feet high. As the Armant factory produced only raw brown sugar, the warehouses stored bagged and bulk brown sugar before it was shipped to the primary refinery at Southdown in Houma, Louisiana. These and other agricultural outbuilding sheds and tractor stalls were commercially constructed buildings of steel beam frames and galvanized sheeting. Commercially designed and fabricated agricultural and industrial outbuildings at Armant were not folk architecture. They did not represent the traditional folk cultures with which other plantation elements could be identified. They did, however, reflect traditional management intentions by their location in an agglomerated outbuilding complex. The older forms could not be expected to remain through two centuries of plantation activity at Armant, but it should be understood that the spatial arrangement and siting of functional structures represented a continuum of plantation traditions dating from the 1790s.

Other Landscape Features

A plantation store and an office were located near the center of the settlement. The store closed early in 1969 because nearby retail stores in Vacherie had taken away much of the business. A new brick office building was built across the road from the older office site. Armant was one of the few plantations that had a church erected on the premises for use by the workers. The church was located upriver between Armant and Saint James Plantations and served both. Armant's white employees described it as a "Negro church" because it was used by African American field and mill workers (Rehder 1971, 273–74).

Adjacent to the habitation and outbuilding complex was a twenty-five-acre tract reserved for bagasse storage. Bagasse, sugarcane stalk pulp once thought of as trash but later considered a significant by-product, was dried and baled at the Armant sugar factory. The bagasse was transported 330 yards to the storage tract, where bales were stacked into warehouse-shaped piles and covered with a loose sheet metal roof. An independently owned bagasse processing plant redried the bagasse and sent it for further processing to the Celotex Corporation in Merrero near New Orleans, where ceiling tiles and fiberboard products were made (Rehder 1971, 274). The bagasse plant was closed along with the Armant sugar factory, and only ruins remain.

Fields and Drainage

Armant still retains long, narrow fields running back from the Mississippi River, just like all other sugar plantations in the delta. Fields, rows, and ditches all conform to this pattern on the backslope of the natural levee to provide proper drainage, preventing root damage. Individual field plots, marked and separated by roads or ditches, average about 2.5 acres each (range, 1 to 7 acres). Lengths of plots range from 143 to 330 yards, and widths vary from 12 to 110 yards. Field cuts, the larger field units, are usually bounded by roads and contain three to nine plots, with an average of five plots each. As in the case of other sugar plantations in the region, Armant field rows are set at a standard measure of six feet from furrow to furrow.

Drainage is made possible by an elaborate grid work of ditches. The smallest are quarter drains, with two or three per field plot. Next are long, narrow lateral ditches that separate plots. Deep cross ditches are placed at right angles to the lateral ditches and join canals that lead to the backswamp. Two main canals, one of which is partially navigable, extend into the backswamp. In the 1890s, W. P. Miles experimented with drain tiles, but they were not feasible because they clogged with sediment so quickly. Field fences were not present at Armant. Barbed wire fences on the artificial levee contained cattle. Small yard fences of wood and chicken wire surrounded house sites in the quarters (Rehder 1971, 274–75).

Land Use

For centuries, sugarcane has been primary crop at Armant; in 1969 only 300 acres were in company-grown cane on site. This figure hardly seemed to be enough sugarcane for a sizable plantation, much less one with a large operating sugar factory. In addition to company cane acreage, 650 acres of tenant cane and other company-owned cane from adjacent plantations were processed at Armant's sugar factory. Soybeans, a secondary commercial crop, accounted for 313 acres. Planting managers throughout the sugar region generally despised soybeans and, were it not for federal controls on sugarcane acreage, planters would have cultivated sugarcane instead of soybeans. Soybeans were not harvested by Armant's plantation workers or equipment; rather, hired harvesting companies went from plantation to plantation to harvest soybeans. Aside from sugarcane and soybeans, no other commercial crops were produced. Fallow land covered 153 acres and was marked into rows ready for the next year's planting. Tenant lands had 280 acres of fallow ground set aside for the next sugar crop. Throughout the plantation at individual house lots, vegetable gardens for home use yielded onions, peppers, beans, okra, tomatoes, and some corn, especially among the tenants (Rehder 1971, 275–76).

Armant Plantation, 1981–1998

Armant as a landscape entity no longer exists; much of that ended about 1981. All or nearly all buildings that were ever on the plantation are now gone. Only greatly modified storage buildings remain, and they are at Toth Chemicals on the former sugar factory site. Yet cane surprisingly continues to be produced in record quantities in all available fields on site. The strategies of the new owners in 1974 were primarily responsible for the demise in the landscape. South Coast Sugars, the mega-corporation masterminded by outside entrepreneurs and following a California model, decided that sugar factories at Armant, Greenwood, Montigut, Georgia (Mathews), and eventually Oaklawn should be closed and if possible sold and all unnecessary buildings should be razed. This CALA model was probably the most destructive force to hit the plantation landscape in some years. It was more destructive to the built environment than a hurricane because it was a complete and permanent removal of a plantation architectural morphology and philosophy. The actions in the CALA model that severely affected Armant Plantation also affected nearly forty other plantations covering approximately 80,000 acres and representing approximately one-third of the sugar industry in the region. Armant Plantation, unfortunately, was caught up in the collection. The two plantations discussed in the following chapters, Ashland and Oaklawn, were also caught up in the CALA model in South Coast Sugars. Although their evolutionary patterns and cultural identities are remarkably different, they met the same fate as Armant.

CHAPTER 6

Ashland Plantation, 1828–1998

Ashland Plantation is located in Terrebonne Parish on Bayou Grand Caillou, about seven miles south of Houma (see map 5.1). Now virtually nonexistent, Ashland had been one of the larger sugar plantations in the state. Two other former plantations have the same name. One is the famed Ashland-Belle Helene, where the mansion built in 1841 is all that remains of the former Duncan Kenner Plantation in Ascension Parish. Another Ashland is a small holding that is no longer identifiable on the landscape in West Baton Rouge Parish, about eight miles north of Baton Rouge. Could it be that plantations named Ashland are jinxed?

Ashland, a bayou-block plantation type, was a vigorous, vibrant agricultural concern covering 1,446 acres, with 652.7 acres in sugarcane in 1969. Ashland Plantation proper was a smaller part of a multiplantation division that had 7,253 acres within the entire 42,000-acre SouthCoast corporation. A sugar factory had operated until the 1930s, but it was closed when the property was purchased by SouthCoast in 1935. In 1969, Ashland's fifty-nine resident workers included fifty agricultural laborers, five overseers, two

FIG. 6.1. *Ashland Plantation was an Anglo-American bayou-block plantation in Terrebonne Parish in 1940. Little had changed between 1940 and my early period of observations in 1967 and 1969.* Sources: U.S. Department of Agriculture, Agricultural Stabilization and Conservation Service aerial photograph, Dec. 17, 1940, image CQC 5A-76; Rehder 1978, 148. Used with permission from the Louisiana State University School of Geoscience

managers, a store clerk, and a plantation clerk. Another fifteen laborers lived off the plantation premises. The plantation was a block-patterned plantation settlement, a bayou-block subtype. Ashland's block settlement pattern and numerous old structures initially attracted me to investigate the property. Armed with the visible evidence of a block settlement in 1969 and the general historical record of nineteenth-century Anglo-American planters settling Terrebonne Parish, I felt certain that the plantation had an Anglo-American origin.

Historical Succession

The initial owner-builder of Ashland Plantation was James Henry Cage from Wilkinson County, Mississippi. Cage arrived in 1828, like other Anglo-Americans, to acquire the land that was so easily obtainable, to establish a plantation, and to grow sugarcane in Terrebonne Parish. The area already had sparse settlements of Acadian and Spanish farmers, trappers, and fishermen. Between 1803 and 1828, much more land became available for claims from the U.S. government or was purchasable from initial claimants for low prices.[1]

Cage purchased his initial tract for Ashland on November 25, 1828, from James Bowie, a land speculator best known for his famous knife design and his later role at the Alamo in Texas. Cage paid $9,300 for a tract on the east bank of Bayou Grand Caillou, measuring 46½ arpents front by 40 arpents depth. By 1832, James Cage apparently had the plantation proceeding along in its development because he was readily purchasing more slaves.[2]

Cage began to increase the size of the Ashland holding on April 25, 1840, when he purchased a 12½- by 40-arpent tract located above his property on both sides of the bayou from Lemuel Tanner and wife for $5,000. Three years later Cage paid $7,500 to the Tanners for a 25- by 40-arpent tract located immediately across the bayou and west from his original tract. Another purchase a week later, on February 28, 1843, brought in two tracts totaling 20 by 40 arpents, located on the west bank below the previously purchased Tanner tract. These two tracts were bought for $6,000 from a Kentuckian named Tobias Gibson.[3] Altogether, Cage had spent an average of six dollars per acre on tracts acquired over the fifteen years between 1828 and 1843.

By 1843, James Cage had completed his land acquisitions for Ashland, and a year later he was producing 965 hogsheads of sugar (Champomier

1844, 7). He also brought in as a partner his brother Harry Cage from Wilkinson County, Mississippi. For $83,333.66 Harry Cage purchased the undivided third of Ashland from a total land area containing 59 by 40 arpents on the east bank of Bayou Grand Caillou and 57½ by 40 arpents on the west, which in total measured 4,660 surficial arpents.[4] Besides land, the sale included "the sugarhouse, engine, dwelling houses, cabins, outhouses, farming tools and utensils, stock of all description, appertaining to said plantation (saddle horses and blooded horses excepted) . . . and fifty male negro slaves, . . . thirty-four female negro slaves, . . . nineteen male slave children, and . . . twenty-two female slave children."[5]

Sometime during the next twelve years, Harry and James Cage divided the landholding. James retained Ashland; Harry received the land at the north end of the property and called it Woodlawn Plantation. By 1850, Ashland had approached its maximum size while owned by James Cage. James Cage entered into another partnership on June 5, 1856, when William S. Mayfield purchased an undivided half of the Ashland property for $50,000. Production at Ashland was a booming success, with figures of 893 hogsheads, 1066 hogsheads, and 880 hogsheads for the years 1845, 1851, and 1855, respectively. Disaster struck the next year when unseasonal frosts and periods of flooding caused production to drop to a scant 50 hogsheads.[6]

Three years after Mayfield became a partner, he relinquished his undivided half of the plantation to Duncan S. Cage for $36,500. The records of this transaction show that the remaining Ashland lands comprised 25 by 40 arpents on the right bank and 37½ by 40 arpents on the left bank of the bayou. Duncan S. Cage was almost as important to the landscape evolution of Ashland as the original owner, James H. Cage. During his thirty years of ownership, Duncan S. Cage managed operations despite problems incurred by the Civil War. Duncan Cage had Ashland producing an average of 354 hogsheads, with the highest production reaching 532 in 1880.[7] Bouchereau's survey of the sugar crop of 1869 reported that the plantation had a wooden sugarhouse with a steam-engine-powered mill and open kettles in the boiling house. These were typical sugar factory equipment and construction for the time. The plantation that year produced 280 hogsheads of sugar, 2,580 bushels of corn, and 20 barrels of rice. Apparently the rice crop was only temporary because after 1869 no rice was recorded in the agricultural land use (Bouchereau and Bouchereau 1869, 60).

On January 9, 1886, another family member, Hugh C. Cage, purchased Ashland plantation for a bid of $18,500. It was the same 25- by 40-arpent

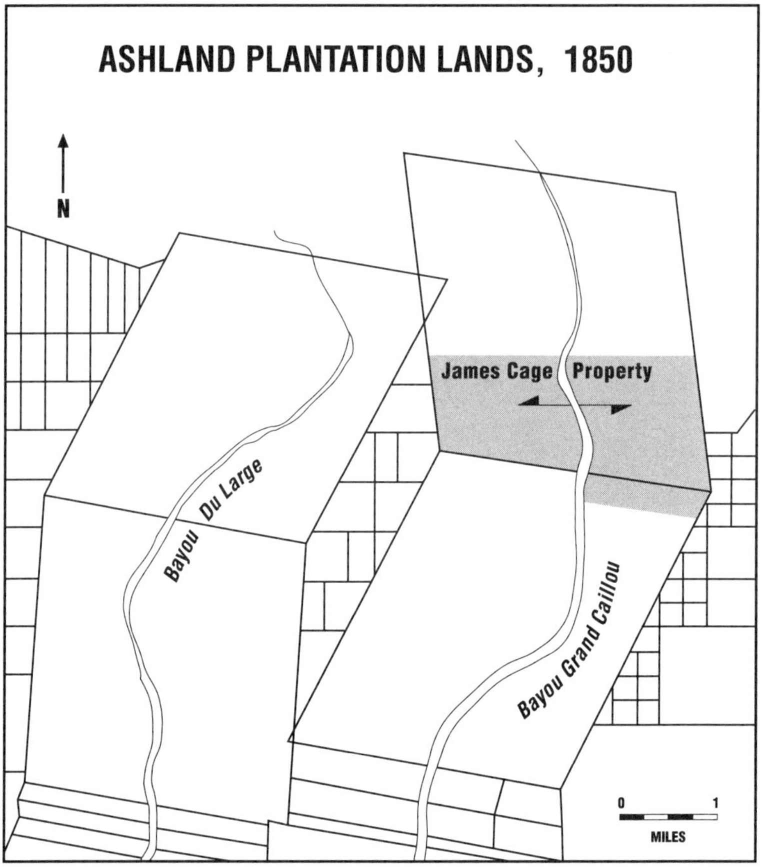

Map 6.1. *By 1850 James Cage, the initial planter at Ashland, had completed his acquisition of plantation lands (shaded area).* Source: Louisiana State Land Office, surveyor's township plats, 1850

west bank and 37½- by 40-arpent east bank property. Bordering lands were, above on both banks, the Woodlawn Plantation owned by Thomas M. and George W. Cage; below on the east bank, the Caillou Grove Plantation of Lafrene and Barrow; and below on the west bank, the Cedar Grove Plantation owned by O. R. Dasfrit. The sale included everything–lands, dwellings, "sugarhouse, mill, engine, machinery, carts, wagons, plows, farming utensils, implements of husbandry and twenty-nine mules." On the same day as that transaction, Hugh C. Cage sold the complete plantation plus thirty-five mules for $23,992.95 to still another Cage, Gayden Cage. This last Cage kept the holding for only a year and one month, and on February 7, 1887, Gayden Cage allowed Ashland Plantation to slip finally from the Cage family for the sum of $23,392.15.[8]

The sale marked a milestone in Ashland history because it ended the era of Cage ownership. The transfer also marked the initiation of great changes to be made in the sugar factory portion of the plantation. J. Norbert Caillovet and Charles B. Maginnis, the new owners, installed double-effect evaporators, vacuum pans, and centrifugals in the wooden boiling house. These modifications and lands already owned by Caillovet brought production for 1890 and 1892 to the equivalent of 1,156 hogsheads (578 tons) and 1,277 hogsheads (638 tons), respectively.[9] By 1900, the wooden sugar factory building needed repairs and was reroofed with metal roofing. In 1906, with J. Norbert Caillovet as president and a relative, J. L. Caillovet, acting as trustee, the Ashland Plantation ownership and name was changed to Ashland Planting and Manufacturing Company Limited. The company-owned lands consisted of the same Ashland tract of 25 and 37½ arpents front and 40 arpents depth on each bank, plus Woodlawn Plantation, located immediately north of Ashland; Ranch Plantation due east of Woodlawn on the next bayou, Bayou Petit Caillou; Caillou Grove Plantation, a 28- by 40-arpent tract south of Ashland; and assorted swampland and smaller tracts throughout the Bayou Grand Caillou area. The total value was $450,000.[10]

Ashland Planting and Manufacturing Company Limited continued successful operation until 1934, with over seventeen hundred acres in sugar and production reaching the equivalent of 9,165 hogsheads (4,582 tons) in 1911. The same sugar factory and its equipment remained at Ashland for many years, and it was only one of four operating sugar factories in all of Terrebonne Parish between 1911 and 1916. Ashland Corporation folded in bankruptcy in 1934 and was eventually sold in a sheriff's sale to Frank Wurzlow for $250 plus mortgages. Wurzlow kept the properties only a few weeks before passing them on to the SouthCoast corporation on February 20, 1935.[11] With this transfer of ownership, Ashland now had become part of the Louisiana plantation corporation model (see chap. 4), owned and operated by SouthCoast in the forty-four years between 1935 and 1979. In spite of turbulent economic conditions during the Great Depression and World War II, the SouthCoast tenure was a time of reasonable landscape stability that would continue through the 1950s and 1960s. But change awaited Ashland Plantation in the 1970s.

In the late 1970s ownership and landscape patterns at Ashland took an entirely different direction. This complicated story follows a sequence of changing corporate management, California and Arkansas entrepreneurs, petroleum prices leading to an oil boom, land and subdivision development schemes, and then a drop in oil prices and a serious decline in landscape morphology. On September 6, 1979, SouthCoast, the huge Louisiana corporation that dated back to the 1930s, was sold in its entirety for $42.5 million to Mid South Mortgage Company, led by Thomas B. Goldsby Jr. from West Memphis, Arkansas, with partners James G. Robbins from Memphis, Tennessee, and Leland J. McCarthy and Richard H. Cook from Fresno, California. Ashland Plantation was a small part of the larger Ashland Division, which covered 7,253 acres. The Ashland Division, in turn, was a part of a much larger package that included a sugar refinery, three sugar factories, and numerous plantations covering 42,000 acres.[12] Although most of the cropland elsewhere in the corporation continued in sugar production, Ashland was to take a different track, with its location so convenient to the rapidly expanding urban development at Houma and oil business near the Houma ship canal and staging area on Bayou Petit Caillou.

By September 18, 1980, Mid South Mortgage, now called Mid South Partnership, had sold 724 acres of the northern part of the Ashland Plantation for four million dollars to Jimmy Don Winemiller from Newport, Arkansas. In January 1983, Winemiller sold a 302-acre parcel in two tracts of 222 acres and 80 acres, respectively, in the northeast quadrant of Ashland Plantation for $3,322,000 to LRT Corporation, headed by Leon Toups.[13] By May 1983, plat maps for a subdivision had been drawn; NASA aircraft imagery taken in 1990 verified subdivision road patterns on the landscape. Under Toups's development plan the property became Ashland North, a subdivision with underground utilities, curved cement-paved streets, streetlights, and a magnificent entrance. Meanwhile, the oil boom failed and so did Ashland North. Only three model homes were ever built. Weeds covered lots and choked ditches, and fading signs pointed to a dream that never became a reality. LRT removed its name from the fading signs, but tax delinquency notices still appeared in the court records in the late 1980s.[14] Most recently, Ashland North subdivision has become a trailer park. Southern Louisiana's coastal marshes are subsiding at such a rate that thirty square miles of marshlands are lost per year. Folks living along the lower, ever narrowing bayous are literally seeing their land sink to the point that some people no longer have enough land to support a mobile home. So they are moving north, up the bayou, to higher ground, to places like Ashland North.

As for the southern part of Ashland, a 184-acre tract called Ashland

FIG. 6.2. *Ashland plantation lands in 1990 had white curving streets in Ashland North, a failed housing subdivision from the 1980s. The settlement core of Ashland Plantation was in the lower left corner of the subdivision pattern and across the road and bayou.* Source: NASA high-altitude photograph, Dec. 8, 1990, roll 4174, image 2264

South sold on January 15, 1981, for $553,260 to Roger Cotton.[15] The property had been platted and subdivided, but as of 1993 the property had not developed into a genuine subdivision. Instead it had at the southwest corner a gleaming new prison called the Terrebonne Parish Criminal Justice Center. Nearby model houses became offices for bail bondsmen. A yellow sign, appearing as an unfortunate epitaph, pointed toward the backswamp, to the Ashland Sanitary Landfill.

These unusual ownership and land use patterns at Ashland developed for several reasons. For forty-four years Ashland had been preserved by old patriarchal management in the SouthCoast corporation. Then it became a part of the CALA model, which put the property in jeopardy, ready for demolition. Ashland's location on wide, dry levees close to Houma opened the opportunity for urban development. The oil boom of the 1970s ignited a land rush in south Louisiana, and Ashland was caught up in it only because Houma had become a land base for offshore drilling companies, suppliers, and distributors. With booming oil prices, the lands at Ashland were just too attractive for developers. Pieces of Ashland were bought and sold at ever-escalating prices to would-be developers until oil prices fell, dashing the hopes and dreams of turning the once-proud plantation into an inelegant housing tract called Ashland North.

Landscape Morphology in 1969

Turning back to the Ashland I knew in 1969, a wonderful though not really complete plantation was visible (Rehder 1971, 340–51). Ashland's sugar factory had met its demise in 1935, but the plantation's other landscape components reflected past functionally significant aspects of a working plantation. Old dwellings, sheds, and warehouses dated to the nineteenth century and indicated that the settlement had not been altered spatially in form and function. In short, the material culture of the plantation had been well preserved, and it was worthy of study.

Ashland Plantation was a classic bayou-block settlement, with buildings on both sides of the bayou. Agriculturally related outbuildings were located on the west bank, and on the east bank were the manager's house, store, office, brick foundations of the sugar factory, former warehouses, and purgery. Ashland no longer had a mansion, but one had been located quite far north of the settlement, very near the Woodlawn plantation line. The site

Fig. 6.3. *The Ashland plantation manager's house, built in the middle to late nineteenth century, was the home of C. L. Denley.* Photograph, 1969

remains in the vicinity of the Cage family cemetery, but the house burned at some undetermined time. In 1969, Mr. C. L. Denley, the manager who had been at Ashland since 1943, told me that he had never seen the mansion, but he knew older people on the place who had seen the house destroyed by fire. The manager's house came closest to being a mansion structure at Ashland in 1969. It was set apart from the other buildings on a separate house site surrounded with pine trees, pecan trees, oaks, magnolias, and carefully tended flower gardens. Incidentally, pines and pecans were culturally Anglo plants in the past century and would have been considered foreign introductions to this part of French Louisiana.

The quarters in 1969 contained fifty-two houses dominated by two types, eighteen shotguns and twenty-three small Creole types. Remaining houses were bungalows and attached-porch houses. All quarter houses had yellow tar-paper siding and galvanized steel roofs. In yards, especially at the manager's house and in part of the quarters, the nonindigenous pine was ubiquitous. Yard fences were far less prevalent at Ashland than at the other

FIG. 6.4. *Of the fifty-two quarter houses at Ashland Plantation, eighteen were shotgun houses.* Photograph, 1969

plantations studied. Even so, some quarter houses had scrap lumber slats or wire fences. Gardens were abundant at Ashland. Plantation manager Denley probably had set a good example with attractive vegetable and flower gardens at his home, so most of the quarter houses had rich vegetable gardens behind them as well.

Outbuildings at Ashland were numerous, variable in size, construction, and age, and all centered on both banks of the bayou nearest the center of the plantation. On the west bank of the bayou were two newly constructed all-metal buildings (built in 1967) used for storing and maintaining sugarcane harvesters, a welding shop, and general maintenance. Both buildings were about all that remained in 1998. A heavy-timbered structure with wooden siding was used for a tractor shed. Another smaller, board-and-batten shed was near the site of the old mule barn, a 400- by 150-foot structure that once sheltered two hundred plantation mules. A very small pump house was on the west side, and it was one of the last older structures still standing in 1996.

FIG. 6.5. *A small Creole quarter house at Ashland dated to the late nineteenth century. The fence was similar to the old* pieux debout *fences of the eighteenth century.* Photograph, 1969

Agricultural and sugar mill activities were segregated; the west side of the bayou was for agricultural buildings and the east side was for sugar factory structures and functions. The sugar factory site was marked by the foundations of the furnace and an intact purgery building. Two wooden buildings were warehouses. An old bunkhouse on this east side of the bayou was used by additional hired hands during grinding season. Nearby were a small shed and a cement mixing area where cement drain pipes were made for this and all other SouthCoast plantations. A weatherboarded plantation store and separate offices were also on the east side of the bayou.

There were two churches on the plantation. The Baptist church was located about one-half mile north of the settlement, and a Catholic one was located on the southern edge of the block settlement. The Baptist church was an older structure for the African American population at Ashland; other folks attended the Catholic church. A family graveyard located near the Woodlawn-Ashland boundary was a good diagnostic culture trait point-

Fig. 6.6. *The main tractor shed at Ashland Plantation had been converted from an earlier wooden barn.* Photograph, 1969

ing to Anglo-American origins, as family graveyards in the South were predominantly associated with Scotch-Irish or Anglo culture (Rehder 1992, 116). The cemetery had been used exclusively by the Cage family in the nineteenth century, but the Baptist church on the plantation later gained usufruct rights to it.

Other Landscape Features and Land Use

Ashland's site and situation differed from those of plantations along the Mississippi River. The natural levees on Bayou Grand Caillou were noticeably small in height and width because the stream was a small distributary from an ancient course of the Mississippi River's Lafourche delta. Arable lands at Ashland extended only one-half to one-quarter mile from levee crest to backswamp on the narrow levees. This alone may explain the immense frontages granted and sold during initial plantation development. A fifty-arpent frontage on a stream like the upper Bayou Lafourche or espe-

FIG. 6.7. *This structure had been the main warehouse to the old sugar factory that closed in 1935. Later the building stored fertilizer and other farm chemicals at Ashland Plantation.* Photograph, 1969

cially on the Mississippi River was unheard of, and a more than fifty-arpent frontage by forty-arpent depth on *both* sides of the stream could be found only on these lower, narrow bayous of Terrebonne and Saint Mary Parishes.

Fields at Ashland were not without drainage problems. The limits of cultivation generally extended a quarter mile to the backswamps, where much of the acreage was under pump. Natural drainage was so inadequate that backlands had to be drained by mechanical means. Mr. Denley told me that the drained land and the limits of good cultivation actually ceased about 1 acre *before* one reached the backswamp. Ashland Plantation's cultivated lands were restricted to only those between the levee crest and the wetlands limits at the backswamp. Field plots averaged 1.5 acres, with full field cuts measuring 13.5 acres.

Few animals were present at Ashland in 1969. Company livestock had been a major part of the landscape until the 1940s, when there were as many as two hundred mules on the plantation. In the 1930s and 1940s, Mr.

FIG. 6.8. *The bunkhouse at Ashland Plantation housed bachelors and seasonal hands and resembled the bunkhouses found on western cattle ranches.* Photograph, 1969

Denley periodically went to Missouri and Tennessee on mule-buying trips; he had to personally inspect the mules before purchase. "Sugar plantation mules must be bigger and have larger hooves to plow the heavy clays of the backlands," he explained. The only livestock I saw were a few chickens and ducks kept by some of the workers in the quarters.

Land use at Ashland was no different from that at any other sugar plantation in southern Louisiana. Sugar was cultivated on 652.7 acres, while soybeans and fallow land took up 500 and 571 acres, respectively. Beans were harvested as oil beans and not turned under for fertilization. Fallow ground was already prepared for the next season's crops of cane and soybeans.

Ashland no longer bears any resemblance to a plantation. All that remains of the buildings are two tractor sheds built in 1967, a small pump house, an open wooden shed, and the foundation brickwork of the sugar factory, de-

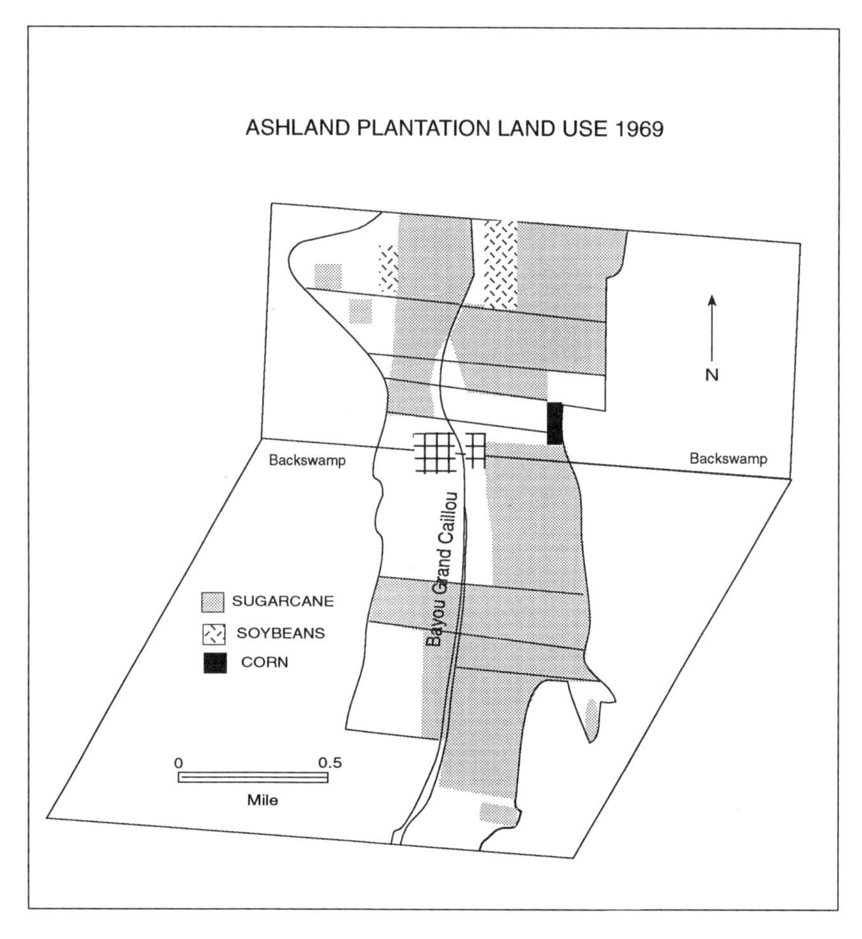

MAP 6.2. *Land use at Ashland Plantation in 1969 included crops of sugarcane, soybeans, and field corn.* Source: Rehder 1971, 350

stroyed long ago. Even most of the fields are gone. Archaeologists might have real trouble finding this one! The how and why of Ashland's evolution and decline began with Mississippi planters named Cage in the 1820s and ended with an oil boom 150 years later. The end began with the CALA model and was completed with land development schemes during the oil boom of the 1970s and early 1980s. We can explain this evolution in terms of progress and development strategies. The American way is to tear down to build up. But in this landscape of limited dry land on which to build, land was and is at a veritable premium. It was too valuable to keep for agriculture. The land was worth much more per acre for housing projects that could fill the need and greed in a local housing shortage. Ashland was already being phased out of the sugar business, but was its demise anything that could have been prevented? Ashland's fate was determined by its location–too close to a growing oil-boom city; too far from the conscious efforts of management to retain an agricultural landscape, especially a plantation one; and too far in time from its past role as a sugar plantation on the North American landscape.

CHAPTER 7

Oaklawn Plantation, 1812–1998

Oaklawn Plantation is located in the Bayou Teche region in Saint Mary Parish six miles east of Franklin and on both banks of Bayou Teche in the Irish Bend district (see map 5.1). In 1969, this bayou-block plantation type was one of the more materially complete plantations in the delta. Since 1979 and especially since 1989, the plantation has been undergoing considerable landscape destruction. Ownership at present is divided and complicated. The finely preserved mansion and a small part of its former 76-acre site belong to Louisiana Governor Mike Foster. The sugarcane lands, razed sugar factory site, and bayou-block settlement (last owned by South Coast Sugars under the name Teche Sugar Company) are now in the hands of the lenders. In 1969, Oaklawn Plantation had 1,716.2 acres of cleared land, with 956.2 acres of it in sugarcane. The mill processed nearly 8,000 acres of cane from nine company-owned plantations in the area. About 112 employees and their families lived at Oaklawn, and during grinding season the figure increased to 150 (Rehder 1971, 351–72). In February 1991, only 4 employees remained at Oaklawn.

Historical Succession

Oaklawn's original owner was a young Irish immigrant named Alexander Porter who first settled in Nashville, Tennessee, in 1801. By 1812, the young lawyer had arrived in Saint Mary Parish and begun buying land along Bayou Teche (Stephenson 1934, 8–11). Between 1812 and 1814, Porter and his partner Isaac Baldwin bought four tracts of land. On November 14, 1812, they purchased a six- by forty-arpent tract on both sides of the Teche for $1,920 and 640 "American acres" on the left bank of the Teche for $1,500. A year later they paid a Mr. Bundick $3,200 for an adjacent tract of ten by forty arpents. Their fourth and final purchase was on November 10, 1814, when they paid John Nopper $100 for his claim of six by forty arpents. Eventually, Porter acquired Baldwin's share of these initial landholdings, but the transfer does not appear in the parish court records.[1]

Porter continued building his estate with other land purchases. In 1823, he paid Ruffin G. Sterling fifteen thousand dollars for a tract of fifteen by forty arpents on both sides of the bayou. Four years later, he acquired a ten- by forty-arpent tract situated on the American Bend of the Teche. The American Bend had been named for the numerous Anglo-Americans who initially settled there; later it was called Irish Bend after the Irishman Porter. He paid Thomas Martin from Tennessee four thousand dollars for this final tract.[2] The completed plantation lands of Porter's Oaklawn consisted of thirty-five arpents frontage on both sides of Bayou Teche by a depth of forty arpents. It was bounded above by the lands of Thomas Martin and below by property owned by heirs of David Smith.

Porter was often absent from the estate because of his legislative and judicial duties. It was not unusual for planters to be absent from their holdings, and New Orleans and Natchez became havens for planters who did not care for all aspects of plantation life. When Porter was away from his plantation, an overseer managed the operations for him. Between 1837 and 1840, Judge Porter returned to set up permanent residence and began construction on the mansion (Stephenson 1934, 116–18). Porter's Oaklawn became a showcase plantation and was visited by writers and travelers. Charles Daubeny of Oxford, England, described Oaklawn as

> a fine and fertile tract of land, extending for nearly a mile on either side of the river Teche, which consists for the most part of what was originally prairie, but now is converted into sugar fields. A belt of wood however bounds the

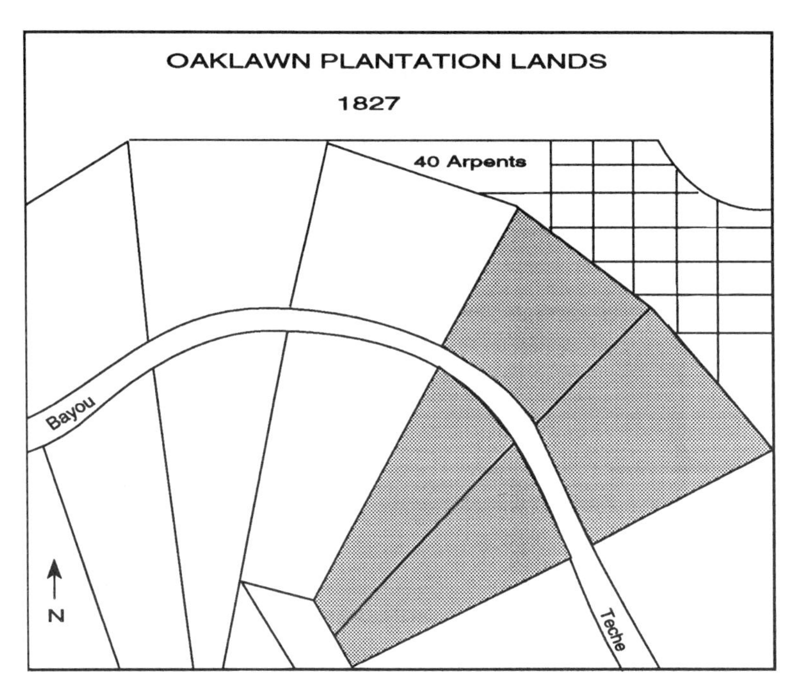

MAP 7.1. *By 1827, Alexander Porter, the initial planter and owner, had completed the acquisition of the Oaklawn plantation lands (*shaded areas*).* Source: Louisiana State Land Office, surveyor's township plats T13S, R10E for 1827

> plantation on either side, and near the borders of the river is an almost regular line of the finest live oaks I have ever seen. He [Porter] has brought nearly 2,000 acres into cultivation, and has a stock of about 160 negroes, 40 horses, and a variety of other cattle. He is at present lodged in a small cottage, but is erecting a handsome and commodious mansion. I should not call it in England very spacious, considered as the residence of one of the largest proprietors in the country, but in Louisiana it is remarkable enough to attract curious persons from considerable distances (Daubeny 1843, 143).

Alexander Porter died in 1844, and Oaklawn was bequeathed to Porter's brother James, who had been a partner in the estate. James kept the operations until his death in 1850, when the estate passed to his wife (Stephenson 1934, 124–25). An inventory in 1850 reflected Oaklawn's nineteenth-century landscape expression and culture traits. Plantation lands and buildings were valued at $120,000 and covered fifty-two arpents front on both banks by forty arpents in depth. Backlands east of the plantation toward Grand

FIG. 7.1. *Many of the landscape features of a complete Louisiana sugar plantation are depicted in these engravings of Oaklawn Plantation in 1864.* Reproduced from *Frank Leslie's Illustrated Newspaper,* Feb. 6, 1864, p. 316

Lake had 1,304.8 acres of backswamps and prairie. The inventory included 320 Negro slaves valued at $90,350. Oaklawn's livestock, some of the best on any plantation at that time, included 80 work horses and mules; 160 head of horned cattle; 10 blooded mares, 7 of which had foals; 5 blooded and 24 ordinary colts and fillies; 4 carriage horses; and 160 sheep. Agricultural implements were 12 horsecarts, 2 farm wagons, 4 ox carts, 60 plows, 150 hoes, 100 axes, and 50 spades and shovels. A sawmill, old engine, and sugar mill were valued at $6,000. All this, added to the furnishings and library in the mansion, buildings, slaves, lands, implements, sugar and molasses in storage, and debts due to the estate totaled $268,054.41.[3]

The Porters were successful sugar planters. Earliest production records date to 1828, when Alexander Porter produced 184 hogsheads to rank

fourth highest in the parish out of seventy growers (Degalos 1892, 67). The sugarhouse burned to the ground in 1837, and production plummeted (Stephenson 1934, 123). In 1844, the year Alexander Porter died, sugar production was 333 hogsheads. By comparison, nine other planters along Bayou Teche surpassed this mark, and three of them produced over 700 hogsheads. Production improved by 1850, when Oaklawn made 420 hogsheads of raw sugar.[4] Mrs. James Porter, with managerial help, kept the plantation operating through the next twenty-four profitable but stormy years. The 1855 Champomier survey recorded a production of 1,140 hogsheads from the steam-powered Oaklawn sugar factory (Champomier 1855, 31). In 1862, while under Mrs. Porter's ownership, production reached 1,138 hogsheads from two sugarhouses. Both brick buildings were slate roofed and had steam-powered mills, and the boiling houses were equipped with open kettles (Bouchereau and Bouchereau 1869, 65). Oaklawn by now had reached the pinnacle of its antebellum landscape expression.

On January 2, 1874, the heirs of Mrs. James Porter sold Oaklawn for $163,343 to a group of New Yorkers, Edward, Stephen, and Charles D. Leverich. The lands included the Oaklawn Estate and that portion of Oaklawn on the east side of the bayou known as the Dogberry Place, all measuring fifty-two by forty arpents on both sides of Bayou Teche, and the backlands of over thirteen hundred acres. All buildings and improvements on the lands, fixtures, implements, and machinery related to the operations of the plantation were included in the sale. Among the buildings and implements were a mansion house and outhouses; two sugar mills, a sawmill, and a grist mill with the engines, implements, tools, and apparatus for sugar making; and carpenter, cooper, and blacksmith shops with the necessary tools and implements. Livestock included seventy-two mules and twenty-four oxen. The number of implements in the sale indicates the approximate probable number of workers on the plantation at the time. A conservative estimate would be about sixty hands based on forty plows, twelve harrows, ten cultivators, twenty large carts, fifty hoes, fifty cane knives, thirty double sets of harness, two corn planters, one set of blacksmith's tools, one set of carpenter's tools, one set of cooper's tools, twenty-four axes, twelve spades, twelve shovels, twenty scythes, and one pair of timber wheels, among other things. Marketable items included 290 hogsheads of sugar and 400 barrels of molasses in storage. Excluded from the sale were all personal belongings and furnishings in the mansion house, the library, two mules, and the milk cows and sheep kept at the mansion house site.[5]

For the next fourteen years, ownership of Oaklawn passed back and forth among Leverich family members. In 1876, original buyers Edward, Stephen, and Charles D. Leverich sold the plantation to Henry S. and Charles P. Leverich for the sum of one dollar. Five years later the holding momentarily passed from the Leveriches to T. J. Foster, but Foster returned the "merchandise" on the very date of purchase and sold the plantation back to A. F. Leverich for ten dollars.[6]

In a more serious transaction, Robert E. Rivers purchased Oaklawn for $45,000 in 1888. While under Rivers's eight-year tenure, Oaklawn underwent extensive alterations, particularly in the sugar factory. When the Leveriches owned the property, the brick and slate sugar factory remained outdated, with open kettle equipment in the boiling room. Robert Rivers installed triple-effect evaporators, vacuum pans, and centrifugals in the brick-walled and slate-roofed sugar factory (Bouchereau and Bouchereau 1875, 1890). In 1896, Rivers sold Oaklawn Plantation and the adjacent Oxford Plantation, which he had purchased in 1892, to a newly formed corporation called the Oaklawn Sugar Company. The corporation, headed by president Eugene W. Weems, bought a bargain that included both Oaklawn and Oxford Plantations, 1,304.8 acres of backlands, a 65-arpent tract across from Oxford, plus all materials for $28,000.[7]

From 1828 until 1925, Oaklawn had remained a complete and singularly owned land unit each time it was sold. But Captain Charles A. Barbour, a great admirer of the mansion, purchased the house and its 76-acre site on February 19, 1925, for $121,000.[8] Barbour's ambition was to restore and furnish the house to its original condition. After long years of painstaking work and many thousands of dollars, he and his wife and their daughter eventually completed the project. In the meantime, the separately owned agricultural and sugar factory portions of Oaklawn Plantation were sold to the SouthCoast corporation in 1934 (Gilmore 1934, 36). Between 1950 and 1964, Mrs. Lucile Barbour Holmes and her husband preserved the mansion. Then Mr. and Mrs. George E. Thompson acquired the mansion and its 76-acre yard and gardens, and the property remained with the Thompsons until 1979.

Mid South Mortgage Company, headed by Thomas B. Goldsby Jr., James G. Robbins, Leland McCarthy, and Richard H. Cook, in 1979 purchased the entire Oaklawn plantation complex of mansion, yard, and 7,150-acre agricultural and sugar-processing enterprise as part of the CALA model that later became South Coast Sugars. For the first time since 1925, the mansion and plantation were back together in a contiguous unit owned by the same people. And for a short time, Thomas B. Goldsby Jr., a land speculator and entrepreneur from Arkansas, lived the life of a Louisiana sugar planter. Apparently, financial burdens in the 1980s forced him to leave Louisiana.[9] Eventually, Oaklawn was separately incorporated and momentarily renamed Teche Sugar Company. The mansion and its 76-acre site were once again separated from the plantation. Between 1979 and 1985 about half of the mansion site yard was subdivided into a residential subdivision. The mansion, now on a much smaller lot, was sold in 1985 to then Louisiana state senator and now Governor Mike Foster. In 1991 moss-draped oaks shaded unsold lots, but by April 1993 several unnatural-looking new houses had been built in the oak groves at Oaklawn.

Landscape Morphology in 1969

Oaklawn illustrated many of the basic plantation components of diagnostic Anglo-American landscape traits for more than 150 years. A block-patterned plantation settlement characterized the main buildings. Just like Ashland Plantation, the Oaklawn settlement had been a bayou-block settlement occupying both banks, or levee crests, of the bayou. The settlement had an interesting spatial arrangement in form and function. Oaklawn's

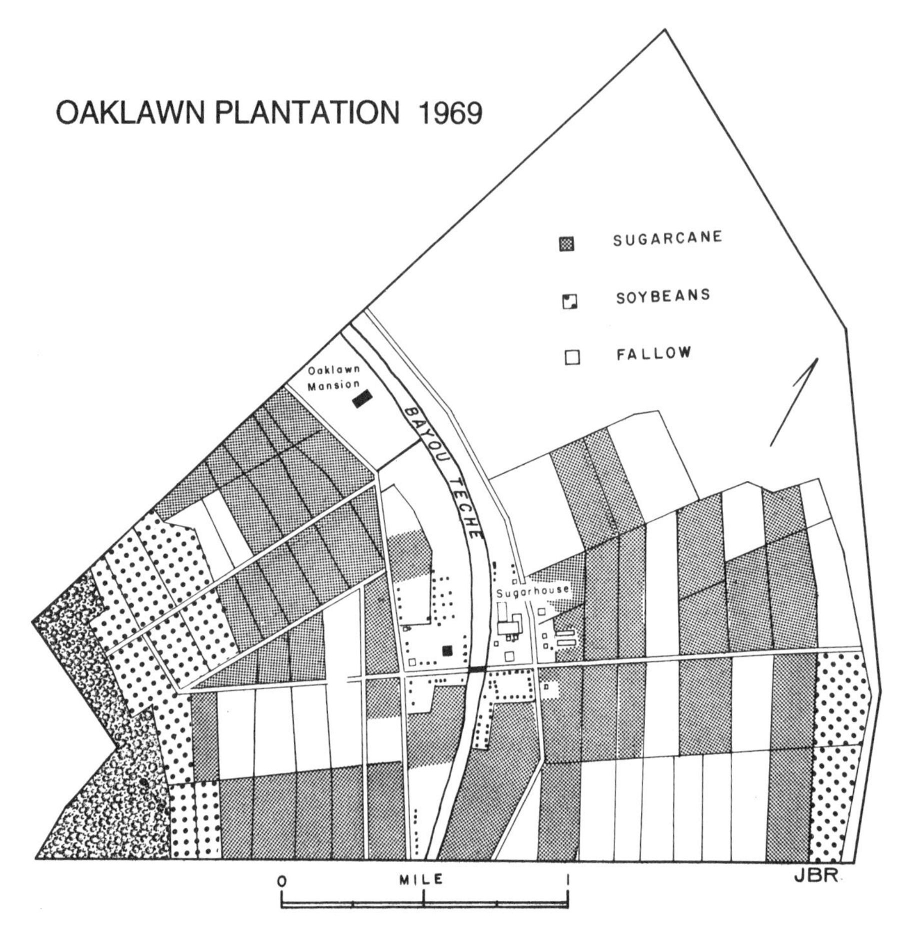

MAP 7.2. *Oaklawn's bayou-block settlement was divided into four quadrants and had sixty-seven quarter houses. Land use in 1969 included 956 acres of sugarcane, 209 acres of soybeans, and 583 acres set aside for fallow ground.* Source: Rehder 1971, 362

block settlement was divided into four units. The southeast quadrant had a quarters with thirty-seven structures. The northeast quadrant had the sugar factory and a large outbuilding complex. The southwest quadrant had dwellings for the field superintendent, several overseers, the sugar factory engineer, a clerk, office personnel, the agriculturalist, some mill workers, the welder, the shop foreman, a dragline operator, the Celotex man, and the night watchman.

The northwest quadrant had a variety of finely separated buildings and functions. On the southernmost side of this quadrangle were more sugar factory workers' houses, the division manager's house (built in the style of the Oaklawn mansion but constructed about 1950), the store and the store clerk's dwelling, and the church. Upstream and north of these structures, separated by a small road, were the remaining plantation quarter houses and a separate unit of cinder-block bungalows for some of the factory personnel (Rehder 1971, 361–72). By 1991, virtually all quarter houses and the store had been razed or were in the process of demolition. By 1995, only the division manager's house, five cinder-block houses, the factory office, and the ruins of the sugar factory across the bayou remained.

Until 1969, the quarters had been located on both banks of the stream. Thirty-seven quarter houses occupied the east bank of the bayou downstream from the sugar factory. Most were small Creole type dwellings. Of the sixty-seven quarter houses in the total settlement, fifty-two were small Creole types; of these, seven were single-family houses. Most of the quarter houses were weatherboarded with cypress lumber, but a few had the popular yellow-brick tar-paper siding. Early during the plantation's demolition in the 1980s, Oaklawn's valuable cypress lumber in the quarter houses was

FIG. 7.2. *The sign out front reads "Oaklawn Store–Grocery & Merchandise." Note the cistern* (left) *at this nineteenth-century store on Oaklawn Plantation.* Photograph, 1969

FIG. 7.3. *Seven of the sixty-seven quarter houses were these single-family units occupied at Oaklawn Plantation. Others were small Creole and bungalow types.* Photograph, 1969

among the first and most sought-after salvage. Oaklawn had been named for its magnificent live oaks that border the banks of Bayou Teche. Huge, moss-covered oak trees shaded not only the mansion and the manager's dwelling, but also the much lesser quality houses in the quarters. In the back yards, outhouses were a natural part of house sites. Fences made of wire and sheet metal enclosed small vegetable gardens.

On the west side of the bayou and clustered in the southwest quadrant were better kept and neater houses for the manager, overseers, and sugar factory workers. Most of these dwellings were bungalows with a variety of construction materials. The newest cinder-block houses were built in the 1950s and were occupied by factory personnel. The remaining bungalows were wood with horizontal weatherboarding for some of the structures and board-and-batten for others. The division manager's dwelling, also built in the 1950s, was larger than most antebellum mansions.

Other Oaklawn dwellings belonged to a miscellaneous category. Located near the outbuilding complex in the northeast quadrant of the plantation

FIG. 7.4. *Pyramidal houses were extremely rare on Louisiana sugar plantations. This dwelling was assigned to one of the sugar factory personnel at Oaklawn Plantation.* Photograph, 1969

were a bunkhouse with board-and-batten construction, a long house with four apartments constructed of board and batten, and a Celotex building that served as office and temporary dwelling for some of the Celotex personnel who worked at the plantation during grinding season.

Oaklawn Plantation maintained a large sugar factory until 1989. The factory served Oaklawn Division's nine plantations, which had a total acreage of 11,631 acres. Sugarcane processed at the mill for the Oaklawn Division plantations in 1966 came from 5,071 acres. Sugar factory buildings and the boiling house were constructed of structural steel with corrugated-galvanized roofing and siding. The vacuum-pan section was reinforced with structural steel and covered and roofed with corrugated asbestos materials. The clarification section, the primary sugar-storage building, and two additional warehouses were wood framed and sided and roofed with galvanized iron. Another warehouse, measuring 80 by 150 feet, had all-steel construction. Total sugar storage capacity at Oaklawn had been 140,000 hundred-pound bags of sugar.

FIG. 7.5. *Two of the five remaining cinder-block quarter houses built in the 1950s are now occupied by caretakers at Oaklawn Plantation.* Photograph, 1995

Functional machinery at the sugar factory had been much the same as in most other Louisiana sugar factories at the time. Three derricks were replaced in 1968 by an overhead crane that loaded two cane carriers to feed canes into the milling apparatus inside. Milling equipment–two crushers and four three-roller mills–gave the mill a grinding capacity of 4,000 tons of cane per day. In 1966 the mill ground 249,663 tons of cane in a 65-day grinding season.[10] Once the canes had passed through the mills, conveyer belts carried bagasse to the baling plant 200 yards across the railroad tracks from the factory. The extracted juices were clarified, evaporated, and processed in the usual manner into raw sugar and molasses. Power came from eight gas-fired steam boilers that generated over 5,000 horsepower.

The factory produced both raw and turbinado sugar, a better grade but not a purely refined sugar. Workers bagged the turbinado sugars, stored them in warehouses, and then packed the bags into railroad cars for shipment. Much of the raw sugar went by conveyor belt to wooden warehouses or directly to the barge dock on Bayou Teche. All raw sugar produced by Oaklawn was shipped by barge to the corporation's Georgia refinery at Mathews, located on Bayou Lafourche. Oaklawn's outbuilding complex had

FIG. 7.6. *The Oaklawn sugar factory closed in 1989, was in ruins by 1993, and after 1995 showed almost no identifiable landscape evidence. Bayou Teche, Saint Mary Parish.* Photograph, 1993

a variety of structures. The former mule barn was an old, large, wooden building that had been converted into an agricultural-equipment warehouse. Across the railroad from the sugar factory were two new tractor sheds, a new cane-harvester shop, an agricultural machine shop, a tire shop, and a blacksmith's shop, all in steel buildings with galvanized metal siding. Some of these agricultural structures remain today because they are still being used on the functional plantation.

Fields and land use at Oaklawn were no different from those on any other plantation. Oaklawn's location on the relatively narrow natural levees of lower Bayou Teche required early acquisition of frontlands along levee crests. Arable levee widths here rarely exceeded 1 mile, compared to the 3- to 5-mile levee widths on the Mississippi River and upper Bayou Lafourche. Arable levee widths at Oaklawn are 1 mile on the east bank and 0.8 mile on the west bank. The company used pumps and artificial levees on the backlands to maintain drainage for sugarcane crops. Field plots measured about 2 acres, and field cuts averaged 25 acres. The 1969 crop included 956.2 acres of sugarcane, 209.7 acres of soybeans, and 583.9 acres of

FIG. 7.7. *The Oaklawn sugar factory was near the peak of its technical capacity in 1969 but illustrated a contrast in technologies with the old, outdated mule barn at* left *and the 1968 overhead crane at* right. Photograph, 1969

fallow ground. Implements on the plantation were seventy-one tractors (three could cultivate three rows, eighteen feet across, at a time), five chemical sprayers, eight mechanical cane harvesters, six field-loading devices, seventy-nine cane carts, and seven mechanical quarter-drain cleaners (Rehder 1971, 361–72). Until the 1960s a mule and plow kept quarter ditches or drains clear of soil, and many plantations still kept a mule or two for this purpose.

Oaklawn's mansion and its 76-acre, oak-covered yard are located one-half mile north of the bayou-block settlement complex, where the mansion site is secluded from the remainder of the settlement. On the natural-levee crest of the west bank of the Teche stands the restored and preserved structure of the mansion built by Alexander Porter between 1837 and 1840. The mansion is an Anglo-American house type with a front-facing gable, inside-end chimneys, a central hallway, and a height of two full stories. An interior stairway ascends to the second floor and beyond to an attic that once was used as a ballroom. The mansion has a simple floor plan. The first floor has

FIG. 7.8. *Oaklawn's mansion, an Anglo-American Tidewater type, dates from 1837, when it was the home of Alexander Porter.* Photograph, 1969; source: Rehder 1978, 142. Used with permission from the Louisiana State University School of Geoscience

four large rooms (two rooms measuring twenty-two by thirty feet, two rooms measuring twenty-two by twenty-two feet) and halls that are sixteen feet wide. The same plan is duplicated on the second floor. The main core of the structure measures sixty by sixty feet and stands fifty-eight feet high. Six columns of Tuscan style support the overhanging roof-porch and pediment at both the front and rear entrances. Beyond the core are eight-foot-wide porches and a kitchen as an appendage.

The mansion in 1925 was unoccupied, in need of paint, barren and yet intact. Shortly after Captain Barbour bought the structure, he photographed the house. Fortunately he had a record of the architecture because fire swept through the structure and destroyed all wooden portions of the building, the porches, roof, and interior ceilings. The brick walls and plaster preserved the original plan so that the mansion could be restored. Throughout its history, destruction, and restoration, Oaklawn Manor remained a classic Anglo mansion, the plantation home of Alexander Porter.

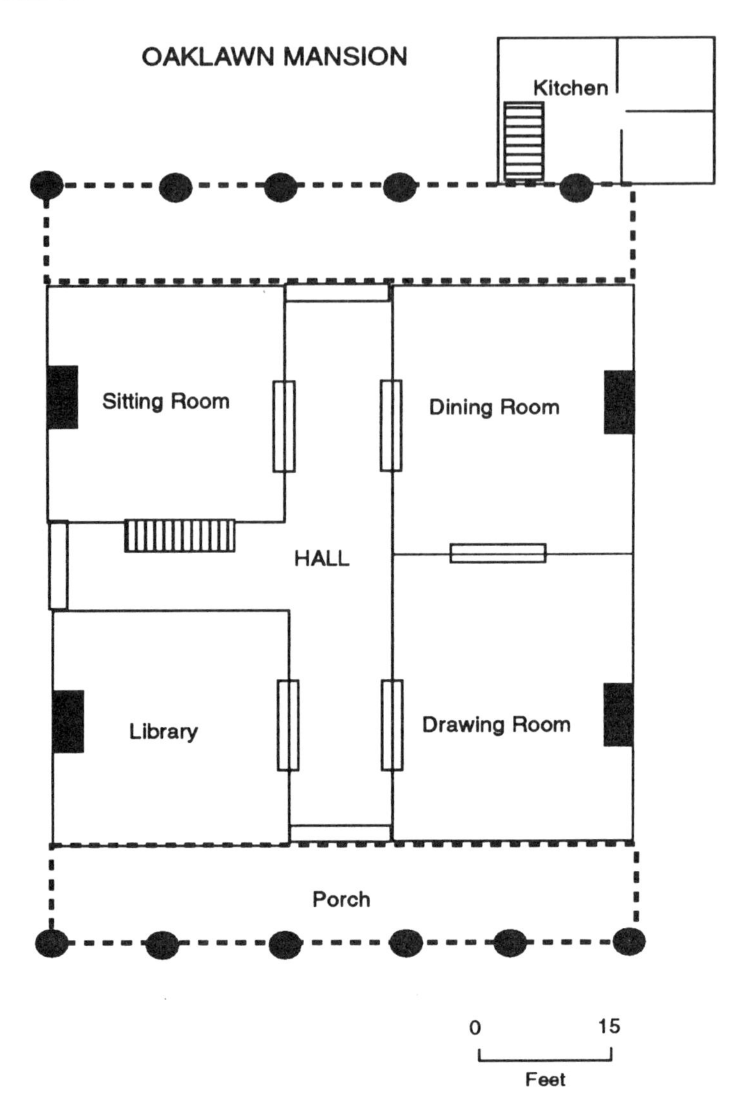

FIG. 7.9. *Floor plan of the Oaklawn mansion exhibits diagnostic Anglo-American Tidewater architectural traits in the front-facing gable, symmetry of four rooms to a floor, central hall, and paired inside-end chimneys.*

Oaklawn, 1989–1998

Illustrations throughout this chapter chronicle an Oaklawn Plantation that has been largely destroyed since the CALA model arrived in 1979. Corporate ownership and management strategies based on a California concept that workers should not live on the premises led to the clearing of diagnostic, culturally significant structures in the settlement. Presently, the mansion remains intact. However, the once-magnificent 76-acre yard filled with moss-draped oaks has been reduced to a subdivision with new streets and several uncharacteristic modern homes deposited in the yard.

The sugar factory closed in 1989 for several reasons. The mill capacity was too large for the amount of cane the company could contract out. "It costs just as much to run a mill at half capacity as it does to run one at full capacity," said Sean McCarthy, assistant chief engineer.[11] In December 1989 a destructive freeze sent temperatures to five degrees Fahrenheit, killing plant canes destined for the next year's crop. Even Bayou Teche was frozen over. A number of sugar land leases also expired in December 1989, so there was no need to renew them with the mill about to close down. Finally, the factory required a major overhaul in areas that contained asbestos materials, so environmentally the structure was ready to be closed. In 1992, Hurricane Andrew blew off much of the roof and further damaged the factory to the point that razing it was the only option. Because of the factory's closing, cane now cultivated at Oaklawn must be trucked seventy miles one way to the corporate headquarters factory at Raceland on Bayou Lafourche.

Sugarcane is still grown on every conceivable acre; even house sites in the quarters have become cane land. The store is gone; so, too, are all but six of the quarter houses. In 1969 there were sixty-seven quarter houses, in 1989 only twenty, in 1991 the figure had dropped to eleven, and in 1995 there were six, with only two of them occupied. As another plantation as big and as once vibrant as Oaklawn vanishes from the landscape, the cumulative effect may one day be that the built environment of the plantation landscape will have finally disappeared.

CHAPTER 8

Cedar Grove Plantation, 1829–1998

The scarce remnants of Cedar Grove Plantation are located on the right bank of the Mississippi River about a mile north of the town of White Castle in Iberville Parish and sixteen miles south of Baton Rouge (see map 5.1). The plantation that I examined in the 1960s had a long, linear settlement pattern suggesting French origin, and, indeed, it proved to be a French linear plantation type. Although under corporate ownership since 1939, Cedar Grove was not part of a large sugar corporation such as Armant, Ashland, and Oaklawn, nor was it a feeder plantation member of a cooperative such as Madewood. In 1969, Cedar Grove was a singular-unit type plantation with its own sugar factory. Cultivated lands measured 1,273 acres, with 726 acres in sugarcane. Approximately thirty-five workers were in the resident labor force, and during grinding season as many as one hundred people were employed. No one lives there now.

Historical Succession

The succession of Cedar Grove ownership is one of the more interesting stories because it reveals much about Louisiana culture. Except for its past two owners, all former owners were French. The first recorded owner of Cedar Grove Plantation was George Deslondes Sr., a Caribbean sugar planter from Sainte Domingue (Haiti), a slave owner, and a free black man who also owned most of his extended family. Earliest information on the Cedar Grove succession dates to December 2, 1829, when conveyance records indicate that George Deslondes, a free man-of-color, purchased a 3- by 80-arpent tract for $7,500 from Joseph Huguet. The vacant land was composed of a first concession of 3 by 40 arpents and a second concession previously purchased by Huguet from the United States in 1822. The depth of the second concession was also 40 arpents "fixée par les arpenteurs des États Unis" [fixed or surveyed by U.S. surveyors]. Deslondes also purchased an additional 2- by 40-arpent tract adjoining the first for $3,000 from Pierre LaCroix, a partner of Joseph Huguet. Except for an 0.8- by 80-arpent tract that belonged to Deslondes originally, these purchases concluded the land acquisition for the contiguous expanse of Cedar Grove at that time.[1]

In October 1836, George Deslondes Sr. died and his estate went to his wife, Mrs. Félicité Deslondes, who was by then a free woman-of-color. The following is a modified inventory of the property at Cedar Grove in 1837.

1. The plantation whereon the widow resides consists of a tract of land established as a cotton and sugar estate containing five arpents in front by forty arpents in depth of the front concession and a double concession on the back and contiguous to the three upper arpents of said land having 180 superficial arpents, $12,000. Included are all buildings and improvements.

2. Another tract containing one-third of 2½ arpents by 80 in depth bound below by the wife of Zacharie Honoré and above by the land of the deceased, $3,000.50.

3. Farm implements:

1 plantation bell (broken)	$3.00
1 barrel of lime	$2.00
1 sugar ladle and 1 skimmer	$2.00
12 weeding hoes	$5.00
8 chopping axes	$8.00

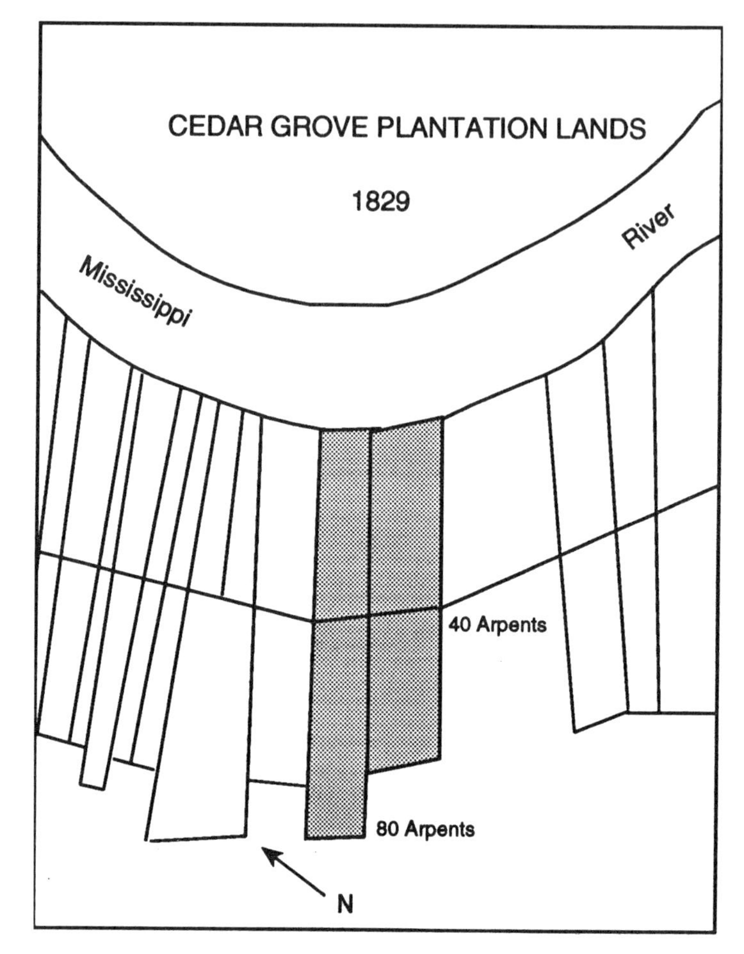

MAP 8.1. *George Deslondes, a planter from Sainte Domingue (Haiti), purchased these tracts to initiate Cedar Grove Plantation in 1829.* Source: Louisiana State Land Office, surveyor's township plats T10S, R13E for 1829

8 cane knives	$2.00
6 sickles	$ .75
5 spades	$2.00
9 ploughs	$30.00
1 lot of horse gear	$6.00
2 harrows	$10.00
4 carts	$12.00
1 grist mill and gears	$25.00

plus numerous other items like carpenter tools, baling rope, tubs, blacksmith tools, etc.

4. Among the household furnishings were furniture pieces of cherry, mahogany, and cypress, plus many other items, including a French academic dictionary.

5. Livestock included 6 horses, 7 work oxen,[2] 10 head of cattle, 24 head of sheep, and 1 lot of hogs.

6. Nineteen slaves valued at $15,000.[3]

In 1837, the Widow Deslondes turned the ownership and operation of the plantation over to her son George Deslondes Jr. The younger Deslondes maintained the plantation for the next thirty years and experienced some disappointing events in the role of sugar planter. He produced 146 hogsheads of sugar in 1844, but frosts in 1850 reduced the yield to only 8 hogsheads. By 1855, partial recovery came with a production of 80 hogsheads, and by 1862 production was 103 hogsheads. The now-aging sugarhouse had been on the property for many years. The Champomier survey in 1844 indicated a steam-powered sugar mill at Cedar Grove owned by George Deslondes Jr. In 1862, Bouchereau recorded a wooden sugarhouse with a steam-powered mill and open kettles in the boiling house.[4]

Cedar Grove was sold in a sheriff's sale in 1867 to Antoine Dubuclet, who also had been a free man-of-color. Informants in the Cedar Grove area today say that the Dubuclet family originally came from Martinique, French West Indies, but court records do not indicate the place of family origin. Dubuclet, who already owned property below and adjacent to Cedar Grove, received for his $25,000 bid the 5¼- by 80-arpent tract, including all buildings, livestock, and agricultural implements.[5]

Dubuclet sold the holding in 1872 to John A. Sigur, who was the owner of an adjacent plantation above the Cedar Grove tract. The purchase price was $12,500, with payment made at $4,000 down and the balance in seven installments. The plantation contained all of the above described with the exception of a ½- by 80-arpent tract on the lower side of the plantation and a grist mill and steam engine.[6]

The Sigurs retained the property for only three years before selling it for $13,488 to Leonce M. Soniat in 1875. The plantation then measured 4¾ arpents wide by 80 arpents deep and included many of the buildings and improvements. All movable property, except the grist mill and steam engine belonging to George Deslondes Jr., were included in the sale.[7] Not mentioned in the records but listed by Bouchereau in 1875 was the wooden sugarhouse with its steam mill and open kettles. This same outdated equipment had been in the sugarhouse since 1862 when Deslondes owned the property.

Soniat maintained ownership of Cedar Grove for forty-five years, one of the longest periods in its history. He made numerous improvements to the plantation by purchasing additional land and re-equipping and modernizing the sugarhouse into a genuine sugar factory. Between 1883 and 1890 Soniat reinforced the shingle-roofed wooden factory with iron beams inside; replaced the old open kettles with double-effect evaporators, vacuum pans, and centrifugals; and with these improvements increased production to 1,388 hogsheads of sugar in 1892. Eight years later he added a triple-effect evaporator and reroofed the sugar factory with composition roofing. By the early twentieth century, Soniat had production soaring to the equivalent of 3,040 hogsheads in 1911, 2,106 in 1914, and 3,021 in 1917.[8]

In 1920, Soniat sold the plantation to the Cedar Grove Sugar Company. Through this transaction evidence of greater modifications and land acquisitions can be seen. The Cedar Grove Plantation that Soniat sold was not the same familiar 4¾- by 80-arpent tract that he had originally bought. Cedar Grove now measured 11 arpents front by about 76 arpents in depth. This larger Cedar Grove had been acquired by Soniat in the following manner: (1) the upper portion, known as Zacharie Plantation, a 4- by 80-arpent tract, had been derived from five different portions, all of which were undivided and purchased between May 1893 and December 1896; (2) the middle portion, known as Cedar Grove, was composed of 4¾ arpents + 30 feet front by 80 arpents depth bought from Mrs. Sigur in February 1875; and (3) the lower portion, comprising the former Augustine Dubuclet tract of ½ arpent + 160 feet front by 80 arpents depth, was acquired through nine different transactions and exchanges from March 1897 to February 1919.[9] The records reveal more about the plantation through reservations and exceptions that Soniat retained in the sale. The provisions were

1. To reserve all present crops, supplies, oil, cooperage, and other articles used in the manufacture of the 1920 crop.

2. To have the use of the sugarhouse and machinery, boarding house, blacksmith shop, and all other equipment needed for preparing the cane crop for market.

3. To have the right to employ laborers from the plantation to manufacture, load, and ship out the marketable crop plus the use of teams of horses and carts, etc.

4. To have the right to storage space for the remaining 1919 crop and the use of one of the larger warehouses for four months.

5. To retain all personal effects, household goods, carriages, cattle, hogs, chickens, one horse, two mules, two carts, and garden implements.

6. To reserve the stock in the store and to have the use of the store and the store clerks' houses for one month after the sale.

7. To have usufruct rights to the residence for M. Soniat and his wife during their lifetime and 25 acres of land at the house site bounded to the north by the river, east by Richland Road, south by the Texas and Pacific Railroad, and west by the fence enclosing the stable lot.

8. To retain the right to use Cedar Grove Road leading to the Mississippi River as well as the use of the steamboat landing in front of the plantation.[10]

Soniat received $300,000 for his sale of Cedar Grove and nearby White Castle Plantations plus 3,134 acres of other scattered tracts behind the property. For the next nine years, the Cedar Grove Sugar Company was managed by Augustin Lasseigne, president, and Charles E. Thibodeaux, secretary and treasurer. Then, in 1929, a bid of $50,000 in a judicial sale placed the holdings back into the hands of the Soniats. In this case the purchaser was Mrs. Leona S. Soniat. The very same plantations that Soniat had previously sold for $300,000 were bought back for $50,000.[11]

Ten years later, on March 17, 1939, the property was auctioned off to William T. Burton Industries for a $150,000 bid. The original purchasing party was Burton and Sutton Oil Company of Lake Charles, Louisiana, which later became William T. Burton Industries. Burton rebuilt the sugar factory in 1939, where it remained intact and operating until 1975. On June 6, 1975, William T. Burton Industries sold the sugarhouse site and its contents, including all buildings and machinery on sixty acres, to Southdown Sugars for $1,250,000. Southdown had just become part of the CALA model under the management strategies of Prudential Southdown Partnership, who bought and closed competing factories, even those as far away as Cedar Grove and Catherine in Iberville Parish. On October 4, 1988, William T. Burton Industries officially became William B. Lawton Industries, and for the time being the ownership succession of Cedar Grove has stabilized.[12]

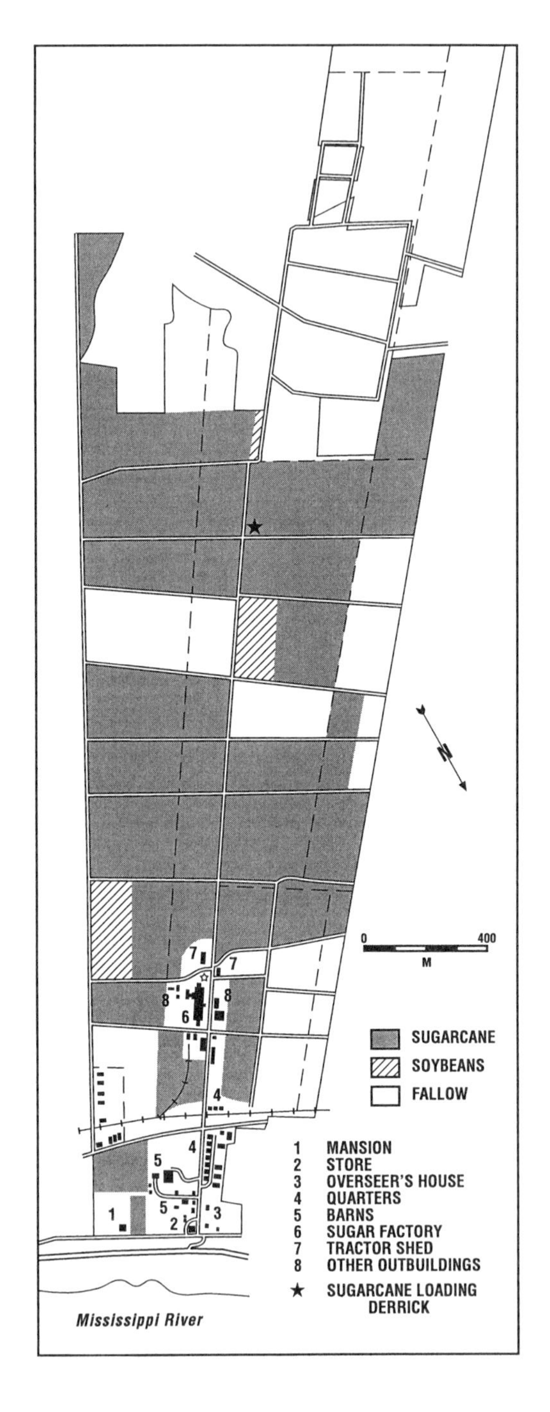

MAP 8.2. *The 1969 Cedar Grove Plantation sustained a complete complement of plantation settlement features and cultivated 726 acres in sugarcane, 70 acres in soybeans, and 176 acres in fallow ground.* Source: Rehder 1971, 305

Landscape Morphology in 1969

For 165 years Cedar Grove has been a long, narrow, flaring landholding measuring between 4¾ and 11⅓ arpents front by 80 arpents deep. With such a narrow frontage, the settlement became a linear plantation almost by necessity. The holding was more than 3 miles long from river to backswamp. From the levee and river road, the succession of buildings proceeded for 0.6 mile from the mansion, store, and overseers' houses, which were along the river road at the levee, to the linear quarters, and ultimately to the sugar factory and outbuilding complex at the rear of the settlement. Cedar Grove was a classic linear settlement with the necessary components of a model linear plantation.

The quarters formed a line of thirty-seven dwellings. All types of quarter houses were represented, but seventeen small Creole type quarter houses dominated the settlement. Other quarter house types included ten attached-porch quarter houses, six shotguns, and one bungalow. Three of the ten attached-porch dwellings were single-family units. Two long multiunit houses and a very small half-sized shotgun house were located near the sugar factory. Garden plots were provided for all households, but only one retiree had a working vegetable garden.

Variable construction details served as time-sequence indicators. The small Creole houses were built with sawn wooden planks weatherboarded in the traditional horizontal manner. Attached-porch dwellings and the two longhouses were board-and-batten construction. Finally in the sequence, a somewhat newer shotgun and the newest bungalow houses were on cinder-block foundations and were sided with asbestos siding. Other dwellings included two larger Creole houses occupied by the field manager and a retired overseer. The plantation welder lived in a small, relatively new non–folk type house.

The empty mansion was neither large nor representative of the early-nineteenth-century plantation mansions found elsewhere in Louisiana. It was a Louisiana bungalow, forty-two feet long like a shotgun, two rooms wide, and with a front-facing gable. It was a type with the appearance of both the shotgun and later bungalows but followed the former and predated the latter. In an urban context like New Orleans, this house type is called a double shotgun. Quite likely, Leonce Soniat had the dwelling built while Cedar Grove was under his forty-five-year tenure because it was the dwelling that he retained when he sold the plantation in 1920. Multiple

Fig. 8.1. *Quarter houses formed the linear plantation settlement pattern in the front section along the main plantation road at Cedar Grove Plantation.* Photograph, 1969

front doors, central chimneys, and a hipped roof are some of the traits of this French dwelling. The dwelling was in evidence when French owners were in tenure at Whitney and Cedar Grove, both French plantations. The distant location of the house and site 0.2 mile from the rest of the settlement, a trait common to French and Anglo plantations alike, indicated the special, exclusive setting for the mansion and its occupants. Several other diagnostic French traits were on site, including crepe myrtle, oak, and willow trees, a cypress cistern, and a surrounding cypress yard fence similar to the French *pieux debout.* In 1982, this small but culturally significant plantation mansion was torn down.

The Sugar Factory and Outbuildings

Like so many of Louisiana's sugar factories, the one at Cedar Grove experienced intermittent operation as well as a rebuilding program since the 1920s. When the mosaic disease wrought havoc on the cane crops of 1927,

FIG. 8.2. *Shotgun and small Creole quarter houses built in the late nineteenth century were representative quarter houses at Cedar Grove Plantation.* Photograph, 1969

1928, and 1929, the Cedar Grove sugar factory was inactive. Operations were restored in 1930 when crops improved with the introduction of new disease-resistant varieties. The factory was rebuilt in 1939 just after the Soniat heirs sold the property to William T. Burton. The 1939 rebuilding program involved first moving a cane shed, which measured 80 by 100 feet, and the old purging-room building, which was 87 by 120 feet, out from the mill site to two new locations some 200 to 400 feet away. The old mill building and boiler house were demolished, and in their place was constructed an all-steel and galvanized sheeted building, the one present on the plantation in 1969 (Gilmore 1940, 78).

The structure had six composite sections: a 45- by 97-foot unit that housed the cane carrier, knives, and cane washing units; a mill section, 53 by 138 feet; an evaporator tower, 12 by 138 feet; a vacuum-pan tower, 40 by 83 feet; the boiler room, 56 by 90 feet; and the purging room and warehouse, which measured 85 by 120 feet. The tallest part of the building, except for the chimney, was a 70-foot-tall vacuum-pan tower. Materials de-

Fig. 8.3. *Cedar Grove's mansion was somewhat uncharacteristic because it was a Louisiana bungalow built about 1875. The house was razed in 1982.* Photograph, 1969

rived from the demolition of older mill buildings were used to construct a hay barn, several storage sheds, and the mechanic's building in the nearby outbuilding complex.

The sugar factory improved considerably after 1939 in milling equipment and grinding capacity. When new roller mills were installed in 1939–40, capacity was at 1,200 tons of cane per day for the eleven-roller mill. In 1967, the capacity rose to 2,000 tons of cane per day ground by fourteen rollers. Another index of the relative size of a sugar factory was the total tonnage ground; at Cedar Grove that went from 99,909 to 80,095 and back up to 97,000 in the years 1965, 1966, and 1967, respectively (Dupy 1967, 60). About 45 percent of the cane ground at the sugar factory came from the company-owned lands on Cedar Grove and White Castle tracts. The remaining cane was bought from about sixty neighboring growers, who delivered the cane to the mill by tractors and trucks.

Adjacent to the sugar factory were several warehouses for bagged-sugar storage. In 1968–69, a new warehouse was erected. Molasses tanks, ma-

FIG. 8.4. *Cedar Grove's sugar factory was at the zenith of its operation and landscape presence in 1969.* Photograph, 1969

chine shops, small storage sheds, and two cane loading derricks completed the sugar factory and outbuilding structures. Other agriculturally related outbuildings included two tractor sheds, an old wooden shed, and a new steel-fabricated shed that remained until 1993. Cane harvesters were stored in an old equipment shed that had been part of the pre-1939 sugar factory. Most outbuildings in the factory area were covered with galvanized sheet metal, except for one shed constructed with wooden board-and-batten materials (Rehder 1971, 303–14).

Cedar Grove had two outbuilding complexes, one located at the sugar factory site and another smaller complex nearer the front of the plantation that was only sparsely used in the 1960s. The latter site had a feedlot, barns, and several rarely used equipment sheds. Apparently, it had been designated for plantation mules and was located nearer the quarters for use by agricultural laborers. The plantation store had been built with a front-facing gable, like most plantation stores in the area, but this store had multiple appendages. It had a credit system and served plantation workers and their families.

Fig. 8.5. *Cedar Grove's plantation store was the center of commercial and social activity. The store closed in the late 1970s and was razed in 1994.* Photograph, 1969

Land Use and Other Landscape Features

Land use at Cedar Grove was much the same as on other sugar plantations in Louisiana in 1969. Cropland had sugarcane on 726 acres, 70 acres in soybeans, and 176.8 acres of fallow ground. Field patterns were long, narrow plots of 2 to 6 acres each that collectively formed numerous large cuts of 7 to 30 acres. The landholding at Cedar Grove had no backswamp because the plantation was favorably located on the upper portion of the White Castle crevasse. Water from fields eventually drained toward a backswamp, but not before passing into drainage canals at Richland Plantation, located immediately behind Cedar Grove.

Other features included a cane loading derrick at the far end of the landholding where cane was hoisted onto trucks and cane carts and hauled to the sugar factory 1.6 miles away. No fences surrounded Cedar Grove fields. However, a barbed wire fence enclosed the feedlot behind the store in the front outbuilding complex, where two pleasure horses were kept. Fences were once located at house sites, where smooth wire fences surrounded the manager's and overseer's dwellings and chicken wire and slat fences were used in the quarters. There were no plantation livestock. The only animals I saw were the two pleasure horses in the feedlot and a few chickens kept in the quarters. One item of more recent appearance on the landscape was a small airstrip used by the manager and by small crop-dusting planes that occasionally landed on the field. The airstrip was not all that unique because private airstrips had been built on other sugar plantations in Louisiana (Rehder 1971, 309–14).

Cedar Grove, 1989–1998

For the past several years, I have had the sad privilege of watching Cedar Grove fall into ruin and all but disappear, being a silent witness to its early deterioration and now wholesale destruction. I had not seen Cedar Grove for twenty years when I returned in 1989 to find a plantation with a closed sugar factory in ruins, a closed store, numerous dilapidated and empty quarter houses, and perhaps four or five quarter houses and two overseers' houses still occupied. But this time the occupants were retired workers, widows, and relatives of the agricultural workers.

I spoke with Leland Overton Jr., a twenty-five-year-old, unemployed black man who shared many poignant thoughts and recollections of key events in the history of Cedar Grove. When I showed him a photograph of the store in 1969, he laughed and said, "Hey man, look at that car! It's Mr. Burton's car! I remember Mr. Burton driving up here to check on things, stopping at the store, talking with folks, giving us candy. Yeah, I remember even though I was probably not even ten years old." Leland Overton Jr. knew that the sugar factory closed in 1975 and that the mill workers were told to move off the plantation. They were no longer welcome to stay. He knew that the agricultural workers and families were allowed to remain in their little quarter houses because some were still employed by William T. Burton Industries to cultivate sugarcane on the plantation. In 1989 Cedar Grove was hanging on by a thread, and Leland Overton Jr. was its corporate memory.

I returned to Cedar Grove in 1991, 1993, 1994, 1995, and 1996. On each visit I saw more and more deterioration. Each time more buildings were abandoned and destroyed. Hurricane Andrew blew the roof off of one of the former overseers' houses. The sugar factory was reduced to skeletal ruins, and the quarters settlement was all but gone. In May 1994, I arrived to find a man named Johnny driving a small bulldozer over a house site. By this time only one quarter house remained, and it was going fast, as the roof had already been removed. Only one overseer's house remained in the front section of the former settlement. Staring blankly at scenes he could not see, Johnny the dozer driver told me that he was "cleaning" up the place and that much of his payment was based on things he could salvage and sell. I tried to show him that I understood. That, yes, nobody lived here anymore, and the buildings were falling down; nobody would want to live in these shacks. We both were thinking this way. It was logical to tear them down, especially if there was something of value that Johnny could sell to a scrap dealer. But I was angry that we both agreed on a form of destruction that was a payoff for him and a permanent loss of landscape features for me. In the outbuilding complex only a few rusting tanks and a huge tractor shed with some missing roof panels remained, but the plantation was still raising cane.

CHAPTER 9

Whitney Plantation, 1790–1998

More than two centuries ago on the right bank of the Mississippi River in an area known as the German Coast, the Haydel family, descendants of German immigrants, initiated a plantation later to be called Whitney (see map 5.1). Traditional French and assimilated Germans, such as the Haydels, were not in the habit of naming their plantations other than family surnames. For some inexplicable reason Anglo-American planters named their plantation properties with colorful appellations such as Oaklawn, Ashland, Madewood, or New Hope. In 1868, a New Yorker named Bradish Johnson became the new owner of the Haydel property and named the plantation Whitney after his grandson.

The plantation property, a French linear plantation type, covers over eighteen hundred acres in Saint John-the-Baptist Parish approximately thirty-three miles upstream from New Orleans. When I observed the plantation in 1967, the resident work force of eight employees included a manager, an overseer, and six agricultural-machine operators. These eight workers and their families plus four households of retirees constituted the

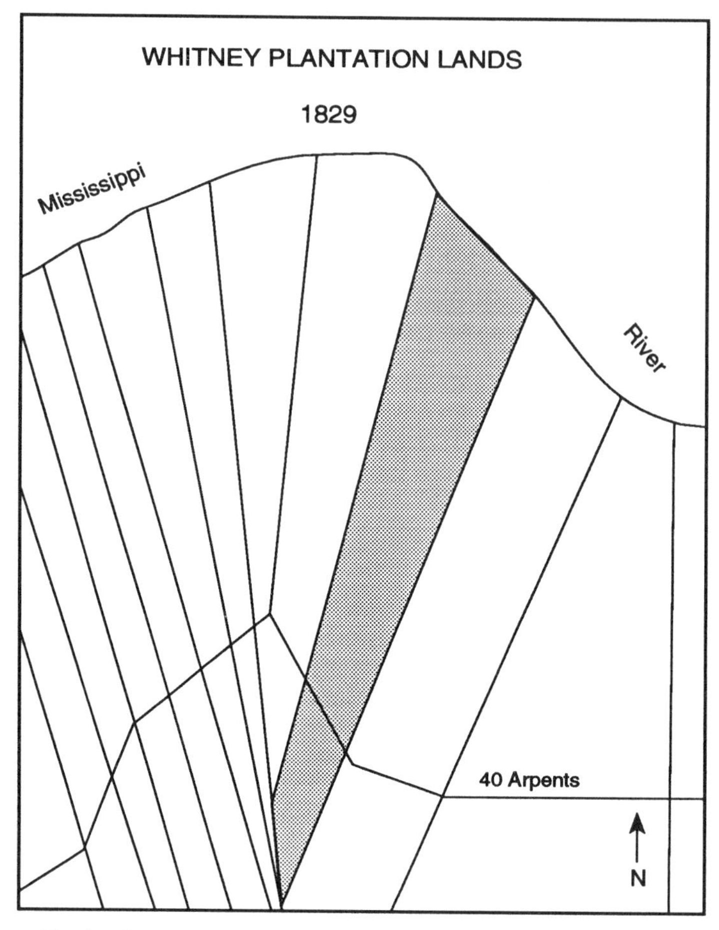

Map 9.1. *The family of Jean-Jacques Haydel, the first effective settlers and planters, developed the plantation lands later known as Whitney (*shaded area*).* Source: Louisiana State Land Office, surveyor's township plats T12S, R18E for 1829

total population at Whitney. The absentee plantation owner was Alfred M. Barnes of New Orleans. Commercial crops were sugarcane, soybeans, and rice, with the latter produced until 1967. Between 1969 and 1990, Whitney experienced the loss of many buildings, a reduction in the resident labor force from eight to none, a deterioration and decline in the landscape, and a critical change in ownership.

Historical Succession

Initial ownership for Whitney Plantation traces to Jean-Jacques Haydel. The earliest recorded evidence indicates that in 1766 Haydel was acquiring land on the west bank of the Mississippi River on the German Coast. By 1781, a tract of land measuring seventeen arpents along the river was in his possession. But the earliest map of plantation evidence appeared much later in 1803.[1] Conveyance records from 1820 for the ownership successions indicate that Jean-Jacques Haydel Sr. sold a well-established plantation to his sons Jean-Jacques Haydel Jr. and Marcelin Haydel. The plantation was described as follows:

> 1. An habitation situated in the Parish of Saint John-the-Baptist measuring 23 *arpents de face* on the River Mississippi, of which 10 arpents have a double concession, the others having only the ordinary depth of 40 arpents, forming nine degrees opening of which 580 [surficial] arpents are cultivated; having 300 in plant cane, 200 in *sources* [cane destined for the next year's

FIG. 9.1. *Whitney Plantation (at* center*) can be seen in relation to adjacent land parcels and within the context of the Chemical Corridor.* Source: NASA high-altitude image, Dec. 8, 1991, roll 4175, image 2452

planting], and 80 in maize; the remaining lands are in wood or in *savanne* [marshes].

2. The buildings, consisting of two *maisons de maître* [master's dwellings] (one is a two-story structure), kitchen, storehouses, mills for rice and maize, a sugarhouse with a steam-driven mill, purgery, *cases à Negres* (Negro cabins), stables, and so forth.

3. Twenty-five horses, and twenty pairs of oxen.

4. Fifty-seven slaves.[2]

The 1820 inventory listed two master's dwellings, or mansions, on the plantation. This would indicate the possible merger of two plantations at an earlier time (Rykels 1991, pt. 2).

Slaves in 1820

The inventory revealed much about places of origin, ethnic patterns, and the sociology of African-born slaves on the plantation. It also illustrated a nineteenth-century value system held by the Haydels. Twenty-one of the fifty-seven slaves were foreign born; the others were creole, meaning American-born, with most of them born in Louisiana. African-born slaves numbered nineteen and came from Mandinka, Bambera, Susu, Timne, Kiamba, and Kongo tribal and linguistic groups. Four older men were Kiamba, an old name for the modern Temba in Dahomey (Curtin 1969, 293). Two Mandinka men were Lubin, aged fifty and valued at 50 piastres, and Achille, aged twenty-two and valued at 1,780 piastres. At this time, a piastre was worth about one dollar. Three men were Susu (35, 50, and 60 years old). Three young men were Bambara, and one more was Timne. All Susu, Bambara, and Timne people came at various times from the West African slave coasts of Senegambia. According to Gwendolyn Hall, two-thirds of the slaves brought by the French slave trade to Louisiana in the eighteenth century were from Senegambia (Hall 1992, 29). The single largest African-born group at Whitney were Kongo–five men and one woman who ranged in age from twenty to thirty-five and were valued from 205 to 2,015 piastres.

In the slave population, the adult group included thirty-three men and twelve women. The slaves' ages ranged from six months to sixty years. Categorically, fifteen slaves were younger than nineteen years; twenty-six were nineteen to forty years old; and fourteen were forty-one to sixty years old. Perhaps most revealing was the value placed on each slave. Much had to do

TABLE 9.1

Slaves at Whitney Plantation in 1820

Name	*Ethnic Origin*	*Gender*	*Age*	*Value (piastres)*	*Comments*
Raphael	Kiamba	M	60	50	
Lubin	Mandinka	M	50	50	has a hernia
Mars	Kiamba	M	60	50	
Augustin	Kiamba	M	50	400	a good negro on the hab.[a]
A grifton[b]		–	–	550	transported to another country
François	creole	M	50	900	carpenter
Alexdre	Bambara	M	30	50	sick at length with miasma
Antoine	creole	M	20	1,600	good cart driver, laborer, attentive with the animals
Hilaire	creole	M	19	1,600	"
Etienne	creole	M	19	1,700	"
Azor	creole	M	19	1,700	"
Joseph	creole	M	20	1,700	"
Robine	creole	M	20	1,600	"
Dick	creole	M	25	1,600	"
Jean-Pierre	creole	M	22	1,600	"
Requine	creole	F	18	1,100	a good negress on the hab.
Catherine	creole	F	16	1,100	a good negress on the hab.
Julien	creole	M	16	1,200	"
René	creole	M	30	1,200	carefully tends the animals
Claire	creole	F	20	1,600	with a child named Ursin (7)
Sam	Susu	M	60		
Marguerite	creole	F	60		wife of Sam
Marie-Joseph	creole	M	50	1,500	
Elenore	creole	F	9		
Marie	creole	F	43	800	cook
Pauline	?	F	10	1,155	a négrite[c]
Honoré	Kongo	M	30	205	good negro, has a hernia
Agathe	creole	F	43	2,115	
(with 3 children ages 18 months to 4 years–Jeanne, Jean, and Jumeaux)					
Rosette	creole	F	40	905	good negress on the hab.
Rose	creole	F	14	1,605	

(continued)

TABLE 9.1 *(continued)*

Name	*Ethnic Origin*	*Gender*	*Age*	*Value (piastres)*	*Comments*
Françoise	creole	F	?	965	
Eugenie	creole	F	24	1,730	
[with 2 children Basile (4) and Syphosien (1)]					
Sile	creole	F	40	1,305	
[with 2 children Toussant (2) and Moliere (6 months)]					
Bernard	Kimba	M	50	210	
Alexis	Bambara	M	20	300	a bit of a sugar maker
Henry	creole/ Jamaican	M	50	1,000	
Flore	creole/Ste. Domingue	F	60	360	
Sophie	Kongo	F	35	625	good negress on the hab.
Barnabé	Bambara	M	30	1,850	sugar maker
Manuel	creole	M	23	600	a thief and complainer
Lucas	Susu	M	35	1,450	good negro on the hab.
Hector	Susu	M	50	355	
Michel	Kongo	M	30	2,015	
Valere	Kongo	M	25	2,015	good cart driver, wood cutter
Achille	Mandinka	M	22	1,780	"
Philipe	Timne	M	30	1,700	"
Isidore	Kongo	M	20	800	"
Gabriel	Kongo	M	25	915	"
Baptiste	creole	M	15	1,605	carefully tends the animals

Source: St. John-the-Baptist Parish 1820, *Original Acts*, book A, p. 1123.
[a] hab., habitation, plantation.
[b] A *grif* was a black-Indian, negroid-mongoloid racial mixture.
[c] *Négrite* or *négritte* was an ancient term that meant "little black girl."

with the person's age, sex, health condition, and skills. Old, sick, or contrary slaves were valued lowest, at 50 piastres. The highest valued individuals were two Kongo men, Valere and Michel, aged twenty-five and thirty, who were appraised at 2,015 piastres each. Mothers with children were evaluated in a lump sum. Agathe, a forty-three-year-old creole, and her three small children together were appraised at 2,115 piastres. These were the extremes; the majority of able-bodied slaves at Whitney in 1820 were young creole men eighteen to twenty-five years old and valued at 1,600 to 1,700 piastres each.[3]

With this slave labor force, Jean-Jacques Haydel Jr. and his brother Marcelin operated Whitney until about 1830, when Marcelin took over the complete operations. The records do not indicate when the transaction took place, but the brothers certainly were at Whitney in 1828 (Degalos 1892, 65–68). While under the singular ownership of Marcelin Haydel, a three- by forty-arpent habitation was purchased for seven thousand dollars in 1835. This addition was not contiguous to the main property; it was a small place consisting of a principal dwelling (*maison de maître*), a kitchen, a storehouse, two pigeonniers, and two cabins for Negroes.[4] During Marcelin Haydel's tenure, the interior walls of the mansion at Whitney were richly decorated with hand-painted plant motifs and monograms with the initials *MH*. Surprisingly, these elaborate decorations survive in the mansion.

In the mid-1840s, after the death of Marcelin, his widow Azelie Haydel remained on the plantation and operated it, apparently with some success. The 1844 production reached 326 hogsheads, but by 1851 and 1852 production dropped to 235 and 260 hogsheads. Production reached its zenith for the antebellum period in 1854 with 390 hogsheads. Disaster came in 1855 when an early frost devastated the sugar region and production fell to 60 hogsheads.[5]

Azelie Haydel also acquired two tracts of swampland near the rear of the Whitney property in 1852 and 1853. These tracts were not immediately adjacent to the plantation and were obtained perhaps because of the low price for one tract and the fact that the other was a patent of property granted by the U.S. government. Regardless of these outside land acquisitions, the Whitney holding had stabilized while under Marcelin Haydel's era of ownership. Mrs. M. Haydel retained the property until 1867, when she sold it to the Bradish Johnson Company.[6]

In the years 1868–69, the Alcee Bouchereau survey described the Whitney Plantation as having a wooden sugarhouse with a steam-powered sugar mill and open kettles in the boiling house section. In 1869, the plantation produced 182 hogsheads of sugar and 3,500 barrels of corn (Bouchereau and Bouchereau 1869).

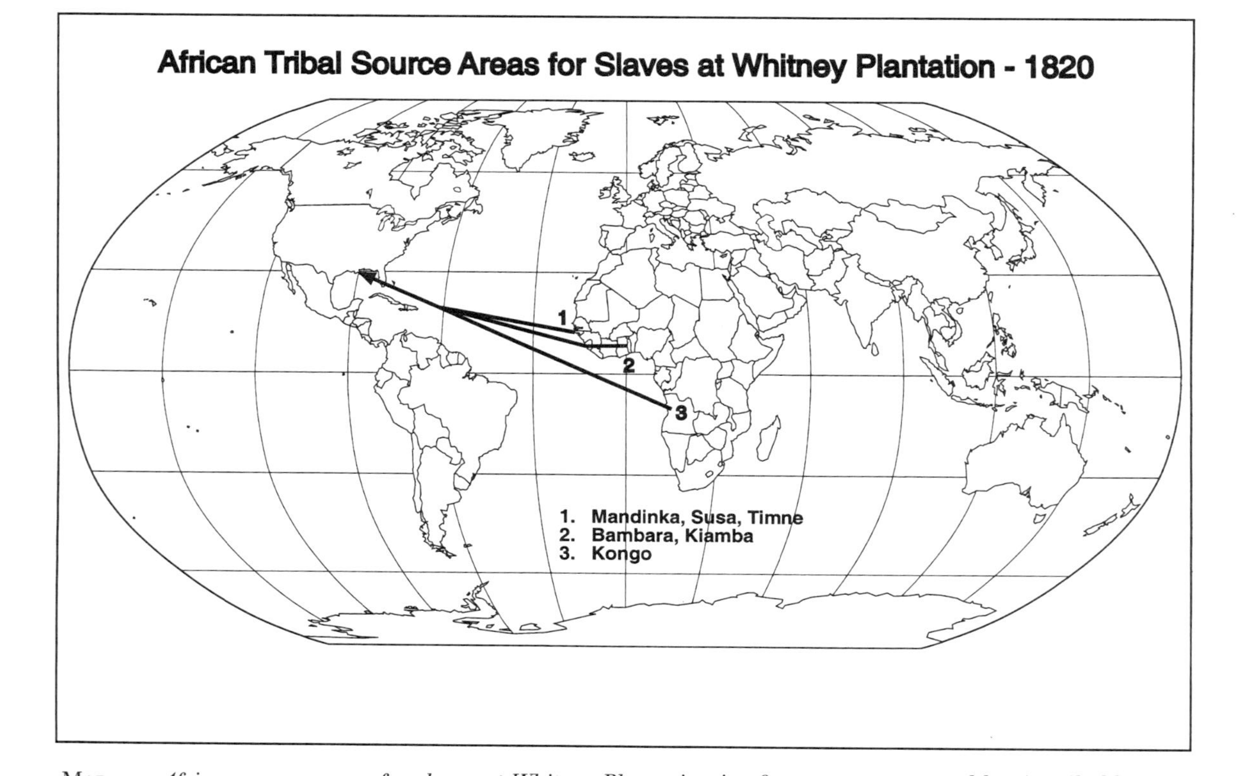

MAP 9.2. *African source areas for slaves at Whitney Plantation in 1820 were represented by six tribal-linguistic groups from West Africa and the Congo.* Data source: Saint John-the-Baptist Parish 1820, *Original Acts,* book A, p. 1123

Wage Laborers in 1868

The paid labor force under Bradish Johnson in 1868 were among the first wage-earning laborers after emancipation (table 9.2). Pay rates ranged from twenty cents to $1.75 per day, but the normal daily average pay was between forty and sixty cents. Men outnumbered women fifty to nineteen and were paid at a higher rate; men received sixty cents compared to the forty cents per day for women. In November 1868, however, the daily rate rose to eighty-five cents for the majority of the workers because it was "grinding season," the hectic harvest time for extremely arduous work (Johnson 1868).

The annual round of agricultural activity varied from season to season, and the number of workers needed varied as well. Under normal conditions during the period January through April, workers were cutting wood and ditching. Activity in May through July centered on normal cultivation practices–hoeing and grubbing. Almost nothing was done in August and September, and none of the agricultural labor staff was paid during this time. From October through December grinding season dominated all activity at the plantation. Residents and neighboring workers were hired at higher daily wages to cut cane by hand, load the stalks onto animal-drawn

TABLE 9.2
Labor Force at Whitney Plantation, Louisiana, 1868 and 1880–1885

	No. of Laborers						
Month	*1868*	*1880*	*1881*	*1882*	*1883*	*1884*	*1885*
Jan.	–	43	29	38	15	12	16
Feb.	–	44	39	33	34	32	17
Mar.	–	38	37	31	25	31	16
Apr.	–	30	30	32	19	22	15
May	55	40	42	25	21	23	22
June	57	28	27	29	21	14	45
July	54	25	30	16	35	9	16
Aug.	54	23	27	21	11	5	16
Sept.	62	26	8	–	–	12	–
Oct.	62	33	33	56	41	23	22
Nov.	67	24	24	–	33	31	23
Dec.	68	118	15	–	26	12	15

Source: Johnson 1868, 1880–85

carts, and transport the cane to the sugarhouse or a neighboring sugar factory for processing. The division of labor of thirty-seven men and four women during grinding season at Whitney in 1868 had two shifts set up to rotate the forty-one workers through the process. Eight workers in each shift carried cane to the mill. Two to three workers tended the actual milling process. Eight in each shift worked the wash process at the kettles where cane juice was ladled from larger kettles to smaller ones. Four more men were to strike the cane syrup into crystallized sugar, a delicate job assigned only to the best sugar makers (Johnson 1868).

While Bradish Johnson operated the concern, production was lower than that before the Civil War. The 1869 crop produced 182 hogsheads, and the following year production dropped to 124. Even by 1875, the sugar crop still had not risen above 185 hogsheads. To make matters worse, disaster struck the Whitney Plantation in the crop year of 1879–80, when fire destroyed the sugarhouse. At the time of the sugarhouse loss, sugarcane and rice were the only commercial crops. In 1880, the rice crop increased to 2,109 barrels while the sugar crop, though undetermined at the time, was probably as small as or smaller than that in previous years. During the next two years, the sugar crop was only 37 and 15 hogsheads (Rykels 1991, pt. 2).

Shortly after the sugarhouse fire, Whitney was sold to Peter Edward St. Martin and Theophile Perret. The Whitney they bought consisted of twenty-three arpents front by seventy deep on the downstream lower line and sixty arpents depth on the upstream upper line. Court records in 1880 stated that the plantation was 15½ leagues above New Orleans. Apparently from the time of Marcelin Haydel's ownership through that of St. Martin and Perret, the contiguous holding had not been altered, nor had the court records been modified to express the distance from New Orleans in miles instead of leagues. The sale included all buildings, improvements, tools, mules, and everything on the plantation except one hundred cords of wood.[7] Immediately after their purchase of Whitney, St. Martin and Perret improved the rice crops while very nearly neglecting the sugar crop. For the years 1881 through 1884, rice production was 3,029 barrels, 3,913 barrels, 8,230 barrels, and 7,000 barrels, respectively (Bouchereau and Bouchereau 1881–84).

Resident wage labor at Whitney continued to change with the seasons in the 1880s. In May 1880, 40 people were on the payroll, 5 of whom were female. Pay scales ranged from 30¢ to $1.00 and averaged between 50¢ and 80¢ per day. However, during grinding season in December, the work force swelled to 118, 15 of whom were female. Daily pay during grinding season ranged from 25¢ to $1.25, but nearly everyone was making $1.00 a day (Johnson 1880–85).

In 1919, St. Martin and Perret purchased from Alovon Granier 260 acres of frontland along the upstream boundary at Aurelia Plantation. According to subsequent sources, this plantation was indeed the former Mialeret Plantation (Rykels 1991, pt. 2). The parcel became the last piece of land acquired in the plantation's land acquisition history. The St. Martin and Perret partnership remained from 1880 until 1938, at which time the ownership title changed to St. Martin and Tassin. This was really a minor change because Mrs. George Tassin was formerly a Perret. In 1946, Alfred M. Barnes of New Orleans bought the Whitney Plantation.[8] Barnes, the plantation's first absentee owner, retained the resident labor force of managers, overseers, and field hands for several years thereafter. When I first investigated the Whitney plantation morphology in 1967 and again in 1969, Mr. Maurice Tassin was the storekeeper, and in earlier years he had managed the entire plantation. The Barnes family used Whitney primarily as a weekend retreat where pleasure horses were kept for riding and a picnic pavilion was constructed for outdoor parties.

In 1990, the Barnes family sold Whitney for $7,998,864 to Taiwanese industrial giant Formosa Chemicals and Fiber Corporation, which intends to build the world's largest rayon plant at Whitney.[9] Considerable litigation and environmental and industrial posturing have taken place since 1990. An environmental impact statement and cultural resources preservation plan were required before permits were granted for Formosa to proceed with the rayon plant. In 1996, construction of the facility was still on hold and preservation of the Whitney complex of historic structures had not proceeded beyond the erection of a chain link fence to protect the deteriorating relic structures from vandalism.

Landscape Morphology in 1969

The linear settlement pattern and French Creole raised mansion at Whitney were diagnostic culture traits indicating a French plantation. The Haydel family progenitors of the plantation had become French by cultural assimilation. Initial German immigrants including the Haydels (originally spelled Haidel) settled the area in the 1720s, but by 1820 most had been assimilated by the dominant French culture, including the Jean-Jacques Haydel family. The linear settlement pattern in the primary quarters contained

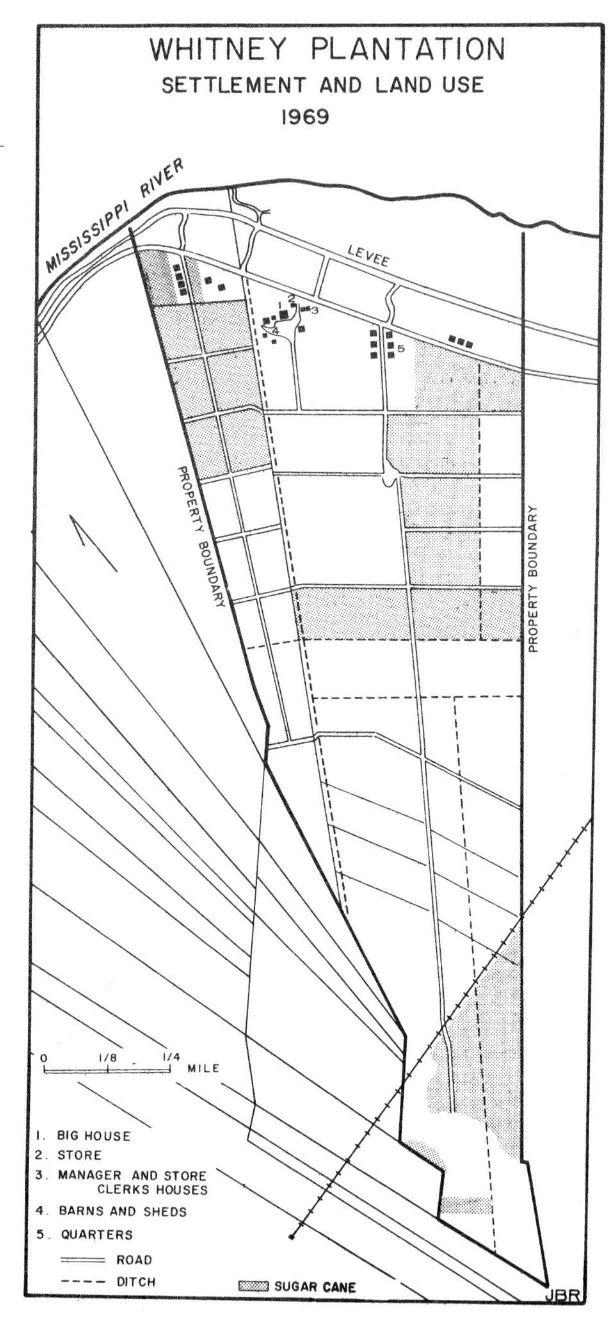

MAP 9.3. *In 1969, the Whitney plantation settlement had three groups of quarter houses. The 1,800-acre plantation had 325 acres in sugarcane, 550 acres in soybeans, and 150 acres in fallow ground. Until 1968, about 450 additional acres had been in rice.* Source: Rehder 1971, 283

six small Creole quarter houses located 0.3 of a mile downstream from the mansion. At the mansion site was a complex of many structures: the store and dwellings for the store clerk, the plantation manager, and the overseer; sheds, agricultural-equipment storage structures, and every barn except one. Four-tenths of a mile upstream from the mansion, a smaller linear quarters on the former Mialeret property contained four small Creole quarter houses. Between this smaller quarters and the mansion were two larger Creole type houses fronting the river road, which also had been part of the Mialeret place. Downriver 0.4 mile from the mansion were three more quarter houses originally built for field hands (Rehder 1971, 282–94).

The Mansion and Other Major Dwellings

The mansion, constructed about 1790 during the Haydel ownership period, is an extremely large and impressive structure. It seems overlarge when compared to the relatively small plantation at Whitney. The structure measures seventy feet long by almost forty-three feet wide. The width mea-

FIG. 9.2. *The core plantation complex–mansion, store, and other houses–illustrates the compact nature of Whitney Plantation. The view is from the levee and shows the large area of frontland sugarcane fields.* Photograph, 1967

FIG. 9.3. *Whitney's French Creole plantation mansion, at the west and south elevations, demonstrates diagnostic traits in the hip roof, central chimneys, and exterior stairs to the loggia, a semiprivate, shuttered rear gallery used as a veranda.* Photograph, 1991

surement includes an eight-foot-wide gallery that extends the entire length of the front of the house. The mansion displays prominent culturally diagnostic French traits such as a hip roof, central chimneys, two full stories with brickwork below and wooden half-timbered colombage above, outside stairs, six front doors, and a full-length gallery incorporated into the roof design. Construction materials and techniques include a wooden roof covered with newer false slate shingles, weatherboarded colombage walls, and a bricked, raised basement ground floor. The woodwork throughout is cypress.

Adjacent to the dwelling in the mansion yard are several other material French culture traits: a brick pigeonnier with a pyramidal roof; three small cypress outbuildings for the former uses of pantry, outhouse, and kitchen; crepe myrtle shrubs; and a fence enclosing the mansion site. The trait of a total fence enclosure appeared at the nearby sites for the manager's, overseer's, and store clerk's dwellings. The overseer's house is a Creole house type, the clerk's house is a bungalow, and the manager's dwelling is a Louisiana bungalow. Yard trees associated with southern French Louisiana flourished, including crepe myrtle, live oak, and magnolia. Many French traits at Whitney had been traditionally and unknowingly preserved by the store manager, Maurice Tassin, grandson of the former owner, Mrs. George Tassin. In Mr. Tassin's vegetable garden were characteristic ethnobotanic

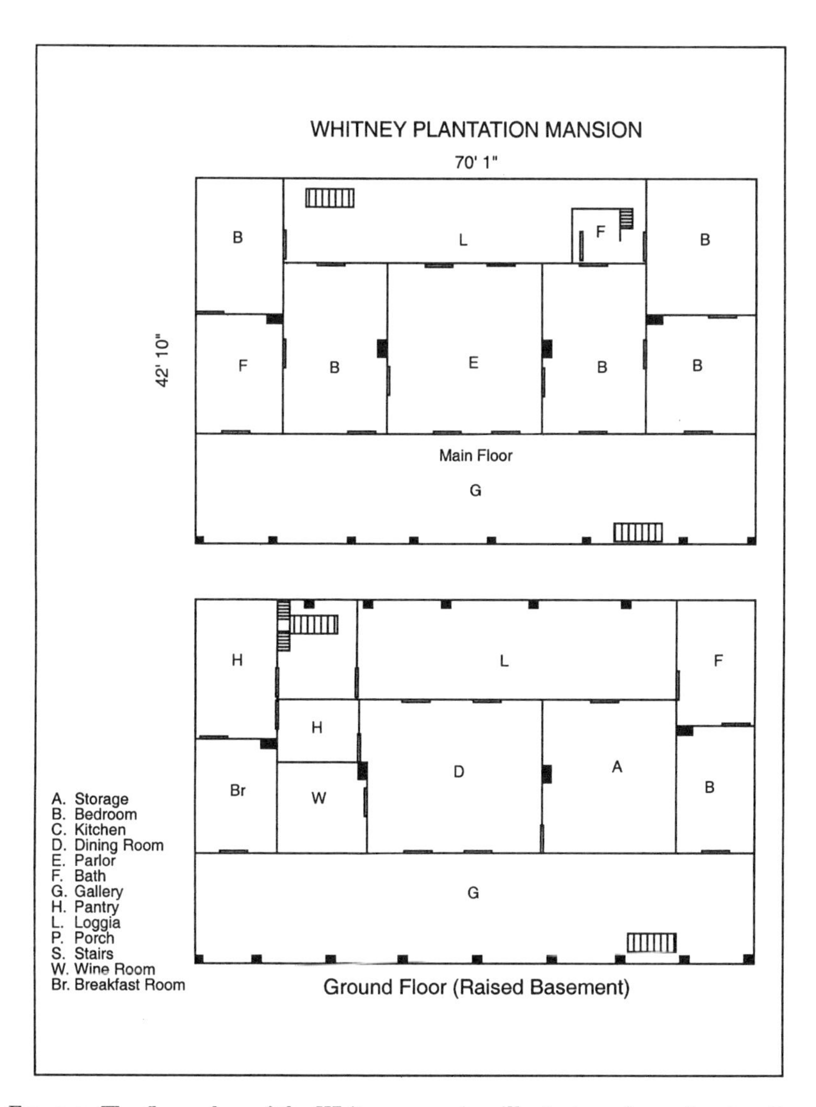

FIG. 9.4. *The floor plan of the Whitney mansion illustrates a large front gallery and rear loggia, multiple doors, central chimneys, six bedrooms, and various functional rooms.* Data source: Cizek, Lehman, and See 1983 in Rykels 1991

Fig. 9.5. *Interior details of Norman roof trusses, rafters, and the central chimney reveal eighteenth-century (circa 1790) French construction techniques that employed mortise and tenon pegged connections for the cypress timbers in the Whitney Plantation mansion attic.* Photograph, 1991

folk foods of okra, tomato, squash, onion, garlic, eggplant, a single row of potatoes, a multitude of dried beans, but no corn because traditionally corn came from the field. Throughout French Louisiana dry beans have been a major staple consumed as red beans or white beans on rice in a starch-on-starch, sometimes colorless, but always flavorful and filling low-cost diet.

Two large Creole dwellings occupy a site on the former Mialeret portion of Whitney, two hundred yards upstream from the mansion. In 1969, one dwelling had two families living in it; the other had one family. The large size of the houses, yards, and former upkeep suggest that they were designated for people with higher status, such as overseers or managers, but by 1969 they housed the plantation's more skilled agricultural-machine operators. Two years earlier, in 1967, people from the quarters were not allowed to live in these large Creole houses. But as a promotion and an enticement to keep skilled workers at Whitney, concessions were made to move these families from the quarters up to the larger Creole dwellings (Rehder 1971, 284–86).

FIG. 9.6. *This exceptionally large pigeonnier, a brick dove cote, was built about 1790 at the same time as the mansion at Whitney Plantation.* Photograph, 1991

The Quarters

Most of the workers lived in two separate quarters settlements located 0.3 and 0.4 mile from the mansion and outbuilding complex. Six identical white-painted, wooden, small Creole type quarter houses with galvanized metal roofs huddled in a cluster. Garden spaces had been provided but were not used. Chinaberry trees, rough wire, and galvanized sheet fences enclosed yard sites. Rarely seen elsewhere in the region were crudely assembled galvanized sheet metal garages at each quarter house. The low-profile garages had hardly enough space for an automobile, and only in rare cases were they used for auto storage; they mostly contained scrap lumber, car parts, and junk. Not everyone living in the quarters worked on the plantation. Of the thirteen usable quarter houses at Whitney, five were occupied by salaried employees, four were vacant, and four were occupied by retirees and persons on welfare who no longer worked on the plantation but who received free housing (Rehder 1971, 287). By 1989, only one house remained occupied–the former store clerk's house then occupied by a caretaker family. Management had fallen to a Mr. Hymel, who lived elsewhere

Fig. 9.7. *The plantation store clerk's bungalow* (left) *was built between 1940 and 1953, and the plantation manager's house* (right) *dates to the 1920s. Both houses were the last to be occupied at Whitney Plantation and then only by caretakers.* Photograph, 1991

and only rented cane acreage on the plantation. No one has lived at Whitney since 1990.

Outbuildings

After the Whitney sugar factory was destroyed in 1880, no attempt was made to rebuild one. Sugarcane crops were milled at the Saint James Cooperative in Saint James Parish. A cane loading derrick, located at the main quarters, facilitated the loading of tractor-trailer trucks that hauled Whitney canes thirteen miles north to the mill. As an outside grower's enterprise in the Saint James Cooperative, Whitney remained an independent agricultural concern that supplied part of the cane milled at the factory. All outbuildings, except a single barn near the geographic center of the landholding, were clustered around the mansion.

A very old hipped roof barn at the mansion site on the Whitney Planta-

FIG. 9.8. *Whitney Plantation's primary quarters in 1967 had six quarter houses built during the Bradish Johnson era of ownership in the 1870s.* Photograph, 1967

tion is thought to be perhaps the oldest surviving barn in Louisiana (see fig. 3.13). The small building is located behind and downstream from the mansion and is thought to date from about 1790, making it the same age as the mansion. The structure measures about fifty-three feet long by thirty-four feet wide. The hip roof was built as a Norman truss roof, and the external roofing material had been cypress shakes (shingles) but is now covered with a relatively newer tin roof installed in the 1960s. The internal construction is a cypress frame all mortised and pegged; externally, the barn is weatherboarded with twelve-inch-wide horizontal cypress planks. In plan, a large internal space is surrounded by four smaller storage rooms. The entire wooden structure is raised four feet off the ground and supported by a thick brick foundation. The building in 1969 was used as a mechanical shop and to store farm supplies and did not appear to have ever been used as a shelter for farm animals. The building is in disrepair but survives.

Other outbuildings were built within the past fifty years and were used for agricultural equipment. Three sheds were used for this purpose; one was located out in the field 0.5 mile from the other two, which are in the ag-

FIG. 9.9. *The Whitney Plantation store dates to the 1870s.* Photograph, 1969

glomerate outbuilding center. Two tool sheds and two open drive-through tractor sheds complete the outbuilding agglomeration. All outbuildings at Whitney were oriented toward the agricultural service of the plantation.

The plantation store was especially needed because there were no other retail stores nearby. The building remained an elongated structure with a front-facing gable, similar to most other plantation stores along the Mississippi, and had batten doors and windows and a familiar white-paint covering trimmed in green. In the 1960s, a credit system at the store accounted for the higher prices on most merchandise. Dry goods, clothing, boots, utensils, and canned foods lined the shelves. A large screened case contained penny candy, cookies, and gum. The greatest volume of sales was from the screened case, as children were constantly coming into the store (Rehder 1971, 288–90).

Other Landscape Features

Whitney had the benefit of being physiographically situated on a point bar, or convex bend, on the right bank of the Mississippi River. The landholding had an unusually wide frontage of twenty-three arpents that sharply narrowed into a pie slice shape toward the backswamp. The location was important for several reasons. With a twenty-three-arpent frontage on a favored high-elevation levee crest, the site was extremely valuable for settlement. The soils on levee crests were better textured and more easily drained so that the twenty-three-arpent front was agriculturally favorable for sugarcane. Because of alluvial deposition the point bar on the meandering river continuously increased its land area in the direction toward which it pointed. A negative factor in the site location was the former requirement for locally owner-built and owner-maintained levees. Before the federal construction of the artificial levees by the U.S. Army Corps of Engineers throughout the Mississippi River Valley, the costly protective levee along a wide frontage was a burden to the landowners of Whitney. Despite its favorably wide frontlands, the landholding lacked the same depth of useful drained land toward the backswamp as compared to the crevasse materials found at Armant. Since a point bar was unlikely to experience crevasses, the extent of arable land here was only 2.0 to 2.5 miles from levee to backswamp.

Commercial agriculture at Whitney Plantation included both sugarcane and rice. Sugarcane had been produced on the property for more than two centuries. Rice cultivation did not become really important until about 1880, although rice apparently had been grown before 1820 because records indicate a rice mill on the property at that time.[10] In 1967, the rice crop covered approximately 450 acres, but it was the last year for rice production at Whitney. Even then Whitney was the only Mississippi River plantation that had both commercial rice and sugarcane on the same property. Cultivation of both crops led to differences in land use as well as materially different field and drainage patterns. The two crops occupied different parts of the plantation as a response to soil moisture differences. Cane required the better drained frontland soils classified as Commerce and Mhoon loam, silt soils. The rice crop occupied the backlands, where thick, clay soils of the Sharkey series prevent subsoil percolation and thereby allow controlled flooding in the fields.

Large field cuts and ditches were oriented linearly from the levee crest toward the backswamp, but major pattern differences appeared among individual plots. Sugarcane plots were long and narrow, whereas rice plots were small block squares numbering many more per field cut. For example, one of the sugarcane field cuts contained 14.7 acres, with eight plots av-

FIG. 9.10. *A cemetery surrounded by a petrochemical plant demonstrates the competition for land in the Chemical Corridor south of Whitney Plantation in Saint Charles Parish.* Photograph, 1993

eraging 1.83 acres each. The average length and width of each cane plot was 957 feet by 90 feet. By comparison, an adjacent rice field cut containing 14.3 acres had fifty-six plots averaging 0.25 acres each. The rice plots averaged 119 by 94 feet. Drainage ditches and the network of rice irrigation canals, though differing in function, for the most part were identical. Water control gates at various points allowed drainage of cane lands and flooding of rice fields. Two pumps, one located on the river side of the artificial levee and one further inland, maintained water control for both flooding and draining the rice lands.

In 1969, agricultural land use at Whitney included 325 acres in sugarcane, 550 acres in soybeans, and 150 acres in fallow land. Rice was no longer cultivated even though the elaborate irrigation and ditching systems, pumps, and rice harvesting combines remained. Landscape features expected at Whitney but missing were field fences, a plantation bell, a church, a sugar factory, and livestock, except for two pleasure horses stabled in front of the mansion. Beef cattle were raised on the plantation until the 1950s. The last of the working mules left in 1948 (Rehder 1971, 290–92).

FIG. 9.11. *Formosa's sign at the Whitney Plantation points to a new era of economic development in a region known as the Chemical Corridor, the American Ruhr, and Cancer Alley.* Photograph, 1995

Mechanization affected the number of resident laborers at Whitney. In 1820 there were forty-nine working slaves at Whitney. By 1868 there were fifty-six hired laborers working for sixty cents to a dollar a day, depending on whether it was grinding season, during which the latter rate applied. In 1880 the number of laborers had dropped to about forty, and wages were nearly the same as in 1868.[11] In 1969 six full-time resident employees received minimum hourly wage rates of $1.75, plus housing and utilities. Six workers with tractors and mechanized sugarcane harvesters could cultivate the same acreage that had required forty or more laborers before mechanization arrived.

Whitney Plantation, 1990–1998

In recent years, controversy has surrounded the plantation. In 1988, Governor Buddy Roemer went to Taiwan on a mission to attract industry to Louisiana. He found that Formosa Chemicals and Fiber Corporation was interested in building a rayon complex in Louisiana. The initial site presented to Formosa was on the right bank of the Mississippi River downstream from Whitney at Willowbend and was owned by Shell Oil Company. What seem to be obscene lease arrangements at one dollar per year were offered but were politely turned down. Formosa wanted to own a site outright. During a helicopter overflight, Formosa officials spotted the eighteen-hundred-acre Whitney site; when they found the price to their liking, they decided to purchase the Whitney Plantation (Rykels 1991, pt. 1). Formosa finalized the purchase of Whitney Plantation on April 30, 1990, from the twelve-member Alfred M. Barnes family for $7,998,864.[12]

Once the word had spread that a $700 million rayon plant was about to come to Saint John-the-Baptist Parish, people began choosing sides. Developers and politicians could easily envision economic development on a grand scale, the kind of growth that mere mortals only dream about. There would be hundreds, maybe thousands of jobs to feed the local economy and build up the parish tax base. Environmental and historic preservation groups joined forces to prevent foreseeable wholesale environmental and cultural landscape destruction. The preservationists had two weapons. First, an environmental impact statement (EIS) was required and a cultural resources inventory in the form of a survey of historical buildings and a restoration/preservation plan were to be put in place before any factory construction could begin. Second, little did anyone know that a select few local citizens in nearby Wallace would refuse to sell adjacent land that Formosa thought it could easily acquire.

In the past forty years, 138 petrochemical plants have been built on the banks of the Mississippi River between Baton Rouge and New Orleans, an area known officially as the Chemical Corridor but locally called Cancer Alley. What difference would one more chemical plant make? Cancer Alley was no misnomer. Too many young people were sick, and many people in the area reportedly were dying from cancer. Local citizens were worried. Could Formosa's rayon plant be an environmental problem? Many believed so because the plant would be dumping fifty-three million gallons of waste water into the Mississippi River daily. Furthermore, the plant's daily wood consumption of eight hundred tons would be emitting steamy odors and fumes into the atmosphere. It had been rumored that Formosa had further plans to build a PVC (polyvinyl chloride) factory adjacent to the rayon plant at Whitney, explaining the requirement for additional land.[13] For the moment Formosa has scaled back its effort to pursue the project, and the future of Whitney plantation remains uncertain.

CHAPTER 10

Madewood Plantation, 1823–1998

Throughout the South, plantation names reflected cultural identities, economic conditions, and sometimes indications of hope for the future. Anglo-Americans liked picturesque plantation names like Oaklawn, Ashland, and Madewood. But some plantations named in the post–Civil War period became New Hope, Home Place, Good Luck, Hard Times, and even Hard Scrabble (Owens 1997, 191–209). Madewood Plantation was named by its initial owner Thomas Pugh in the first quarter of the nineteenth century. Madewood, a nodal-block plantation type, is a small plantation with a majestic mansion located on the left bank of Bayou Lafourche two miles below Napoleonville in Assumption Parish (see map 5.1). The plantation site is favorably situated on one of the widest sections of alluvial natural levees on Bayou Lafourche. In approximately A.D. 1100, the Mississippi River flowed along the course that the Lafourche now takes and built up wide Mississippi-sized levees (Fisk 1952). The contemporary Bayou Lafourche is but a small distributary that branches from the Mississippi River, but its oversized natural levees are two to three miles wide here in the upper reaches between Donaldsonville and Thibodaux. In 1969, Madewood had 676 acres in cropland and eight resident laborers. It was the smallest of the case-study plantations. Presently, Madewood operates on about seven hundred acres with three or four machine operators, an overseer, and two owner/managers, the father-and-son team of John and Steve Thibaut.

When I returned to the plantation landscape in 1989, I was surprised to find Madewood surviving rather well. Logic would have us believe that, over time, small plantations would not survive the vagaries of environmental and economic upheaval. It was once thought that survival in the sugar industry depended on the size of the operation. One expected that the larger the plantation, the greater chance for economic survival. Such is not the case in the six plantations analyzed in chapters five through ten. Armant, Ashland, and Oaklawn were three large, once vibrant plantations. When they became part of the CALA model, they closed their sugar factories, eliminated their quarters, and so had little remaining landscape evidence to show that they were ever plantations. The huge South Coast Sugars (in the CALA model) was reported to have suffered financial difficulty in 1993, and ownership reverted to the lenders. Size is not everything. Of the six plantations, Madewood is the only one that maintains a balance of landscape survival and economic stability, resting largely on a continuum of family-owner management.

Historical Succession

Madewood's first owner and primary landscape builder was Thomas Pugh, a young man of Welsh ancestry from Bertie County in the English Tidewater culture region of eastern North Carolina. Pugh and his two brothers arrived in 1818 and for two years tried to settle in the Bayou Teche area near Franklin before they moved to the Bayou Lafourche area (Lathrop 1945, 11–15). Levee crest frontlands on Bayou Lafourche had been settled earlier by small Acadian and Spanish farmers, so Pugh's land acquisitions were typical of other Anglo planters who entered the area. Nineteenth-century planters bought small access tracts and any other available frontlands from the small farmers, but the Anglos' primary acquisitions were large-acreage back concession lands located behind the forty-arpent line purchased from the U.S. government.

The earliest recorded land purchase by Thomas Pugh for Madewood

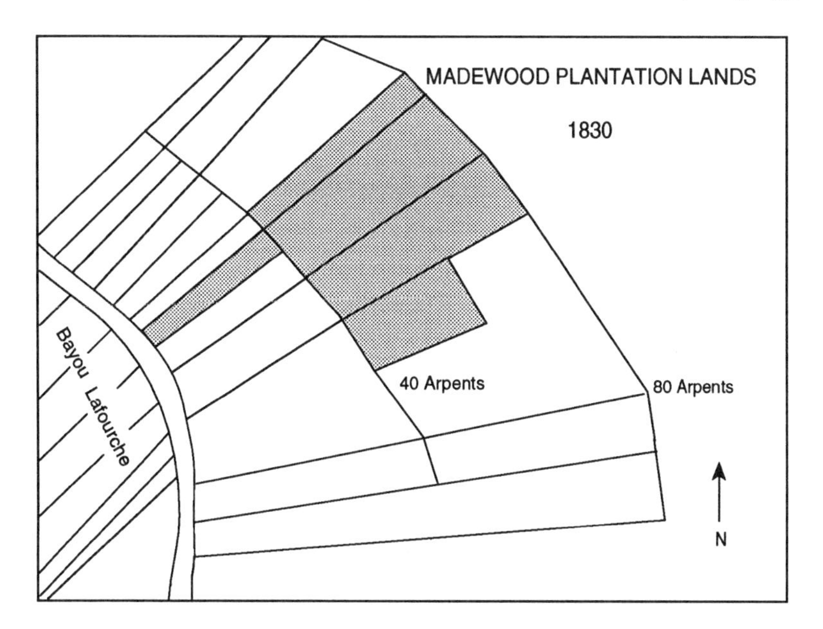

MAP 10.1. *In 1830, Thomas Pugh was acquiring large tracts of backlands beyond the forty-arpent line and small access tracts of frontlands from* petit habitants *(small farmers) to form Madewood Plantation along Bayou Lafourche.* Source: Louisiana State Land Office, surveyor's township plats for 1830

dates to November 1823, when he purchased from Pierre Aubert a 5- by 40-arpent tract for nine thousand dollars. This first tract was to become the core of the plantation. Other frontland acquisitions included 50 acres acquired in 1830, a 2- by 40-arpent tract, and a 1- by 40-arpent tract. In 1836, a tract of 5⅔ arpents by 40 arpents was purchased, and another 79-acre tract was acquired in 1838. Pugh received back concessions from the U.S. Land Office in March 1835, March 1836, and June 1836, for a total of 993 acres of partially arable lands and backswamps.[1] Pugh gained more land in 1844, 1845, and 1848 to total 1,076 acres of swampland beyond the back concessions, finalizing the Madewood tracts (Lathrop 1945, 457–58). Much of this swampland was unfit for agriculture, but the cypress timber became valuable construction and fuel materials. By 1848 Pugh had completed acquisition of the Madewood contiguous landholding. The total plantation measured over 1,000 acres of arable land and contained 1,700 acres of backswamp land. It remains the same size and shape today.

The Madewood Plantation initiated and developed by Thomas Pugh was the best groomed of the Pugh properties and came nearest to an agriculturalist's dream. Aside from its mansion, the plantation differed little from many of the sugar plantations in Louisiana. Solon Robinson in 1849 described Madewood as

> one of the best [plantations] in the state. Not the largest, though quite so to satisfy any man of moderate desires as the value of the annual crop is from $30,000 to $40,000. Mr. P. owns here about 3000 arpents–1000 cleared, 550 in cane, 250 in corn, and 200 in pasture, yards, gardens, etc. Of the first named crop, 440 arpents made 700 hogsheads of sugar, and about 60 gallons of molasses to the hogshead. The remainder of the cane was reserved for seed planting. One acre of cane is required to plant five acres. Mr. P. has 100 working hands, producing about seven hogsheads of sugar each (Kellar 1936, 197–201).

Like other sugar planters, Pugh experienced years of crop fluctuations primarily caused by bad weather and floods. Production at the budding Madewood was 108 hogsheads in 1828, but in the following year it was only 53 hogsheads (Degalos 1892, 67). In later years production revealed a growing expansion of the plantation, with an 1844 production of 756 hogsheads (Champomier 1844, 6). The maximum crop production of 1,050 hogsheads came in 1853, the same year that Thomas Pugh died. Madewood Plantation was then valued at $273,995 with its 251 slaves.[2]

Pugh's widow successfully operated the enterprise until 1883. The 1860 production was 765 hogsheads of sugar and 9,000 barrels of corn. Even after the Civil War in 1869, production was 340 hogsheads of sugar and 2,500 bushels of corn.[3] Mrs. Pugh sold an undivided half of the Madewood holding in 1883 for fifty thousand dollars to her son Robert:

> The undivided half of a sugar plantation named the Madewood Plantation . . . containing sixteen arpents front on said bayou [Lafourche] by a depth of forty arpents and having a surficial area of one thousand arpents. . . . Also the undivided half of the double concession thereunto belonging and attached, containing thirteen hundred arpents, and the one undivided half of some lands in the rear of the said double concession and forming part of said Madewood Plantation containing nine hundred and twenty acres, together with the undivided half of the dwelling house, of the sugarhouse, of all plows, and other farming implements, and utensils, of all the building materials and all the machinists', engineers', carpenter's, and cooper's tools, on said plantation.[4]

The other half of the property was in litigation for the next three years, but eventually Robert Pugh purchased it in a public sale. Robert Pugh introduced technological improvements such as vacuum pans and centrifugals in the sugar factory and a steam train in the fields before donating Madewood, including all debts, to his nephew Llewellyn Pugh.[5]

The Pugh family finally relinquished Madewood in 1896 for thirty thousand dollars when Leon Godchaux bought it.[6] Leon Godchaux was the former peddler turned sugar magnate and founder of Godchaux Sugars, and he simply added Madewood to his collection of innumerable properties in a sugar empire throughout southern Louisiana. Madewood's sugar factory had been destroyed by fire, necessitating the sale and lower price. Excluded from the sale was the family graveyard at the mansion site, a significant culture trait identified with Anglo plantations here in particular and with the South in general (Rehder 1992, 116).

Godchaux kept Madewood for fourteen years and on February 18, 1910, sold it to Henry Delaune. Afterward, Madewood changed hands with great rapidity. On the same day of the sale to Henry Delaune, the latter sold the property for the same price of $97,500 to Alcee F. Delaune, who was probably a relative. This Delaune retained the property for six years and then sold it on March 20, 1916, for $47,000 to Emile Sundberry, a speculator who realized a quick profit when on the next day he sold Madewood for $50,000 to Robert L. Baker.[7]

Baker and his heirs retained the property until 1946. During his tenure, Baker operated the plantation by cultivating sugarcane on nearly nine hundred acres of cropland. Baker's family continued to operate the plantation until April 3, 1946, when they sold it to D. Bronier Thibaut, the father of the president of Madewood Incorporated, John Thibaut.[8] Since that time, the Thibauts have maintained the plantation as an agricultural unit feeding the larger Glenwood Cooperative sugar factory.

One last transaction was the sale of the Madewood mansion and its six-acre site. This finely preserved mansion now belongs to the family of Mr. and Mrs. Harold K. Marshall of New Orleans. Mrs. Naomi Marshall personally purchased the mansion and site in November 1964.[9] The Marshall's careful and loving maintenance of the mansion and the Thibaut's trusted operation of the remainder of the plantation have preserved Madewood's form and function for more than three decades on the Louisiana plantation landscape.

Landscape Morphology in 1969

Madewood is emblematic of an Anglo-American plantation. It had Anglo owners from the time of its initial construction through its development until 1896, as evidenced by ownership successions. Material culture traits clearly reveal an Anglo identity: (1) a block-patterned settlement; (2) a Tidewater mansion with a front-facing gable, a central hallway, end chimneys, Georgian symmetry, and Greek revival ornamentation; (3) a family graveyard; and (4) pine and pecan yard trees. The latter two traits may appear irrelevant, but they support Anglo identification (Rehder 1971, 321–22).

The block-shaped settlement located one-half mile behind the mansion toward the backswamp, contained a quarters with shotgun and bungalow house types. The former overseer's bungalow and all of the outbuildings associated with the agricultural operations were also in the block settlement. Few quarter houses were needed for the eight full-time laborers who then lived at Madewood. Of the eleven quarter-house dwellings in the settlement, four were shotguns, five were bungalows, and two were small Creole quarter houses. All had wooden construction, but several were sided with tar paper in the yellow brick pattern; two had asbestos siding, and one house was weatherboarded and painted white. Many dwellings had smooth wire fences enclosing vegetable gardens. Two of the laborer's families kept hogs and chickens in their back yards. Each quarter house had a functional outhouse. Yard trees of chinaberry, poplar, and pecan shaded the quarters. An overseer's bungalow in the middle of the block measured twenty-five by forty-two feet and was weatherboarded and painted dark green. In the center of the block and outbuilding complex were a new all-metal building for cane harvesters, a wooden tractor shed, a wooden warehouse twenty-five by thirty-five feet formerly used for corn storage but later used to store equipment, a mechanic's and former blacksmith's shop with board-and-batten construction, and the plantation bell. Although the sugar factory was destroyed in 1896, Madewood Plantation continued producing successful crops that were sent to various nearby sugar factories for processing. Currently, all Madewood sugarcane is processed at the Glenwood Cooperative sugar factory, located about four miles upstream on Bayou Lafourche (Rehder 1971, 322–24). Management is simple on this plantation. The overseer visits the Madewood operation daily, and the manager periodically checks on Madewood's agricultural operations. Neither overseer nor manager resides on site at Madewood.

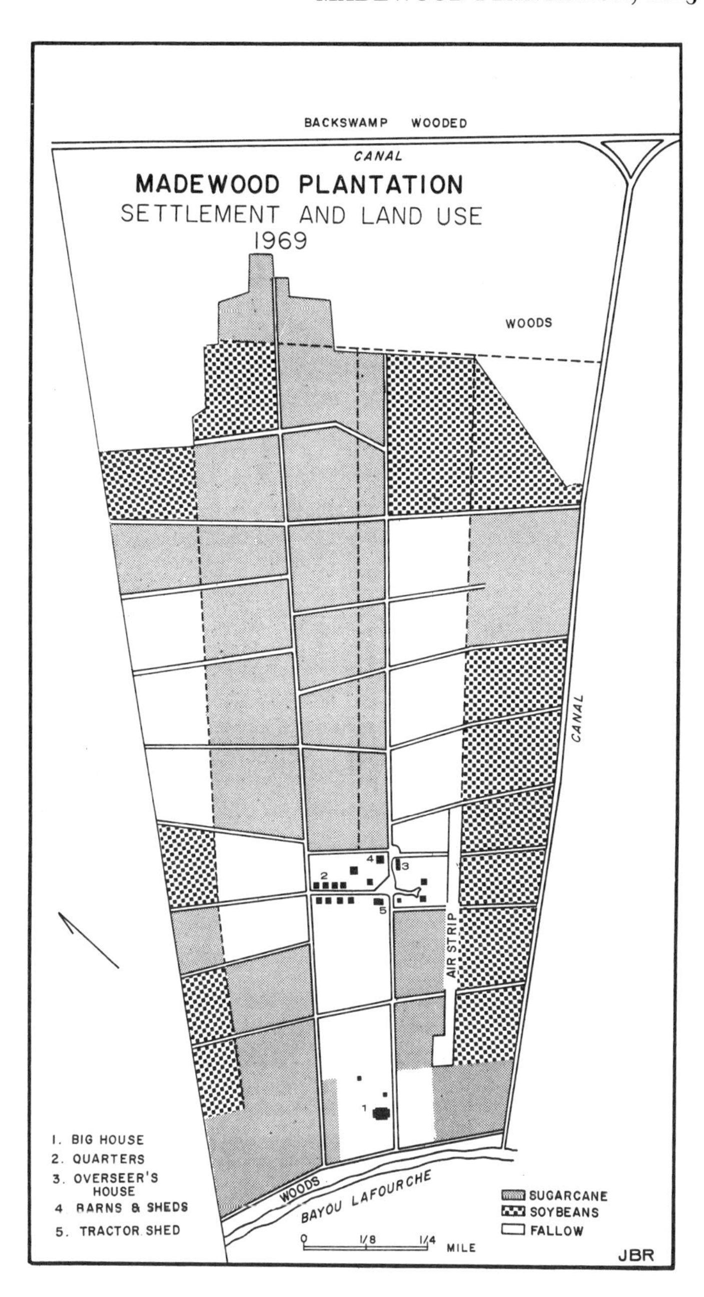

MAP 10.2. *The Madewood plantation settlement in 1969 had the Tidewater mansion nearest the bayou and the quarters and outbuilding complex a half mile away in back. Madewood's cropland included 347 acres in sugarcane, 177 acres in soybeans, and 153 acres in fallow ground.* Source: Rehder 1971, 327

FIG. 10.1. *The overseer's house was a green-painted bungalow from the early twentieth century. Madewood Plantation's work bell in the nineteenth century called field hands to their tasks at dawn and signaled breaks at noon and dusk. Later, the bell signaled lunch breaks and emergencies.* Photograph, 1969

Like other Lafourche plantations, Madewood characteristically had two separate yet related building agglomerations in the form of the block settlement and the mansion site. Located adjacent to the levee road, the mansion site retained an aloofness demonstrated by its half-mile separation from the block settlement. A large servant's house, a brick carriage house, and the family graveyard were clustered at the mansion site. The surrounding yard and gardens featured moss-draped oaks, magnolias, old pines, and pecan trees.

The impressive mansion was built between 1840 and 1848 (some date it at 1846), and it remains one of the best examples of Anglo Tidewater house types in Louisiana. The mansion was designed by Henry Howard, an Irish architect from Cork, then living in New Orleans. The building measures sixty by sixty-eight feet and has front-facing gables, four inside-end chimneys, and a central hallway eighteen feet wide that extends the full length of the main house. On the main floor, two rooms measure twenty-one by thirty-four feet and two rooms measure twenty-one by twenty-five feet. On

Fig. 10.2. *The tractor shed and mechanic's shop are two of the three functioning outbuildings remaining at Madewood Plantation.* Photograph, 1991

the second floor are four large bedrooms that are each twenty-one feet square. The facade, characterized by Greek Revival ornamentation, has six Ionic pillars supporting an overhanging portico and pediment. Georgian symmetry is reflected by two matched wings flanking the main structure. Each wing has end chimneys and rooms for special functions. One wing has a music and study room, a ballroom that measures twenty-four by forty-eight feet, a pantry, and the kitchen. The other wing is used as an apartment by the owners.

Madewood's construction reflects Anglo techniques and materials, which differed considerably from French colombage half-timbered construction. Madewood was built with more than 600,000 bricks to form walls that are eighteen to twenty-four inches thick in the main structure and wings (Kellar 1936, 200). Even the columns internally are brick. Most of the house is plastered so that bricks are not visible. For interior walls, a lath-and-plaster construction was used. Flooring, porches, shutters, and the roof support are all cypress. The roof, once covered with slate, has been replaced with a newer lightweight roof material.

Fields on this Anglo plantation were no different from those on other sugarcane plantations because drainage ran from levee crest to backswamp, and fields, rows, and ditches all formed linear patterns in the same direction. Field plots in 1969 ranged from 0.5 to 4.6 acres, with an average of 2.2 acres. Field cuts varied from 21.5 acres to 6 acres. A 3,500-foot airstrip was used by crop-dusting planes and particularly by the corporation's flying president, John Thibaut (Rehder 1971, 324–31). Drainage on the plantation was accomplished by two lateral ditches, several cross ditches, and a large canal on the south flank of the plantation. The canal had been on the property since 1829, when it was called the Carlos Canal.[10] Part of Madewood's excess runoff drains into Carlos Canal, which drains into the Baker Canal, a larger backswamp canal named for Madewood's former owner Robert L. Baker.

In 1969, land use on the plantation included 347.2 acres of sugarcane, 177 acres of soybeans, and 153.8 acres of fallow ground. The latter were put

FIG. 10.3. *Madewood Plantation mansion, a Tidewater Anglo-American house type, was built for Thomas Pugh between 1840 and 1848.* Photograph, 1967; source: Rehder 1978, 147. Used with permission from the Louisiana State University School of Geoscience

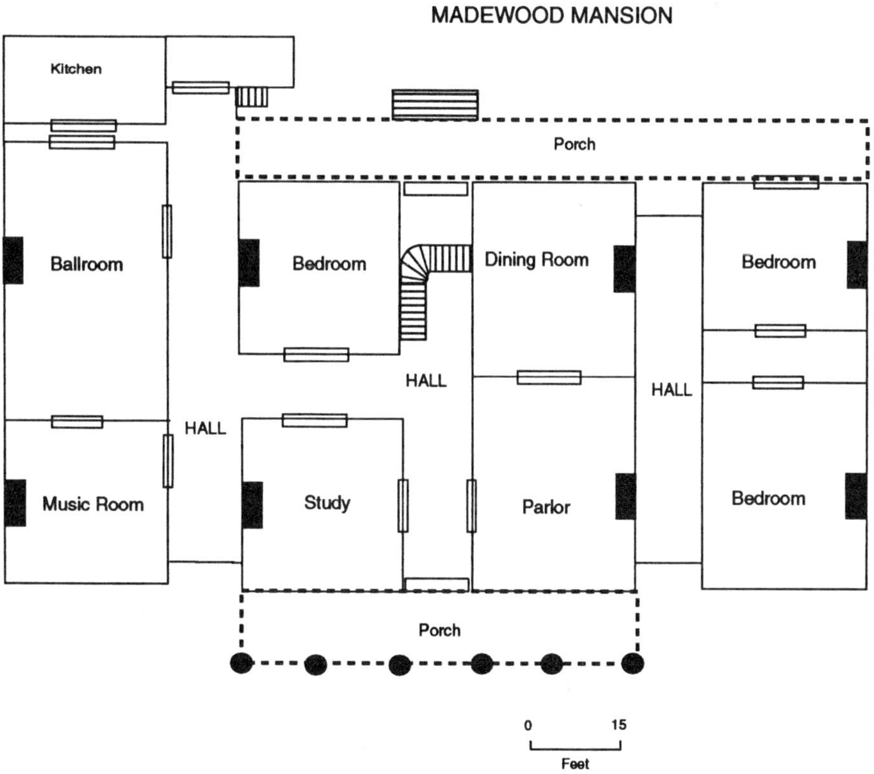

FIG. 10.4. *The floor plan of Madewood's mansion reveals the typical symmetry of paired wings, paired rooms separated by a central hall, and paired inside-end chimneys.*

into sugarcane cultivation for the next year. The plantation also had crawfish ponds in its extensive backswamp. Crawfish are a delicacy in Louisiana and seasonally provide a lucrative business. While most crawfish are still harvested in their natural swamp habitat, enterprising fellows about forty years ago began diking portions of backswamp lands to impound them to control, cultivate, and harvest crawfish (Rehder 1971, 330–31).

Madewood, 1989–1998

One of the most pleasant surprises on my return to Louisiana in 1989 was the discovery that Madewood was not only still on the landscape, but also carefully preserved in both form and function. Certainly, the quarters

FIG. 10.5. *The quarters at Madewood is smaller but still survives. The center bungalow is one of two newer cinder-block quarter houses built in the 1970s in hopes of retaining skilled laborers on the plantation.* Photograph, 1995

had lost houses, but two new ones had been added. John Thibaut built new cinder-block quarter houses in hopes that improved housing would convince his better machine operators to stay on the plantation. It worked, at least for a little while. In the outbuilding complex, a new tractor building and mechanic's shop replaced the older wooden tractor shed and blacksmith's shop that had been torn down. The overseer's house remained but was empty. The plantation bell was now gone, but it was preserved at Thibaut's house in Napoleonville. Up at the mansion site, no noticeable changes had taken place until 1992, when Hurricane Andrew blew the roof off the servant's quarters behind the mansion. Otherwise, Madewood has stood the test of time.

What makes a plantation persist in the face of the destructive landscape changes taking place elsewhere? I believe that Madewood endures because its owners want it to endure. They recognized early on that the preservation of this small plantation made sense. It especially made sense to the Marshall family, who have maintained the mansion and its six-acre gardens for tourists since 1964. Naomi Marshall originally bought the mansion in 1964, but her son Keith Marshall carefully manages the mansion today. It has also made sense to continue to produce sugarcane at Madewood even though it remains but a small part of the multiplantation operation of the Thibaut family. I emphasize *families* because so much goes back to the family farm, the traditional plantation, your place, my place, granddaddy's place, which speak of topophilia, the love of land, that helps this plantation endure. Perhaps it is more than a love of land; it may be a love of life summed up in a business card that declares *John Thibaut–Cajun Entrepreneur.* It is incredibly ironic that the big entrepreneurs in Louisiana's delta sugar in the CALA model failed and that a Cajun named Thibaut had the presence of mind to preserve a way of life on a landscape that had the mark of Thomas Pugh, a nineteenth-century Anglo planter from the English Tidewater region of North Carolina.

Each time I visit the plantation landscape I go with the hope that I will see no further deterioration and perhaps be surprised to find higher levels of preservation. On May 25, 1994, I was pleasantly surprised to find that the one-time vacant and thoroughly dilapidated Laura Plantation home built in 1805 at Vacherie was undergoing extensive restoration by the Norman Marmillion family with the help of architect Eugene Cizek. To be sure, this plantation will be saved.

CHAPTER 11

Agents of Change

The Louisiana sugar plantation is woven into the fabric of regional and national policies that affect the operational nature of the sugar industry in the United States. Since the late nineteenth century, policymakers have controlled the destiny of the sugarcane industry; major changes since 1974 have influenced the landscape and threatened its future. The regional and national events that initiated trouble for the U.S. sugar industry, invited a California agribusiness model, and led to a transformation of Louisiana's sugar corporations began in 1974, which marked the beginning of one of the most dynamic, catalytic, catastrophic periods in the past three decades of the Louisiana sugar industry. Sugar prices fluctuated from thirteen to forty cents and back to fifteen cents per pound. Energy costs escalated. Environmental regulations began to affect the sugar industry. Petrochemical industries rapidly surfaced along the Mississippi River between Baton Rouge and New Orleans. While labor costs rose, skilled, reliable labor became more difficult to find or retain. Even Southdown, one of Louisiana's largest sugar corporations, was looking for ways out of the sugar business.

Changes since the 1970s

Sugar Prices

In 1974, world shortages of sugar and speculative buying caused sharp rises in retail prices (Hawaiian Sugar Planters' Association 1990, 17). Prices for raw sugar in Louisiana fluctuated from thirteen cents in 1973 to forty cents in 1974 and then instantly fell to fifteen cents per pound in 1975 (Campbell 1977, 8). The rising price of sugar early in 1974 attracted considerable attention from potential investors. When Southdown put the then-largest available sugar corporation on the market at a reasonable price of $11 million, outside investors were interested. Little did anyone realize that the high price of sugar would go down as fast as it had gone up. Anderson and others from the U.S. Federal Trade Commission explained that "there appears to be little evidence that the explosion of raw sugar prices in late 1974 was due to anything other than natural market factors" (Anderson, Lynch, and Ogur 1975, vii). But prices really were out of control, and 71.6 percent of the raw sugar received by U.S. sugar refineries in 1974 was imported.[1]

Government Controls

Between 1934 and 1974, the U.S. Sugar Act controlled sugar production, wages, prices, and other aspects of the industry (Hodson 1980). After the defeat of the U.S. Sugar Act in 1974, the United States faced seven chaotic years without a national sugar policy. Ironically, a British regulatory device called the Commonwealth Sugar Agreement was also rescinded in 1974 (Abbot 1989, 343–60). World sugar production soared; subsidized sugar imports entered the United States and threatened the survival of the U.S. sugar industry (Hawaiian Sugar Planters' Association 1990, 17). Moreover, the soft drink industry turned from cane sugar to high fructose corn syrup to sweeten beverages at a time when the market for soft drinks was expanding. Cane sugar was in trouble.

Raw sugar prices were once again federally controlled under the Food and Security Act of 1985. The act established safeguards against low sugar prices, controls on imports of raw sugar into the United States, and ways to deal with any surpluses (Shuker, Heagler, and Chapman 1986, 2–3). Although price controls were back in place, acreage quotas were not, so sugarcane acreage and production in Louisiana expanded. Cane sugar ostensibly was out of trouble.

The Energy Crisis and Industrial Competition for Land

The energy crisis of the 1970s created additional problems for the Louisiana sugar industry. More than one sugarcane grower complained about fuel prices, and one explained to me that increased fuel costs initiated the downfall of the Southdown corporation. The crisis also became a two-edged sword as skilled workers were drawn to higher paying jobs in Louisiana's offshore oil fields and to petrochemical industries along the Mississippi River. After World War II and especially after 1965, the region between Baton Rouge and New Orleans along both banks of the Mississippi River rapidly grew into what became known as the Chemical Corridor, the American Ruhr, or, as locals referred to it, Cancer Alley. In a distance of less than one hundred river miles, 138 chemical plants, mostly petrochemical ones, now occupied lands that had once been sugar plantations. The resources that a sugar plantation could provide for a chemical plant site were perfect: large tracts of land, much of it on the dry levee crest adjacent to North America's largest river waterway, adjacent rail lines and highways, ample water supply, oil and gas lines, and a state government that heavily encouraged industrial expansion at apparently any cost. Little wonder that on the banks of the Mississippi one now saw grotesquely contorted pipes, tanks, and metal structures with vents spewing noxious fumes, flames, and steam and smoke clouds into Louisiana skies. Some 350 legally permitted outfall pipes would discharge untold thousands of gallons of waste fluids into the river. Industry meant money. It was the way progress in Louisiana would be measured, and it meant that a plantation landscape dominance along this hundred-mile stretch of the Mississippi River would be no more.

Environmental Protection

The Environmental Protection Agency began to control Louisiana's sugar industry in 1975. Federal regulations, written expressly for Louisiana raw cane sugar-processing factories, set pollution controls for effluent temperatures and waste water content from sugar factories (Office of the Federal Register 1990). Owners either complied with water and air quality regulations or considered closing their sugar factories.

Soybeans

In the 1970s, soybean production expanded enormously throughout the southern United States. From Missouri to Louisiana, the Mississippi River floodplain grew into a major national soybean region.[2] Even within the sugar parishes of Louisiana, soybean production increased by 82 percent between 1969 and 1978 (Fielder 1981, table 16). Posing as soybean experts, two Arkansas land speculators embarked on a plantation-buying spree in southern Louisiana. Their largest acquisition, South Coast, was acquired in 1979 with California entrepreneurs and financially backed by Prudential and Northwest Mutual Life Insurance Companies.[3] In spite of the proposal to turn southern Louisiana into another soybean empire, financial troubles drove one of the entrepreneurs from the state, and any talk of soybean expansion seemed to have ceased.

Meteorological Disasters

In 1984 and again in 1989, severe freezes devastated the Louisiana sugar crops. Production dropped from 603,000 tons in 1983 to 452,000 tons in 1984. A more destructive freeze in December 1989 caused production to plummet from 844,000 to 438,000 tons.[4] Predictably, more trouble was headed for the Louisiana sugarcane industry, but this time in the form of damage from Hurricane Andrew in August 1992. The storm could not have come at a more inopportune time because the immature cane crop contained low levels of sucrose and many tall cane stalks were bent and broken into useless tangles. Regionwide losses were estimated at 20 to 25 percent. Sugar parishes in the direct path of the eye of the storm were estimated to have losses of 50 percent. Revised data later revealed the losses to be not quite as high as originally estimated. The 1992–93 crop had been expected to yield one million tons of sugar but actually produced 876,000 tons, or about 13 percent less than expected.[5]

Sugarcane Expansion

Changes in crop quotas after 1985 meant that sugarcane could not only make a comeback, but also expand. Figure 11.1 indicates trends in the changing numbers of sugar-producing farms in the delta. Between 1987 and 1992, the number of sugar farms increased from 687 to 755, a net gain

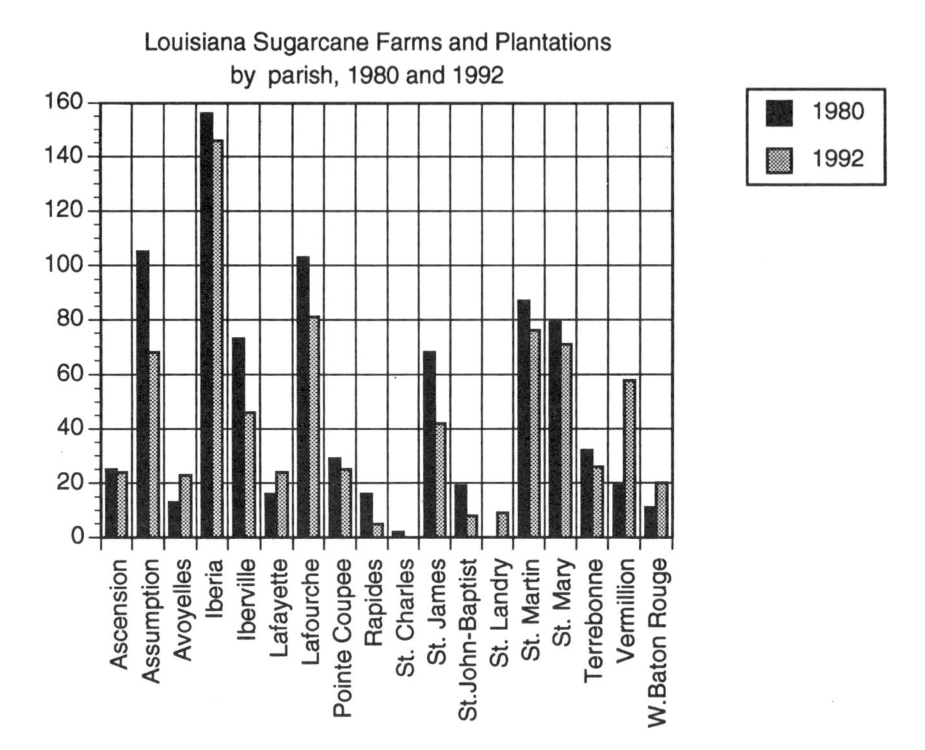

FIG. 11.1. *Trends in numbers of sugarcane farms and plantations by parish between 1980 and 1992 indicate that thirteen parishes are losing farms. Parishes gaining farms are Avoyelles, Lafayette, Saint Landry, Vermillion, and West Baton Rouge. Largest gains are in parishes on the western and northwestern margins of the sugar region.* Data source: U.S. Department of Commerce, Bureau of the Census, 1992 Census of Agriculture

of 98 farms.[6] This ran counter to what seemed to be happening in American agriculture, where farm consolidation has been the norm. Where were the ninety-eight farms coming from or expanding to, and were any parishes losing sugar farms, as one might expect? The old traditional sugar parishes along the Mississippi River indeed were losing sugarcane farms to consolidation and especially to urban and industrial expansion in the Chemical Corridor. However, new sugarcane growth areas were emerging in parishes to the north and west of the old sugarcane region. Vermillion Parish added twenty-nine more farms in just five years; Avoyelles had twenty-three; Lafayette had eighteen more; Saint Landry had nine; Pointe Coupee increased by eight; Rapides had five; and West Baton Rouge Parish added three farms. According to a U.S. Department of Agriculture June 1997 report, "a new growing area in western Louisiana of about 6,000 acres will be harvested this fall for seed cane to expand plantings which would be harvested in 1998. The new area may expand to 30,000 acres within a few years."[7] It will be interesting to observe this spatial trend of new sugarcane producers expanding into the western parishes, where soils historically did not favor sugarcane, and to the north, where meteorological freezes certainly have placed limits on expansion throughout the history of commercial sugarcane in Louisiana.

Economies of Size: Sugar Factories and Cane Growers

As the number and distribution of sugar factories declined, sugarcane production increased dramatically between 1969 and 1995. Improved efficiency through technology in sugar factories, measured by grinding capacity in tons per day, indicated an enormous change in the way sugarcane was processed. Large factories became larger; smaller and less efficient ones closed.[8] All but one of the sugar factories in the CALA model were deliberately closed to consolidate milling functions and to control more of the processing industry. Between 1969 and 1989, fewer cooperative factories closed as compared to private and large corporate sugar factories. Some private sugar factories, like Billeaud and Valentine, did not reinvest profits made in the 1960s and 1970s into upgrading and meeting environmental regulations and therefore closed as obsolete units.[9] Cooperative sugar factories such as Glenwood, owned and operated by associated groups of grower-members, survived largely because their grower-owners knew how to weather the vagaries of the sugar business, conservatively managing in good times and bad.

At the grower's level, as farm size changed management requirements changed as well. An economic analysis of 250 cane farms for optimum resource allocations recommended that farms with up to three thousand acres should have the range of resources shown in table 11.1. Small farms had to produce with high efficiency by maintaining lower operating costs. Larger farms, obviously needing more labor and equipment, required exceptional management skill (Hart 1991a, 66–67). By the late 1970s, 75 percent of Louisiana's sugarcane farms were larger than three hundred acres (Campbell 1977, 1). A minimum acreage was necessary to support the high costs of farm equipment. At what point should the grower decide to contract

TABLE 11.1
Optimum Resource Allocation for Sugarcane Operations

	No. Needed			
Resource	*500 acres*	*1,000 acres*	*2,000 acres*	*3,000 acres*
Hired labor	4	8	17	26
Tractor	3	6	11	17
Cane harvester	1	2	4	5
Cane loader	1	1	2	5
Subshank plow	1	1	2	3
Chisel plow	1	1	1	1
Disk	1	1	1	1
3-row chopper	1	2	4	6
Row opener	1	1	1	2
Row marker	1	1	1	2
3-row flat roller	1	1	1	2
Mech. planter	2	3	5	8
Liq. fert. applicator	1	1	2	3
Cultivator	1	2	4	5
Cane wagons	3	5	9	13
Herbicide rig	1	1	2	2
Drain cleaner	1	1	2	2
Burning unit	1	1	1	1
Rotary mower	1	1	2	3
2-row plow	1	1	2	3

Source: Shuker, Heagler, and Chapman 1986, 46–49

for outside hired custom harvesting? For farms up to three hundred acres, custom harvesting was advised because it was more efficient to rent the service at fifty dollars per acre than to buy and maintain an expensive mechanical harvester (Shuker, Heagler, and Chapman 1986, 46–49). It is unknown how many cane farmers followed the advice of agricultural economic experts. But for some reason, possibly through the symbiotic relationship with co-op factories and their traditionally conservative methods for survival and because cane farming is a way of life, growers not only continued but also expanded operations.

Contemporary Cane Operators and Labor in the 1990s

How does the contemporary 1990s cane operation compare with the functional plantation of the 1960s? Without retilling the entire field, I will outline the fundamental elements and the social and ethnic content. Terms in parentheses were used in the 1960s and in the nineteenth century. Approximately 80 percent of Louisiana's entire sugarcane acreage is now on leased land (American Sugar Cane League 1996, 4).

A contemporary sugarcane operation (plantation) functions with four elements: management, leased land, leased equipment, and nonresident labor. The typical cane operator (planter) leases several thousand acres of sugarcane land scattered among small plantations and cane farms. He leases most if not all land and farm equipment. He especially leases the most expensive farm equipment in terms of tractors, cane harvesters, and long-haul cane trucks. For some operations, the agricultural equipment is not sheltered in barns or under sheds but is left out in the open.

Labor reflects a social, ethnic, and racial hierarchy in the following pattern: (1) white cane operators (planters); (2) white skilled equipment operators (overseers/drivers); (3) black skilled equipment operators (mule/tractor drivers); and (4) unskilled Hispanic migrant laborers (slaves/field hands). Labor is hired on a job by job basis. The operator (planter) hires local skilled or semiskilled laborers to operate mobile equipment–tractors, harvesters, trucks, and any other equipment that is likely to be driven on the roads. The most valuable equipment items in tractors but especially cane harvesters are driven especially by white and some black skilled labor. Black skilled laborers for the most part drive other mobile farm equipment such as tractors, older cane harvesters, cane loaders, and trucks. During grinding season, unskilled migrant Mexican workers, who cannot or are not allowed to drive mobile equipment, perform low-tech and clean-up work like setting field fires, much like field hands did in the past century.

The operation is fundamentally similar to the old categories of a labor hierarchy on plantations in the past; however, it differs considerably by its lack of permanence. Clearly, the functional cane operation on a lease basis is at all times temporary and impersonal, and it requires nothing more of the landscape than management, land, equipment, and labor. Under such management, the resulting landscape retains little of the morphological composition of the old plantation patterns, and the result is a surface that neither needs nor wants mansions, quarters, outbuildings, stores, churches, fences, or much of what we have examined on the plantation landscape. It is hard to imagine a more temporary condition, yet this is the way cane is raised in Louisiana.

Epilogue: Preservation and Progress

Louisiana's sugar plantation landscape has experienced steady decline over the past century. The dramatic drop in plantation numbers in the past thirty years can be attributed, in part, to two things: (1) a changing philosophy toward landscape preservation and (2) progress brought on by land development and technology. Many planter practitioners on the functional landscape have little interest in preserving artifacts that can no longer bring a profit to the sugar industry. Old buildings are perceived to be a liability to corporate agriculture. And quarter houses throughout the delta are a known taxable liability. We saw the result of the CALA model where landscapes at Armant, Ashland, and Oaklawn Plantations were bulldozed in the name of progress and efficiency. We most recently witnessed the decline of Cedar Grove, where all dwellings are now gone, perhaps in the name of progress.

Why does the word *progress* inflame so many people? The answer is not as simple as saying developers want it and preservationists only want to stop it. Ashland Plantation fell victim to the progress brought on by Houma's phenomenal urban expansion during the oil boom of the 1980s. If the boom had continued, as many certainly expected, Ashland North now would be full of subdivision-style houses. Progress was perceived to be good as long as it was associated with profit. Even though progress was halted, nothing could bring back the Ashland plantation.

We have seen the trap that Formosa Chemical and Fibers Corporation fell into with the Whitney plantation property. Formosa chose to purchase an 1,800-acre plantation that just happened to have historic structures on it. Formosa found itself in the awkward position of being forced by Louisiana's cultural resource regulations to take responsibility for preserving historic buildings on site. Reliable sources have told me that Formosa was informed that it would have been wise for the company to bulldoze all the old buildings before state authorities had a chance to stop them. Thanks be that Formosa had the good sense to work the preservation issue in their favor by funding a preservation planning project through Louisiana State University. Still, some might argue that Formosa was trapped into putting up money for the preservation process when they could have avoided the issue by quickly clearing the landscape shortly after they bought it. Alternatively, Formosa could have acquired the Willow Bend site from Shell Oil and already be building their $700 million rayon facility. Was landscape preservation the only issue? Certainly not, because small landowners in the Wallace community adjacent to Whitney had their own issue of reluctance to sell to Formosa. Their worry was far more personal; it had to do with their health and quality of life in the Chemical Corridor. Their resistance to selling coupled with the preservation issue kept Whitney in a state of limbo, with no progress toward either long-term preservation of the Whitney buildings or rayon plant construction.

At Madewood few things changed and, when they did, it was usually for the betterment of the property. A preservation strength resides in the hearts of the owners of Madewood's mansion, for obvious reasons. On the functional landscape at Madewood, the Thibauts have kept some of the landscape preserved simply because that is the way they have always done it, simply out of family tradition.

Technology precipitated the disappearance of plantation morphology. As I explored plantations in the 1960s, technological forces were already at work dismantling the landscape. The mule barn is a good example. Mules left the plantation in great numbers in the 1940s and 1950s, when tractor technology took over. Naturally, with animal power no longer needed, mule barns began to disappear and in their place tractor sheds were erected. If we take this logic several years forward, technology had improved to the point that fewer workers were needed to operate a plantation. Empty, surplus, and deteriorating quarter houses were no longer needed for anything, except for some nostalgic professor like me to wax poetically about them. Not only were quarter houses unnecessary, they were a liability, and to some people the buildings were an embarrassing eyesore. For a few folks who had once lived in the quarters, the buildings may have been a reminder of the days of slavery. For these people leaving the plantation was twentieth-century emancipation. For others, the changing ownership simply forced people off the property. Better paying jobs at petrochemical plants in the Chemical Corridor and in offshore oil fields drew skilled workers from the plantation landscape so that the only people left back at the old plantation were people too old or too poor to live anywhere else. Government-subsidized housing projects in nearby small Louisiana towns became a major drawing card for people to find better housing conditions

FIG. 11.2. *For some things, to change is to vanish. One of the last uses of mules in Louisiana sugar production was for quarter drain work at Green Oaks Plantation in the 1960s.* Photograph, 1967

and physically to leave the plantation. Others moved to the North. If everyone left the plantation, where would today's sugar operators obtain workers? They do much like one sugar operator in Saint Charles Parish told me. "When I need somebody to work, I just take that sombitch truck into town and hire as many men as I need for the job that day."

The Louisiana sugar industry is represented by a resilient body of cane growers and entrepreneurs who manage plantations, corporations, and cooperatives in an area where commercial sugarcane cultivation seems improbable. Marginal at best, the business survives largely because of an expert knowledge base, government assistance, and inertia. The models and case studies in this story represent a framework with which the changes over the past three decades can be observed. Factors related to change help to explain the events that occurred after 1969 and suggest further speculation. In a "what if" interrogation, one might understand forces that were under human control and might even suggest other possibilities. What if sugar prices had not fluctuated so much in 1974? What if the U.S. Sugar Act had not been rescinded in 1974? What if the energy crisis, environmental regulations, and soybeans had not entered the equation? Would the answers to these questions alone cover the issue?

Traditional explanations of geographic phenomena follow the laws of nature, physics, or some other generic form. In a 1988 paper, Warf came to the conclusion that the role of the individual had legitimate explanatory value in regional analysis (Warf 1988, 326–46). Rarely has the role of the individual carried the weight of explanation so much as in the story of the changing sugar industry in Louisiana. After 1974 as much as 30 percent of the Louisiana raw sugar industry came under the controlled design and management strategies of two and initially four individual entrepreneurs. The story has been and continues to be one of cane growers and entrepreneurs, push and pull factors, detrimental and positive consequences in the plantation transformations into corporate forms. However, one of the most cogent explanations for twenty-five years of landscape change in the Louisiana sugar industry is the influence of the diffusion of ideas and the entrepreneurial actions of gentlemen from California and Arkansas who introduced the CALA model to the delta.

The inevitability of change is always with us. As the plantation landscape in *Delta Sugar* permanently vanishes, we can look to examples like Madewood and to the great preservation efforts made at Laurel Valley Plantation near Thibodaux as models for survival. Mansion preservation is not new, but mansions like Oaklawn, Destrehan, Ormond, San Francisco, Evergreen, Nottoway, Houmas House, Oak Alley, and now Laura are evidence of the restorative powers of philanthropic interests in architectural history. My hope is that twenty years from now we will be able to observe a few of the still intact persistent plantations in *Delta Sugar*.

Notes

CHAPTER 1: *Plantation Evolution*

1. Gray [1932] 1958, 333; Surrey 1916, 163–64; Du Pratz 1758, 201–8; "Indigo" 1844, 92

2. Mitchell 1936. The myth continues, as plans have been made to build replicas of Tara and Twelve Oaks mansions in a proposed theme park in Villa Rica, Ga. (*Knoxville News Sentinel,* November 20, 1994).

3. Prunty 1955, 460; Gregor 1965.

4. Sitterson 1953; Heitmann 1987.

5. Moody, Martin, and Byrne 1982, 222–65; Dunaway 1944, 4–15, 28; McWhiney 1988, xxi–xlii.

6. For sugarcane's continuous cultivation, see Avequin 1857, Dart 1935, and Shugg 1937. For its spatial stability, see Hilliard 1979 and 1990 and Hart 1991a.

7. U.S. Department of Agriculture, Economic Research Service [hereafter USDA ERS] 1993; USDA ERS 1997, 8–9; Conrad 1997, personal communication.

8. Taggart 1933, 69–70; Lytle 1968, 11.

9. Barnes 1964, 18–29; Blackburn 1984, 30–42.

10. Alexander 1973, 54–56; Matherne 1969, 1056–61; Barnes 1964, 32, 39–40, 49.

11. Hall 1992, 34–35, 58–60; Curtin 1969, 76–78.

12. For settlement near New Orleans, see Giraud 1996, 24–30; Hirsch and Logsdon 1992; Gayarre 1851, 239; Cable 1884; and Oszuscik 1992a, 136–38. For *concessions,* see Cruzat 1918, 87–153, and Rehder 1971, 29–40. For the delta in colonial times, see Blume 1956 and Carpenter 1847.

13. For Acadian settlement, see Post 1962; Brasseaux 1987, 1992; Conrad 1983; Ancelet, Edwards, and Pitre 1991; and Comeaux 1992. For Cajun population growth, see LeBlanc 1996, 117–24, and U.S. Department of Commerce, Bureau of the Census 1994.

14. Robertson 1911, 2:248, 2:263, 2:292; Brasseaux, Conrad, and Cheramie 1992.

15. Le Conte 1996, 31–43; Conrad 1981; Blume 1956.

16. Sitterson 1953, 23–26; *American State Papers* 1834, 3:67–77, 3:119–21; Degalos 1892, 67; Becnel 1989; Newton 1967.

17. Hall 1996ab, 44–48, 72–102; Hall 1992; Conrad 1981; St. John-the-Baptist Parish 1820.

18. Brandes and Satoris 1936; Artschwanger and Brandes 1958.

19. Artschwanger and Brandes 1958; Alexander 1973, 6–11.

20. Artschwanger and Brandes 1958; Barnes 1964, 2; Blume 1985, 21–24.

21. Watson 1983, 103–19, 190–200; Galloway 1989, 23.

22. Galloway 1989, 38–39; Deerr 1949, 1:7, 2:535; von Lippmann 1929, 338.

23. Blackburn 1984, 1; Galloway 1989, 47.

24. Le Roy Ladurie 1971; Blume 1985, 25; Blume 1967; Galloway 1989, 46.

25. Blume 1985, 28; Galloway 1989, 43–47.

26. Galloway 1989, 77–80; Blume 1985, 167–69.

27. Dunn 1972, 118–24; Gaspar 1993, 101–23; Pitman [1917] 1967; Pulsipher 1991.

28. Merrill 1958, 69; Galloway 1989, 89.

29. Galloway 1989, 88–89; Edwards 1793–1801, 2:255.

30. Galloway 1989, 89; Merrill 1958, 69.

31. Galloway 1989, 105–12; Blume 1985, 175–82.

32. Pusser's Ltd., n.d.; Hill 1995, 144, 259.

33. Labat 1742, 3:449–58; Lasserre 1961, 2:354–55.

34. Lasserre 1961, 2:355; Labat 1742, 3:416.

35. Du Terrage 1903, 198; Robertson 1911, 1:152–53; Martin 1827, 1:320; DeBow 1850, 35.

36. Sitterson 1953, 6–7; Conrad and Lucas 1995, 3–7.

37. Surrey 1916, 157–59; Gray [1932] 1958, 302.

38. Robin [1807] 1966, 108–9; Delanglez 1935, 390.

39. Pares 1963, 327–28, 506; Sitterson 1953, 18–98; DeBow 1850, 15.

40. Gray [1932] 1958, 744. Prices fluctuated from 8½¢ per pound in February 1815 to 13¢ in December 1815 to 16¢ in May 1816. By comparison, when sugar was selling at 4½¢ per pound and cotton at 6¢ per pound, the profits were considered about equal.

41. DeBow 1847b, 391–400; DeBow 1853; DeBow 1855, 688–93; Sitterson 1953, 169–70, 188–93; Lewis 1976, 32–37.

42. *American State Papers* 1843, 67–77, 119–21; Degalos 1892, 67.

43. Bond 1912, 3–12; Hackett 1946, 4:ix, 4:214; Rehder 1973.

44. Mitcham 1986; Ginn 1940, 559–60.

45. For cotton cultivation, see Darby 1817, 78, and Scroggs 1914, 268. For capital needed for a sugar plantation, see *An Account of Louisiana* 1803, 6–7; Robin [1807] 1966, 114–15; and Darby 1817, 222.

46. DeBow 1846, 442; Gray [1932] 1958, 744–48, 1027.

47. Byrnside and Sturgis 1958, 13–16; Peevy and Brupbacher 1962, 12–13.

48. Carrigan 1851, 616; Johnson 1933, 539–53. Bayou Manchac is located about ten miles south of Baton Rouge and runs west to east off the east bank of the Mississippi River.

49. Champomier 1860, 3. One hogshead equals 1,000 pounds of sugar. Louisiana's total sugar production for 1855 was 346,635 hogsheads.

50. Vlach 1993, 7–8; Woodman 1966, 15.

51. Hilliard 1978, 92–95; Clifton 1973, 366–68.

52. Thomas 1997, 570–71, 649; Gray [1932] 1958, 661–63.

53. Vlach 1993, 187–88; Oszuscik 1992b, 170; Oakes 1982, 185.

54. DeBow 1856, 196, 275; Deerr 1928, 11–12.

55. Benjamin 1846, 334–43; Rillieux 1848, 285–88, 291–93.

56. Alexander 1973, 54–56; Barnes 1964, 32, 39–40, 49; Matherne 1969, 1056–61.

57. Zeichner 1939, 22–32; Barrow 1881, 832–33; Aiken 1998.

58. Shugg 1937; Voorhies and Grayson 1939, 514–23; Hawaiian Sugar Planters' Association 1990, 17; Shuker, Heagler, and Chapman 1986, 2–3.

CHAPTER 2: *Culture and Form*

1. Jordan 1966, 26–28; Schlüter 1899, 65; Sauer 1931, 621–23; Kniffen 1965, 549–53; Evans [1942] 1963, 1967; Erixon 1938.

2. Smith 1941, 214; Oszuscik 1997, personal communication; Overdyke 1965.

3. Glassie 1968, 54, 110–11; Neiman 1986, 293; Smith 1941, 214; Oszuscik 1997, personal communication.

4. Wilson 1987, 98, 370–73; Daspit 1996, 154–56; Doré 1989, 39.

5. Oszuscik 1992b, 145–54; Daspit 1996, 26, 31, 44, 53, 98, 147, 152, 155.

6. Vlach 1986, 71–77; Oszuscik 1992a, 164.

7. Wilson 1987, 98, 370–73; Rehder 1971, 258, 277; Wilson 1980, 51–82.

8. Smith 1941, 171–75; Hamlin 1944, 223.

9. Forman 1957, 33; Kniffen and Glassie 1966.

10. Pillsbury 1989, 533–41; Oszuscik 1994, 23; Jordan 1994.

11. Glassie 1968, 54, 101–11; Glassie 1989, 394–402; Neiman 1986, 292–314; Upton 1986, 315–35.

12. Oszuscik 1997, personal communication; Glassie 1968, 54, 110–11.

13. Edwards 1988, 24–25; Oszuscik 1994, 23–31.

14. For houses of the pen tradition, see Kniffen 1965, 561, and Rehder 1992, 104–11. For the Georgian plan, see Glassie 1968, 54, 101–11; Glassie 1989, 400; Upton 1986, 317; and Neiman 1989, 311.

15. Rehder 1992, 109; Jordan and Kaups 1989, 179–96.

16. Pillsbury 1989, 533–41; Glassie 1968, 54, 110–12; Oszuscik 1994, 23; Jordan 1994.

17. Oszuscik 1994, 25–31; Oszuscik 1997, personal communication.

18. Oszuscik 1992b, 144; Vlach 1993, 42; Noble and Cleek 1995, 142, 144.

19. Daspit 1996, 152, 219; Noble and Cleek 1995, 139.

20. Taylor 1950, 271–81; Hall 1970.

21. Carter 1940, 461–62; Moore, n.d., 106; Downs 1960; Hall 1970.

22. Linder 1995; Prunty 1955, 464; Coulter 1955, 21; Lewis 1985, 35–65.

23. Barrow 1881, 883; Prunty 1955, 466–82; Woofter et al. 1936; Aiken 1998.

24. Watson 1848; Blassingame 1979, 254–56; Genovese 1976, 524–31; Joyner 1984; Owens 1976, 23–24.

25. Rehder 1971, 162; Kniffen 1936, 182–83.

26. Russell 1863, 98, 104; Olmsted 1861, 320, 327.

27. Vlach 1986, 62–64; Oszuscik 1992a, 159–66.

28. Vlach 1976, 47–70; Vlach 1986, 58–77.

29. Vernon 1993, 168–71; Oszuscik 1992a, 172–73.

30. Russell 1863, 104; Whelan 1983, 113–26; McDonald 1993, 275–99; Genovese 1976, 535–40; Berlin and Morgan 1993.

31. Hoffsommer 1940, 1941; Sitterson 1948; Laws 1902; Lenoir and Smith 1937; Parenton 1938.

CHAPTER 3: *The Morphology of the Functional Plantation Landscape*

1. Bouchereau and Bouchereau 1868–69 to 1900–1901; Gilmore 1922; Gilmore 1930.

2. USDA ERS 1987, 25; 1994, 13; 1997, 35.

3. Lasserre 1961, 1:352–53; Labat 1742, 3:318, 3:436.

4. Degalos 1892, 65–68; Champomier 1860, 39; Favrot 1844.

5. USDA ERS 1994, 14; 1997, 8–9.

6. Dumont de Montigny 1753, 65; Prichard 1938, 987.

7. Kniffen and Glassie 1966, 46, 48; Russell 1863, 98.

8. Robin [1807] 1966, 124; Olmsted 1861, 319; Comeaux 1992, 189.

9. Kellar 1936, 177, 180, 182; DeBow 1847a, 66–80; Wilkinson 1847, 229; Begnaud 1980, 34–35, 43.

10. Conrad and Lucas 1995, 70; Burrows and Shlomowitz 1992, 64–73.

11. "Sugarcane Loading Devices" 1906, 105–7, 165, 294.

12. Kellar 1936, 65; Surrey 1916, 93; Du Terrage 1920, 129.

13. Cayton 1881; Olmsted 1861, 317.

CHAPTER 4: *A Prescription for Landscape Decline*

1. Gregor 1982; Galloway 1989; Wilkinson 1989; Hilliard 1992.

2. Coastal Environments 1989; Orser 1988; Singleton 1985.

3. Prunty 1955; Babin 1974.

4. Whittlesey 1929; Broek 1932; Sestini [1947] 1962; McIntire 1958.

5. Rehder 1978, 135–40; Rehder 1989b, 554–55.

6. Bouchereau and Bouchereau 1868–69 to 1900–1901; Gilmore 1930.

7. USDA ERS 1987, 25; 1990, 28; 1997, 35.

8. Rehder 1971, 253; St. James Parish 1809, 1860, 1883, 1889.

9. Moody's 1990, 4239; Terrebonne Parish 1977, book 67, p. 523.

10. Gregor 1962, 1974, 1982; Stanley 1979; Scheuring 1983.

11. For sugar factories, see Iberville Parish 1975, book 227, pp. 88–94. For tenants, see Hilliard 1990 and Wall 1981.

12. For Madewood's growth during 1823–48, see Rehder 1971, 314–31, and

Assumption Parish 1823–36, books 2–5. For the 1896 purchase, see Rehder 1971, 318–20, and Assumption Parish 1896, book 42, p. 281. For twentieth-century owners, see Rehder 1971, 321–30, and Assumption Parish 1910, 1916, 1946.

13. Rehder 1971, 321–30; Assumption Parish 1964, book 97, p. 135.

14. Gray [1932] 1958, 317–22; Barrow 1881, 832–33; Prunty 1955, 464.

CHAPTER 5: *Armant Plantation, 1796–1998*

1. For the earliest map evidence, see *Caires' Southern Louisiana Antebellum Plantation Map* 1796. For land acquisition between 1809 and 1818, see St. James Parish 1809, book 1, p. 631; book 4, pp. 671, 952; book 5, p. 571; book 6, p. 98; and book 7, p. 1; for acquisitions between 1820 and 1830, see St. James Parish 1827, book 10, p. 29.

2. Quotation is from St. James Parish 1840, book 18, p. 366. For the 1847 donation, see St. James Parish 1845, book 24, p. 78.

3. For the 1860 sale, see St. James Parish 1860, book 35, p. 23. For the 1883 transfer, see St. James Parish 1883, book 50, p. 398, and 1886, book 51, p. 441.

4. For Miles's modernization of the sugar factory, see Bouchereau and Bouchereau 1884–92 for St. James Parish. For his use of tenant labor, see Sitterson 1953, 264, 272, 261, and Miles, 26 January 1896. For other tenant labor see McCall 1899.

5. St. James Parish 1934, book 67, p. 452; 1948, book 86, p. 284.

6. Terrebonne Parish 1977, book 692, p. 647; book 697, p. 523; book 768, p. 45; Terrebonne Parish 1979, books 744, 749, 769.

7. St. James Parish 1981, book 235, p. 370.

8. St. James Parish 1934, book 67, p. 456.

CHAPTER 6: *Ashland Plantation, 1828–1998*

1. Watkins 1939; Becnel 1989.

2. For Cage's land purchase, see Terrebonne Parish 1828, book D, p. 427. For his purchase of slaves, see Terrebonne Parish 1832, book G, p. 57.

3. Terrebonne Parish 1840, book G, p. 331; Terrebonne Parish 1843, book I, pp. 396 and 411.

4. Terrebonne Parish 1844, book J, p. 166. Linear arpents measured 192 feet, but areal arpents varied between 0.84 and 1.28 acres, with the variance depending on time and place of survey.

5. Ibid., book J, p. 181.

6. For the land division between Harry and James Cage, see Terrebonne Parish 1859, book T, p. 78. For Mayfield's purchase, see Terrebonne Parish 1856, book R, p. 253. For Ashland's production, see Champomier 1845, 8; 1851, 30; 1855, 29.

7. For the 1859 transaction, see Terrebonne Parish 1859, book T, p. 78. For Duncan Cage's sugar production, see Bouchereau and Bouchereau 1869–83 for Terrebonne Parish.

8. For Hugh Cage's purchase and sale of Ashland, see Terrebonne Parish 1886, book JJ, pp. 235 and 337. For Gayden Cage's sale, see Terrebonne Parish 1887, book KK, p. 87.

9. Terrebonne Parish 1888, book LL, p. 177; Bouchereau and Bouchereau 1890, 1892.

10. Terrebonne Parish 1906, book SS, p. 135.

11. For sugar factories in Terrebonne Parish, see Bouchereau and Bouchereau 1911–16. For the sale to Wurzlow, see Terrebonne Parish 1934, book 103, p. 500; for SouthCoast's purchase, see Terrebonne Parish 1935, book 104, p. 215.

12. Terrebonne Parish 1979, book 768, p. 65.

13. For Winemiller's purchase, see Terrebonne Parish 1980, book 880, p. 208. For the sale to LRT Corp., see Terrebonne Parish 1983, book 912, p. 101.

14. Terrebonne Parish 1987, book 1095, p. 84.

15. Terrebonne Parish 1981, book 821, p. 882.

CHAPTER 7: *Oaklawn Plantation, 1812–1998*

1. St. Mary Parish 1812, book BA, entries 131–32; 1813, book F, entry 5163; 1814, book BA, entry 312.

2. For Porter's purchase from Sterling, see St. Mary Parish 1823, book B4, entry 358; for his purchase from Martin, see Stephenson 1934, 8–11, and St. Mary Parish 1827, book B4, entry 556. For Irish Bend district plantations, see Foscue 1936.

3. St. Mary Parish 1849, book G344, entry 6466.

4. Champomier 1844, 7; 1850, 32.

5. St. Mary Parish 1874, book R, p. 418.

6. For the 1876 sale, see St. Mary Parish 1876, book S, p. 405. For the Foster-Leverich transfer, see St. Mary Parish 1881, book U, pp. 253 and 257.

7. For the Rivers purchase, see St. Mary Parish 1888, book X, p. 749. For the 1896 purchase, see St. Mary Parish 1896, book FF, p. 305.

8. St. Mary Parish 1925, book 4K, p. 55.

9. For SouthCoast's purchase of Oaklawn, see Terrebonne Parish 1979, book 768, p. 176. For Goldsby's finances, see St. Mary Parish 1981, book 7, p. 403, and Terrebonne Parish 1985, book 1036, p. 74, and book 1040, pp. 381–82, 418.

10. Dupy 1967, 155–58; Gilmore 1967.

11. Interview, December 1991, with Sean McCarthy, who supervised Oaklawn, then called Teche Sugar Co. Inc., in its last days of operation and closing. He is the son of Leland McCarthy, who was one of the key people in South Coast Sugars Inc., headquartered in Raceland, La.

CHAPTER 8: *Cedar Grove Plantation, 1829–1998*

1. Iberville Parish 1829, book M, pp. 53–54.

2. French planters preferred horses and oxen over mules for draft animals.

3. Iberville Parish 1837, Probate Records book 9, p. 3.

4. Champomier 1844, 2; 1850, 11; Bouchereau and Bouchereau 1862, 19.

5. Iberville Parish 1867, book 8, p. 420.

6. Iberville Parish 1872, book 11, p. 158.

7. Iberville Parish 1875, book 12, entry 212.

8. Bouchereau and Bouchereau 1883, 1890, 1892, 1911, 1914, 1917 for Iberville Parish.

9. Iberville Parish 1920, book 45, entry 365.

10. Ibid.

11. Ibid.

12. Iberville Parish 1939, book 67, 480; Iberville Parish 1975, book 227, 88; Iberville Parish 1988, book 1159, 142.

CHAPTER 9: *Whitney Plantation, 1790–1998*

1. Conrad 1981, 9; Rykels 1991, pt. 2.

2. St. John-the-Baptist Parish 1820, Conveyance Records, book C, p. 120.

3. St. John-the-Baptist Parish 1820, Original Acts, book A, p. 1123.

4. St. John-the-Baptist Parish 1835, Conveyance Records, book M, p. 296.

5. Champomier 1844, 1851, 1852, 1855, 1856.

6. St. John-the-Baptist Parish 1852–53, book Z, pp. 32, 224; St. John-the-Baptist Parish 1867, book ANS, p. 285.

7. St. John-the-Baptist Parish 1946, book 13, p. 43.

8. Ibid.

9. St. John-the-Baptist Parish 1990, book 264, p. 522.

10. St. John-the-Baptist Parish 1820, Conveyance Records book C, p. 120.

11. Johnson 1868 and 1880, time books.

12. St. John-the-Baptist Parish 1990, book 264, p. 522.

13. Rykels 1991, pt. 1; Cheakalos 1992, B4–5.

CHAPTER 10: *Madewood Plantation, 1823–1998*

1. Assumption Parish 1823, book 2, p. 197; 1830, book 2, p. 571; 1831, book 3, p. 113; 1836, book 5, p. 80.

2. Assumption Parish 1852, *Acts of 1852*, vol. 2.

3. Bouchereau and Bouchereau 1863–70 for Assumption Parish.

4. Assumption Parish 1883, book 35, p. 585.

5. Assumption Parish 1896, book 37, p. 638.

6. Ibid., book 42, p. 281.

7. Assumption Parish 1910, book 53, p. 277; Assumption Parish 1916, book 56, p. 162.

8. Assumption Parish 1946, book 77, p. 307.

9. Assumption Parish 1964, book 97, p. 135.

10. Surveyor General 1829–30, map of southeast district T13S, R15E.

CHAPTER 11: *Agents of Change*

1. U.S. Department of Agriculture, Agricultural Stabilization and Conservation Service 1975, 9.

2. Gregor 1982; Hart 1991a.

3. Terrebonne Parish 1979, book 768, p. 45.

4. USDA ERS 1993, 18–44.

5. USDA ERS 1994.

6. U.S. Department of Commerce, Bureau of the Census 1992, table 27.

7. USDA ERS 1997, 8.

8. Heagler 1991; Christy, Chapman, and Heagler 1990; Auty 1976.

9. USDA ERS 1993, 18–44.

Glossary

Acadians–French people driven from Acadia (Nova Scotia) and other parts of eighteenth-century maritime Canada who entered Louisiana as immigrant settlers between 1755 and 1785. Best known today as *Cajuns,* their colorful way of life is expressed in music, rich spicy food (they are said to eat about anything that comes from the water), traditional French language, camps, celebrations, and many other distinctive culture traits. Their culture hearth is concentrated in the parishes surrounding Lafayette, but they are also found concentrated along Bayou Lafourche, in Saint James and Ascension Parishes, and scattered elsewhere throughout southern Louisiana.

Anglo-Americans–people of Scotch-Irish and English ancestry who settled Louisiana after the Louisiana Purchase in 1803. The majority of Anglos settled northern Louisiana to develop a cotton economy in what was to become part of the Bible Belt South. Some Anglo planters made their way to southern French Louisiana and established distinctive nineteenth-century sugar plantations that reflected their Anglo material culture.

arpent (Latin *arapennis* or *arepennis*)–an old French land measure that equals 192 linear feet or, in area, 0.84 acre. The French Arpent Survey System was widely used to survey lands throughout southern Louisiana, where lands were measured at a standard forty arpents depth but at narrower, variable widths along stream banks. The result was "long lot" land holdings.

backslope–the broad surface of a natural levee, with an almost imperceptible slope gradient, located between the levee crest at the stream bank and the backswamp.

backswamp–the swamp that commences at the lowest, farthest point of a backslope on a natural levee. Backswamps mark the limit of cultivation in the floodplain. In southern Louisiana, backswamps and marshes occupy broad expanses in interlevee basins.

bagasse (Sp. *bagazo*)–the fibrous cane pulp remaining after cane juice has been pressed from sugarcane.

barreaux (Fr. plural of *barre*–bar, rod)–small wooden rods wedged between heavy vertical and diagonal timbers in half-timbered colombage construction. Barreaux, also called *rabbits,* are set horizontally in the interstices between timbers to hold mud and moss filling, or nogging, in place.

batture–a stretch of narrow natural levee land situated between the stream bank at the water's edge and the man-made artificial levee.

big house–a colloquial term referring to the large, usually pretentious plantation mansion that is the home of a planter or plantation owner.

block plantation–an agglomeration of plantation buildings based on the arrangement of quarter houses in the shape of a block square or rectangle. Block plantations are identified with Anglo-American planters in the nineteenth-century occupance of the delta region.

bousillage–In half-timbered construction, the interstices (spaces between timbers) are filled with mud, Spanish moss, and small sticks collectively called *bousillage.*

briquette entre poteaux (Fr.–bricked between the posts)–in half-timbered construction, the interstices (spaces between timbers) are filled with bricks.

colombage (Fr.)–half-timbering construction. Walls are constructed of heavy timbers. Interstices are filled with mud and moss (*bousillage*) or with brick (*briquette entre poteaux*).

Creole–in the sixteenth through nineteenth centuries, the term in the Caribbean and in Louisiana identified a person born in the Americas. Later, the term applied to people of mixed racial or cultural heritage. Creole today identifies a house type, a tomato variety, a pony, styles of food, language, and other varied cultural phenomena.

cyprière (Fr. *cypres*–cypress)–a cypress swamp in the backswamp of a plantation or settlement from which cypress timbers, planks, and other lumber were cut.

dogtrot–a small, southern folk house type characterized by an open passage between two pens, or room units. Originally built of logs, this double-pen house type is widespread from Tennessee to Missouri, Arkansas, Alabama, Mississippi, northern Louisiana, and well into Texas and is associated with Scotch-Irish and English settlers.

engages (Fr. *engager*–to engage, pledge)–French indentured servants who entered Louisiana between 1710 and 1740 as agricultural laborers and initial settlers.

façade (Fr.–front)–the front of a dwelling; the face of the building that faces the street.

Florida Parishes–Louisiana's parishes north of Lake Pontchartain and east of the Mississippi River, formerly a part of West Florida when under British rule in the eighteenth century. The Florida Parishes today are East Baton Rouge, West Feliciana, East Feliciana, Saint Helena, Saint Tammany, Livingston, Tangipoa, and Washington.

frontlands–the sandy, silt loam lands that begin at the levee crest and extend partially down the backslope of a natural levee. These well-drained sites are favored for settlement and especially for agriculture.

gable–on the short side of a rectangular building, the area enclosed by or masking the end of a roof that slopes downward from a central ridge. Most buildings have sideward-facing gables; however, structures with a front-facing gable are shotgun houses, bungalows, plantation stores, and Tidewater Anglo plantation mansions.

galerie (Fr.–gallery)–a large, overhanging porch found on French Creole mansions. When accompanied by a hip roof, a galerie may completely surround the structure.

garçonnière (Fr. *garçon*–boy, bachelor)–a bachelor's quarters separated from the plantation mansion but located on the mansion site grounds. The garçonnière was reserved for the planter's sons but could be used as a guest house for overnight bachelor guests.

granulation–the point at which cane syrup begins to crystallize. Granulation can occur only with mature or nearly mature sugarcane with a high sucrose content.

hogshead–a wooden barrel used as a shipping container for raw sugar. Eighteenth- and nineteenth-century hogsheads held a standard one thousand pounds of raw sugar.

lath and plaster–a building material and technique; small wooden strips called *laths* are attached to the inner wall, and plaster is applied to laths.

levee crest–the highest point on a natural levee. A levee crest is the top of the bluff at the immediate stream bank.

levees, artificial (Fr. *lever*–to raise)–man-made embankments or dikes built along a stream to prevent flooding.

levees, natural (Fr. *lever*–to raise)–broad, slightly elevated areas of alluvial deposition formed on the flanks of streams in an alluvial floodplain. Nearly all major inhabited and cultivated areas in the Lower Mississippi River delta are located on natural levees.

linear plantation–an agglomeration of plantation buildings based on the arrangement of quarter houses in a line or linear pattern. Linear plantations are identified with French planters in the initial occupance of the sugar region in the eighteenth and early nineteenth centuries.

massecuite–a mixture of granulated sugar crystals and molasses that forms late in the sugar making process. After massecuite forms, the molasses is purged, or removed by separation processes, from the raw sugar crystals.

mosaic disease–a sugarcane virus that attacks the leaves of sugarcane plants. Plants weakened by mosaic disease are susceptible to insects and other sugarcane diseases. In the 1920s, mosaic disease devastated the Louisiana sugarcane crops for several years; some plantations recovered slowly, and others, not at all.

nogging–material made of mud and moss (*bousillage*) or bricks (*briquette entre poteaux*) that is used to fill the interstices between timbers in half-timbered construction.

open kettles–iron kettles used in early sugar making. Varying in size from thirty-six to seventy-two inches in diameter, the kettles were named from largest to smallest: *grande, propre, flambeau, sirop, batterie.*

outbuildings–barns, storage and repair sheds, and implement sheds associated with the agricultural and manufacturing functions of a plantation.

parish–an administrative unit in Louisiana known elsewhere in the United States as a county.

petit habitant (Fr. *petit*–small; *habitant*–inhabitant)–Acadian, German, and Spanish settlers who became peasant farmers, fishermen, and trappers. *Petit* *habitants* in southern Louisiana rarely became large plantation owners or large landowners.

pieux à travers (Fr. plural *pieu*–stake, slab, or post; *à traverse*–through)–a field fence in French Louisiana constructed of vertical posts driven into the ground, with horizontal rails fitted into holes cut through the posts. Elsewhere the fence is known as a post-and-rail fence.

pieux debout (Fr. plural *pieu*–stake, slab; *debout*–upright)–an eighteenth-century fence constructed of upright stakes set into the ground to give the appearance of a picket fence. The *pieux debout* was used to enclose house yards and small gardens.

pieux en terre (Fr. plural *pieu*–stake, slab; *en terre*–in the ground)–one of the earliest wall construction techniques in French Louisiana circa 1700–1730. Long cypress stakes were driven two or more feet into the ground to form walls for outbuildings and temporary houses.

pigeonnier–a pigeon house, or dove cote, located on the mansion site grounds.

poteaux en terre (Fr. plural *poteaux*–posts; *en terre*–in the ground)–an eighteenth-century construction technique used in French Louisiana whereby large posts were driven into the ground to form permanent walls. The interstices between the posts would be filled with mud and moss and whitewashed to complete the wall.

poteaux sur sol (Fr. plural *poteaux*–posts; *sur sol*–on a sill)–a method of eighteenth-century French construction in which vertical posts were attached to a horizontal wooden foundation sill.

purgery–a room or a separate building in early sugar making where molasses was purged from raw sugar. Wooden barrels containing massecuite, a mixture of molasses and raw sugar, were placed in storage in the purgery, where the molasses would drain from holes in the bottoms of the barrels into vats.

quarter house–originally a slave's house; one of several laborer's dwellings that comprise a village settlement called the *quarters*. Types of quarter houses on Louisiana sugar plantations are the small Creole, attached porch, shotgun, and bungalow.

quarters–a village settlement of nearly identical houses numbering from four to about sixty-five dwellings where the resident labor force resides on the plantation. The quarters, a traditional settlement concept, was occupied by slaves from about 1720 to 1865 in Louisiana and was retained to house resident wage laborers and their families after slavery was abolished.

ratoon (Fr. *rejeton*–to sprout)–the regrowth, or sprouting, of fresh sugarcane stalks from the stubble that remains after harvest. Each ratooning crop can be harvested the following year, but sucrose content diminishes with each cycle. In Louisiana, sugarcane ratoons for two to four cycles; in parts of the Caribbean, some canes are allowed to ratoon for as many as ten cycles. The biological maximum is about twenty.

Rillieux apparatus–nineteenth-century sugar factory technology that replaced open kettle methods of making sugar. Invented in Louisiana by Norbert Rillieux, the device had two vacuum pans connected so that the vapor given off by the evaporation of juices in the first pan provided heat for boiling syrup in the second pan.

shotgun house–a long, narrow house with a front-facing gable, one room wide and several rooms long. So named because of its long, slender appearance. According to folk wisdom, a shot fired through the front door would pass through all the open doors and out the back without damaging the structure.

striking–the precise point at which boiling sugarcane syrup begins to crystalize and at which time the mass must be removed from heat and placed in cooling tanks or vats to allow for further granulation.

sugarcane–a tall, thick-stemmed, perennial tropical grass of the genus *Saccharum*. Under average conditions sugarcane is 72 percent water, 10 percent fiber, and 18 percent sugar.

sugar factory–also called sugarhouse, boiling house, sugar mill, central, ingenio. In this largest building on a plantation, grinding mills and boiling equipment process sugarcane into raw sugar, molasses, and bagasse.

vacuum pan–a vessel used in the boiling process in nineteenth-century sugar making; in the vacuum pan the final stages of boiling were done in a vacuum. The invention resulted in a substantial improvement over the old open kettle method.

Bibliography

Abbot, G. C. 1989. "The Commonwealth Sugar Agreement as a Model of New Sugar Protocol." *Development Policy Review* 7:343–60.

An Account of Louisiana, Being an Abstract of the Departments of State and Treasury. Prepared for President Thomas Jefferson. 1803. Washington, D.C.: Duane Printer.

Addy, Sidney O. 1898. *The Evolution of the English House.* London: Swan Sonnenschein.

Agel, Jerome, and Walter D. Glanze. 1987. *Pearls of Wisdom.* New York: Harper & Row.

Aiken, Charles S. 1978. "The Decline of Sharecropping in the Lower Mississippi River Valley." In *Geoscience and Man,* vol. 19, *Man and Environment in the Lower Mississippi Valley.* Edited by Sam B. Hilliard. Baton Rouge: School of Geoscience, Louisiana State University.

Aiken, Charles S. 1998. *The Cotton Plantation South Since the Civil War.* Baltimore: The Johns Hopkins University Press.

Aime, Valcour. 1878. *The Plantation Diary of the Late Valcour Aime.* New Orleans: Clark & Hofeline Printers.

Alexander, Alex G. 1973. *Sugarcane Physiology.* Amsterdam: Elsevier Scientific Publishing.

Allain, Robert L. 1980. "Louisiana Sugar during the 1970s." In *Green Fields: Two Hundred Years of Louisiana Sugar.* Lafayette: Center for Louisiana Studies, University of Southwestern Louisiana.

American State Papers: Documents, Legislative and Executive of the Congress of the United States, Class III, Public Lands. 1834. 38 vols. Washington, D.C.: Gates & Seaton.

American Sugar Cane League. 1991–96. *The Louisiana Sugar Industry.* Thibodaux, La.: American Sugar Cane League.

Ancelet, Barry Jean, Jay Edwards, and Glen Pitre. 1991. *Cajun Country.* Jackson: University Press of Mississippi.

Anderson, Keith B., Michael P. Lynch, and Jonathan D. Ogur. 1975. *The U.S. Sugar Industry: Economic Report to the Federal Trade Commission.* Washington, D.C.: Government Printing Office.

Arena, C. Richard. 1955. "Land Holding and Political Power in Spanish Louisiana." *Louisiana Historical Quarterly* 38:27–39.

Artschwanger, E., and E. W. Brandes. 1958. *Sugarcane.* Agriculture Handbook no. 122. Washington, D.C.: Government Printing Office.

Assumption Parish. 1852. *Acts of 1852.* November, vol. 2. Napoleonville, La.

———. 1823–1964. *Conveyance Records.* Napoleonville, La.

Auty, R. M. 1976. "Caribbean Sugar Factory Size and Survival." *Annals of the Association of American Geographers* 66:76–88.

Avequin, J. B. 1857. "Sugarcane in Louisiana." *DeBow's Review* 22:615–18.

Babin, Edward C. 1974. "A Functional Typology of Louisiana Sugar Cane Production Units in 1969." Ph.D. diss., University of Georgia.

Barnes, A. C. 1964. *The Sugar Cane.* London: Leonard Hill.

Barrow, David C., Jr. 1881. "A Georgia Plantation." *Scribner's Monthly* 21:830–36.

Barry, John M. 1997. *Rising Tide: The Great Mississippi Flood of 1927 and How It Changed America.* New York: Simon & Schuster.

Becnel, Thomas A. 1989. *The Barrow Family and the Barataria and Lafourche Canal: The Transportation Revolution in Louisiana, 1829–1925.* Baton Rouge: Louisiana State University Press.

Begnaud, Allen. 1980. "The Louisiana Sugar Cane Industry." In *Green Fields: Two Hundred Years of Louisiana Sugar.* Lafayette: Center for Louisiana Studies, University of Southwestern Louisiana.

Benjamin, J. P. 1846. "Louisiana Sugar." *DeBow's Review* 2:322–45.

Berlin, Ira, and Philip D. Morgan. 1993. *Cultivation and Culture: Labor and the Shaping of Slave Life in the Americas.* Charlottesville: University Press of Virginia.

Berquin-Duvallon. 1806. *Travels in Louisiana and the Floridas in the Year 1802.* Translated by John Davis. New York: I. Riley.

Birkett, Harold S. 1980. "Improvements in the Technology of Processing Sugarcane in Louisiana since 1960." In *Green Fields: Two Hundred Years of Louisiana Sugar.* Lafayette: Center for Louisiana Studies, University of Southwestern Louisiana.

Blackburn, Frank. 1984. *Sugar-cane.* London: Longman.

Blassingame, J. W., ed. 1977. *Slave Testimony.* Baton Rouge: Louisiana State University Press.

———, ed. 1979. *The Slave Community: Plantation Life in the Antebellum South.* New York: Oxford University Press.

Blume, Helmut. 1956. "Landschaft und Wirtschaft in Louisiana unter Franzosicher Kolonialverwaltung." *Erdkunde* 10:176–85.

———. 1967. "Zuckerrohr und Zuckerrübe im Subtropischen Trockengürtel der Alten Welt." *Erdkunde* 21:111–32.

———. 1985. *Geography of Sugar Cane: Environmental, Structural, and Economic Aspects of Cane Sugar Production.* Berlin: Verlag Dr. Albert Bartens.

Bond, Frank. 1912. *Historical Sketch of Louisiana and the Louisiana Purchase.* General Land Office, Department of Interior. Washington, D.C.: Government Printing Office.

Bouchereau, Louis, and Alcee Bouchereau. 1862–1917. *Statement of Sugar and Rice Crops Made in Louisiana.* New Orleans: Louis and Alcee Bouchereau.

Brandes, E. W., and G. B. Satoris. 1936. "Sugarcane: Its Origin and Improvement." In *U.S. Department of Agriculture Yearbook of Agriculture, 1936,* 561–611. Washington, D.C.: Government Printing Office.

Brasseaux, Carl. 1987. *The Founding of New Acadia: The Beginnings of Acadian Life in Louisiana.* Baton Rouge: Louisiana State University Press.

———. 1992. *Acadian to Cajun: Transformation of a People, 1803–1877.* Jackson: University Press of Mississippi.

Brasseaux, Carl A., ed. 1996. *A Refuge for All Ages: Immigration in Louisiana History.* The Louisiana Purchase Series in Louisiana History, vol. 10. Lafayette: Center for Louisiana Studies, University of Southwestern Louisiana.

Brasseaux, Carl, and Glenn Conrad, eds., and David Cheramie, trans. 1992. *The Road to Louisiana: The Saint Domingue Refugees, 1792–1809*. Lafayette: Center for Louisiana Studies, University of Southwestern Louisiana.

Bringier, Louis A. Collection. Baton Rouge: Louisiana State University Library, Special Collections.

Broek, Jan O. M. 1932. *The Santa Clara Valley, California: A Study in Landscape Changes*. Utrecht, The Netherlands: University of Utrecht.

Browne, C. A. 1938. "Development of the Sugarcane Industry in Louisiana and the Southern United States." Presented at the Sixth Congress of the International Society of Sugarcane Technologists, Baton Rouge, La., Oct. 24–Nov. 5.

Burrows, Geoff, and Ralph Shlomowitz. 1992. "The Lag in the Mechanization of the Sugarcane Harvest: Some Comparative Perspectives." *Agricultural History* 66:61–75.

Buzzanell, Peter J. 1993. "The Louisiana Sugar Industry: Its Evolution, Current Situation, and Prospects." In U.S. Department of Agriculture. Economic Research Service. *Sugar and Sweetener Situation and Outlook Yearbook* 18, no. 2: 18–44.

Byrnside, D. S., Jr., and M. B. Sturgis. 1958. *Soil Phosphorus and Its Fractions as Related to Response of Sugarcane to Fertilizer Phosphorus*. Louisiana Agricultural Experiment Station Bulletin 513. Baton Rouge: Louisiana State University.

Cable, George Washington. 1884. *The Creoles of Louisiana*. New York: C. Scribner's Sons.

Caires' Southern Louisiana Antebellum Plantation Map. 1796. Louisiana State University Library, Special Collections.

Cameco Industries, Inc. 1996. *Cane Combine Harvester*. Thibodaux, La.: Cameco Industries.

Campbell, Joe R. 1977. *Break-Even Farm and Mill Costs for Producing Raw Sugar in Louisiana, 1937–75*. DAE Research Report 522. Baton Rouge: Department of Agricultural Economics and Agribusiness, Louisiana State University.

Carpenter, Caleb. 1847. "The Mississippi River in Olden Time: A Genuine Account of the Present State of the Mississippi and the Lands on Its Banks from Its Mouth to the River Yasous, 1776." *DeBow's Review* 3:115–23.

Carrigan, Judge. 1851. "Statistical and Historical Sketches of Louisiana: Baton Rouge." *DeBow's Review* 11:611–17.

Carter, Clarence E., ed. 1940. *The Territorial Papers of the United States*, vol. 9,

The Territory of Orleans, 1803–1812. Washington, D.C.: Government Printing Office.

Cayton, Frank M. 1881. *Landings on All the Western and Southern Rivers and Bayous, Showing Location, Post Offices, Distances, Etc.* St. Louis: Woodward, Tierman & Hale.

Center for Louisiana Studies. 1980. *Greenfields: Two Hundred Years of Louisiana Sugar*. Lafayette: Center for Louisiana Studies, University of Southwestern Louisiana.

Champomier, P. A. 1844–60. *Statement of Sugar Made in Louisiana.* New Orleans: P. A. Champomier.

Chapman, Brian A., Arthur M. Heagler, and Kenneth W. Paxton. 1987. *Projected Costs and Returns on Sugarcane, Louisiana.* DAE Research Report 667. Baton Rouge: Department of Agricultural Economics and Agribusiness, Louisiana State University.

Chardon, Roland E., ed. 1983. *Plantation Traits in the New World.* Baton Rouge: Geoscience Publications, Louisiana State University.

———. 1984. "Sugar Plantations in the Dominican Republic." *Geographical Review* 74:441–54.

Chaturvedi, H. S. 1951. "The Sugar Industry of India." *Sugar Journal* 13, no. 3.

Cheakalos, Christina. 1992. "A Thin Green Line: Louisiana Town Fights a Chemical Giant." *Atlanta Journal/Atlanta Constitution,* 23 February.

Christy, Ralph D., Brian A. Chapman, and Arthur M. Heagler. 1990. *Structural Change and Economic Efficiency within Louisiana's Sugarcane Processing Industry.* DAE Research Report 685. Baton Rouge: Department of Agricultural Economics and Agribusiness, Louisiana State University.

Cizak, Eugene D., Bernard Leman, and A. B. See. 1983. Whitney Plantation: Historic American Buildings Survey. In Rykels, Brenda Barger. 1991. "Our River Road Heritage: The Politics of Land Use. A Study of the History and Preservation Process at Whitney Plantation, St. John-the-Baptist Parish, Louisiana." Undergraduate thesis, Department of Landscape Architecture, Louisiana State University.

Clifton, James M. 1973. "Golden Grains of White: Rice Planting on the Lower Cape Fear." *North Carolina Historical Review* 50, no. 4: 366–93.

Coastal Environments, Inc. 1989. *A Tongue of Land near LaFourche: The Archaeology and History of Golden Ranch Plantation, LaFourche Parish, Louisiana.* Report to the Division of Archaeology, Louisiana Department of Culture, Recreation, and Tourism. Baton Rouge: Coastal Environments.

Coleman, R. E. 1952. "Studies on the Keeping Quality of Sugarcane Damaged by

Freezing Temperatures during the Harvest Season 1951–52." *Sugar Bulletin* 30:342–80.

Comeaux, Malcolm L. 1989. "The Cajun Barn." *Geographical Review* 79, no. 1.

———. 1992. "Cajuns in Louisiana." In *To Build in a New Land: Ethnic Landscapes in North America.* Edited by Allen G. Noble. Baltimore: Johns Hopkins University Press.

Conrad, Glenn R. 1981. *The German Coast: Abstracts of the Civil Records of Saint Charles and Saint John-the-Baptist Parishes, 1804–1812.* Lafayette: Center for Louisiana Studies, University of Southwestern Louisiana.

Conrad, Glenn R. 1983. *The Cajuns: Essays on their History and Culture.* Lafayette: Center for Louisiana Studies, University of Southwestern Louisiana.

Conrad, Glenn R., and Ray F. Lucas. 1995. *White Gold: A Brief History of the Louisiana Sugar Industry, 1795–1995.* Lafayette: Center for Louisiana Studies, University of Southwestern Louisiana.

Coulter, Ellis M. 1955. *Wormsloe: Two Centuries of a Georgia Plantation.* Athens: University of Georgia Press.

Cruzat, Heloise H. 1918. "Sidelights on Louisiana History." *Louisiana Historical Quarterly* 1:87–153.

———, trans. 1925. "Documents Concerning the Sale of Chaouachas Plantation." *Louisiana Historical Quarterly* 8:594–664.

Cry, G. W. 1968. *Freeze Probabilities in Louisiana.* Baton Rouge: Louisiana State University, Louisiana Cooperative Extension Service.

Curtin, Philip D. 1969. *The Atlantic Slave Trade: A Census.* Madison: University of Wisconsin Press.

Darby, William. 1817. *A Geographical Description of the State of Louisiana,* 2d ed. New York: James Olmstead.

Dart, Henry P. 1926. "A Louisiana Indigo Plantation on Bayou Teche." *Louisiana Historical Quarterly* 9:565–89.

———. 1935. "The Career of Dubreuil in French Louisiana." *Louisiana Historical Quarterly* 18:286.

Daspit, Fred. 1996. *Louisiana Architecture, 1714–1830.* Lafayette: Center for Louisiana Studies, University of Southwestern Louisiana.

Daubeny, Charles G. B. 1843. *Journal of a Tour through the United States and Canada, Made during the Year 1837–1838.* Oxford, England: T. Combe.

DeBow, J. D. B., ed. 1846. "Extension of the Louisiana Sugar Region." *DeBow's Review* 2:442.

———. 1847a. "Drainage." *DeBow's Review* 3:66–80.

———, ed. 1847b. "Commerce of American Cities: New Orleans." *DeBow's Review* 4:391–400.

———, ed. 1848. "Improved Husbandry, Implements, Etc." *DeBow's Review* 6:131–32.

———, ed. 1849. "Sugar: Its Cultivation, Manufacture, and Commerce." *DeBow's Review* 5:157–59.

———. 1850. "Louisiana and Her Industry." *DeBow's Review* 8:35.

———, ed. 1851. "Louisiana: Her Public Lands and Levees." *DeBow's Review* 10:530–34.

———. 1853. *The Industrial Resources of the Southern and Western States.* New Orleans: J. D. B. DeBow.

———, ed. 1855. "The Present and Future of New Orleans." *DeBow's Review* 19:688–93.

———. 1856. *The Southern States.* Washington, D.C., and New Orleans: J. D. B. DeBow.

Deerr, Noel. 1928. "The Identity of the Creole Cane of the West Indies." *International Sugar Journal* 30:11–12.

———. 1949. *The History of Sugar.* 2 vols. London: Chapman & Hall.

Degalos, Pierre A. 1892. "Statement of Sugar Made in 1828 and 1829." *Louisiana Planter and Sugar Manufacturer* 9:65–68.

Deiler, J. Hanno. 1909. *The Settlement of the German Coast or Louisiana and the Creoles of German Descent.* Philadelphia: Americana Germanica Press.

Delanglez, Jean. 1935. *The French Jesuits in Lower Louisiana, 1700–1763.* Washington, D.C.: Catholic University.

De Laussat, Pierre Clement. 1831. *Memoire sur ma vie pendant les annees 1802–1803.* Pau, France.

Delavigne, J. C. 1848. "Production of Sugar in the United States." *DeBow's Review* 5:144.

Dickinson, Robert E. 1949. "Rural Settlements in German Lands." *Annals of the Association of American Geographers* 39:239–63.

Doran, Edwin, Jr. 1962. "The West Indian Hip-Roof Cottage." *California Geographer* 3:97.

Doré, Susan Cole. 1989. *The Pelican Guide to Plantation Homes in Louisiana.* Gretna, La.: Pelican Publishing.

Dowdey, Clifford. 1939. *Gambrel's Hundred.* Boston: Little, Brown.

Downs, R. H. 1960. "Public Lands and Private Claims in Louisiana in 1803–1820." Master's thesis, Louisiana State University.

Dumont de Montigny, Louis François Benjamin. 1753. *Memoires historiques sur Louisiane.* Paris: C. J. B. Bauche.

Dunaway, Wayland P. 1944. *The Scotch-Irish of Colonial Pennsylvania.* Chapel Hill: University of North Carolina Press.

Dunn, Richard S. 1972. *Sugar and Slaves: The Rise of the Planter Class in the English West Indies, 1624–1713.* Chapel Hill: University of North Carolina Press.

Du Pratz, Antoine Simon Le Page. 1758. *Histoire de la Louisiane.* 2 vols. Paris: De Bure.

Dupy, Colquitt P., ed. 1967. *The Gilmore Louisiana-Florida Sugar Manual, 1967.* New Orleans: Hauser Printing.

Durbin, Gilbert J. 1980. "Changes in the Louisiana Sugar Industry in the Nineteen Thirties, Forties, and Fifties." In *Green Fields: Two Hundred Years of Louisiana Sugar.* Lafayette: Center for Louisiana Studies, University of Southwestern Louisiana.

Du Terrage, Marc de Villers. 1903. *Les dernières années de la Louisiane Française.* Paris: Librarie Orientale & Americaine.

———. 1920. "A History of the Foundation of New Orleans." *Louisiana Historical Quarterly* 3:157–214.

Earle, F. S. 1928. *Sugar Cane and Its Culture.* New York: J. Wiley & Sons.

Edwards, Bryan. 1793–1801. *The History, Civil and Commercial, of the British Colonies in the West Indies.* 3 vols. London: John Stockdale.

Edwards, Jay. 1988. *Louisiana's Remarkable French Vernacular Architecture, 1700–1900.* Baton Rouge: Department of Geography and Anthropology, Louisiana State University.

Erixon, Sigurd. 1938. "West European Connections and Cultural Relations." *Folk Liv* 2:137–72.

Evans, E. Estyn. [1942] 1963. *Irish Heritage.* Dundalk: Dundalgan Press.

———. 1967. *Mourne Country: Landscape and Life in South Down,* 2d ed. Dundalk: Dundalgan Press.

Farr, Whitlock & Co. 1960. *Manual of Sugar Companies.* New York: Farr, Whitlock.

Favrot, H. L. 1844. "One Hundred Years of Sugar Making in Louisiana." *Harper's Weekly* 38:755, 758, 760.

Fielder, Lonnie L., Jr. 1981. *Changes in Louisiana Agriculture, by Parishes and by Types of Farming Areas with Projections to 1990.* AEA Information Series 51. Baton Rouge: Department of Agricultural Economics and Agribusiness, Louisiana State University.

Fisk, H. N. 1952. *Geologic Investigation of the Atchafalaya Basin and the Problem of the Mississippi River Diversion.* Vicksburg: Mississippi River Commission.

Forman, Henry C. 1957. *Virginia Architecture in the Seventeenth Century.* Williamsburg, Va.: 350th Anniversary Celebration.

———. 1967. *Old Buildings, Gardens, and Furniture in Tidewater Maryland.* Cambridge, Md.: Tidewater Publishers.

Foscue, Edwin J. 1936. "Sugar Plantations of the Irish Bend District, Louisiana." *Economic Geographer* 12:373–80.

French, Benjamin Franklin, ed. 1850–53. *Historical Collections of Louisiana,* vol. 2, 1850, Philadelphia: Daniels & Smith; vol. 3, 1851, New York: Appleton; vol. 5, 1853, New York: Lamport Blakeman Law.

Frink, Chris. 1995. "The End of an Era." *Baton Rouge Advocate Magazine,* September 17, 1995, pp. 11–16.

Galloway, J. H. 1989. *The Sugar Cane Industry: An Historical Geography from Its Origins to 1914.* Cambridge: Cambridge University Press.

Gaspar, David Barry. 1993. "Sugar Cultivation and Slave Life in Antigua before 1800." In *Cultivation and Culture: Labor and the Shaping of Slave Life in the Americas.* Edited by Ira Berlin and Philip D. Morgan. Charlottesville: University Press of Virginia.

Gayarre, Charles. 1851. *Louisiana: Its Colonial History and Romance.* New York: Harper & Brothers.

———. 1889. "A Louisiana Sugar Plantation in the Old Regime." *Harper's Magazine* 74:607.

Genovese, Eugene G. 1976. *Roll Jordan Roll: The World the Slaves Made.* New York: Random House.

Gilmore, A. B. 1922. *Directory of Louisiana Sugar Planters, 1922.* New Orleans: A. B. Gilmore.

———, ed. 1930–86. *Gilmore's Louisiana Sugar Manual.* New Orleans: A. B. Gilmore, 1922–41; Fargo, N.D.: Sugar Publications, Gilmore Sugar Manual Division, 1969–86.

Ginn, Mildred Kelly. 1940. "A History of Rice Production in Louisiana to 1896." *Louisiana Historical Quarterly* 23:544–88.

Giraud, Marcel. 1996. "Emigration and Colonial Society." In *A Refuge for All Ages: Immigration in Louisiana History.* The Louisiana Purchase Bicentennial Series in Louisiana History, vol. 10. Edited by Carl A. Brasseaux. Lafayette: Center for Louisiana Studies, University of Southwestern Louisiana.

Glassie, Henry. 1968. *Pattern in the Material Folk Culture of the Eastern United States.* Philadelphia: University of Pennsylvania Press.

———. 1989. "Eighteenth-Century Cultural Process in Delaware Folk Building." In *Common Places: Readings in American Vernacular Architecture.* Edited by Dell Upton and John Michael Vlach. Athens: University of Georgia Press.

Gray, Lewis C. [1932] 1958. *History of Agriculture in the Southern United States to 1860.* 2 vols. Gloucester, Mass.: Peter Smith.

Gregor, Howard F. 1962. "The Plantation in California." *Professional Geographer* 14:1–14.

———. 1965. "The Changing Plantation." *Annals of the Association of American Geographers* 55:221–38.

———. 1974. *An Agricultural Typology of California.* Budapest: Akademiai Kaido.

———. 1982. *Industrialization of U.S. Agriculture: An Interpretative Atlas.* Boulder: Westview Press.

Guerra, R. 1964. *Sugar and Slavery in the Caribbean.* New Haven: Yale University Press.

Hackett, Charles Wilson, ed. and trans. 1946. *Picardo's Treatise on the Limits of Louisiana and Texas.* 4 vols. Austin: University of Texas Press.

Hall, Gwendolyn Midlo. 1992. *Africans in Colonial Louisiana: The Development of Afro-Creole Culture in the Eighteenth Century.* Baton Rouge: Louisiana State University Press.

———. 1996a. "Senegambia during the French Slave Trade." In *A Refuge for All Ages: Immigration in Louisiana History.* The Louisiana Purchase Bicentennial Series in Louisiana History, vol. 10. Edited by Carl A. Brasseaux. Lafayette: Center for Louisiana Studies, University of Southwestern Louisiana.

———. 1996b. "Death and Revolt." In *A Refuge for All Ages: Immigration in Louisiana History.* The Louisiana Purchase Bicentennial Series in Louisiana History, vol. 10. Edited by Carl A. Brasseaux. Lafayette: Center for Louisiana Studies, University of Southwestern Louisiana.

Hall, John W. 1970. "Louisiana Survey Systems: Their Antecedents, Distribution, and Characteristics." Ph.D. diss., Louisiana State University.

Hamlin, Talbot. 1944. *Greek Revival Architecture in America.* New York: Oxford University Press.

Hart, John F. 1982. "The Role of the Plantation in Southern Agriculture." *Proceedings, Tall Timbers Ecology and Management Conference* 16:1–19.

———. 1991a. *The Land That Feeds Us.* New York: W. W. Norton.

———. 1991b. "Part-ownership and Farm Enlargement in the Midwest." *Annals of the Association of American Geographers* 81:66–79.

Hawaiian Sugar Planters' Association. 1990. *Sugar Manual, 1990: A Handbook of Statistical Information.* Aiea, Hawaii: Hawaiian Sugar Planters' Association.

Heagler, Arthur M. 1991. *Projected Costs and Returns, Sugarcane: Louisiana, 1991.* AEA Information Series 91. Baton Rouge: Department of Agricultural Economics and Agribusiness, Louisiana State University.

Hebert, L. P. 1964. *Culture of Sugarcane for Sugar Production in Louisiana.* Agriculture Handbook No. 262. Agriculture Research Service, Department of Agriculture. Washington, D.C.: Government Printing Office.

Heitmann, John A. 1987. *The Modernization of the Louisiana Sugar Industry, 1830–1910.* Baton Rouge: Louisiana State University.

Hill, J. R., ed. 1995. *The Oxford Illustrated History of the Royal Navy.* Oxford: Oxford University Press.

Hilliard, Sam B. 1978. "Antebellum Tidewater Rice Culture in South Carolina and Georgia." In *European Settlement and Development in North America: Essays on Geographical Change in Honour and Memory of Andrew Hill Clark.* Edited by James R. Gibson. Toronto: University of Toronto Press.

———. 1979. "Site Characteristics and Spatial Stability of the Louisiana Sugar Industry." *Agricultural History* 53:254–69.

———. 1984. *Atlas of Antebellum Agriculture.* Baton Rouge: Louisiana State University Press.

———. 1990. "Plantation and the Moulding of the Southern Landscape." In *The Making of the American Landscape.* Edited by Michael P. Conzen. London: Harper Collins.

Hilliard, Sam Bowers. 1992. *The South Revisited: Forty Years of Change.* New Brunswick: Rutgers University Press.

Hirsch, Arnold R., and Joseph Logsdon, eds. 1992. *Creole New Orleans: Race and Americanization.* Baton Rouge: Louisiana State University Press.

Hodson, R. Charles, Jr. 1980. "U.S. Sugar Policy since the 1930s." In *Green Fields: Two Hundred Years of Louisiana Sugar.* Lafayette: Center for Louisiana Studies, University of Southwestern Louisiana.

Hoffsommer, Harold. 1940. *The Sugar Cane Farm: A Social Study of Labor and Tenancy.* Louisiana Agricultural Experiment Station Bulletin 320. Baton Rouge: Louisiana State University.

———. 1941. *The Resident Laborer on the Sugar Cane Farm.* Louisiana Agricultural Experiment Station Bulletin 334. Baton Rouge: Louisiana State University.

Homans, I. Smith, and William B. Danna, eds. 1861. "Tariffs of the United States for 1842, 1846, 1857, 1861." *Hunt's Merchants' Magazine and Commercial Review* 44:459–518.

Hughes, Louis. [1897] 1969. *Thirty Years a Slave: From Bondage to Freedom.* New York: Negro Universities Press.

Iberville Parish. 1829–1988. *Conveyance Records.* Plaquemine, La.

———. 1837. *Probate Record 586.* Plaquemine, La.

"Indigo in Louisiana." 1844. *American Agriculturist* 9:92.

Johnson, Bradish. 1868, 1880–85. Plantation Records. Baton Rouge: Louisiana State University Library, Special Collections.

Johnson, Cecil. 1933. "The Distribution of Land in British West Florida." *Louisiana Historical Review* 16:539–53.

Jordan, Terry G. 1966. "On the Nature of Settlement Geography." *Professional Geographer* 18:26–28.

———. 1994. "The Creole Coast: Romano-Caribbean Ethnicity and the Shaping of the American South." Paper presented at the annual meeting of the Association of American Geographers, San Francisco, Calif., March 29–April 2.

Jordan, Terry G., and Mati Kaups. 1989. *The American Backwoods Frontier: An Ethnic and Ecological Interpretation.* Baltimore: Johns Hopkins University Press.

Joyner, Charles. 1984. *Down by the Riverside: A South Carolina Slave Community.* Urbana: University of Illinois Press.

Kellar, Herbert Anthony, ed. 1936. *Solon Robinson, Pioneer and Agriculturalist: Selected Writings.* Indiana Historical Collection, vol. 22. Indianapolis: Indiana Historical Bureau.

Kirby, Jack T. 1987. *Rural Worlds Lost: The American South, 1920–1960.* Baton Rouge: Louisiana State University Press.

Kniffen, Fred B. 1936. "Louisiana House Types." *Annals of the Association of American Geographers* 26:179–53.

———. 1963. "The Physiognomy of Rural Louisiana." *Louisiana History* 4:291–99.

———. 1965. "Folk Housing: Key to Diffusion." *Annals of the Association of American Geographers* 55:549–77.

Kniffen, Fred B., and Henry H. Glassie III. 1966. "Building in Wood in the Eastern United States: A Time-Place Perspective." *Geographical Review* 56:40–56.

Knipmeyer, William B. 1956. "Settlement Succession in Eastern French Louisiana." Ph.D. diss., Louisiana State University.

Kramer, Mark. 1977. *Three Farms: Making Milk, Meat, and Money from the American Soil.* New York: Bantam Books.

Labat, J.-B. 1742. *Nouveaux voyage aux isles de l'Amerique.* 8 vols. Paris: T. Le Gras.

Lancaster, Clay. 1986. "The American Bungalow." In *Common Places: Readings in American Vernacular Architecture.* Edited by Dell Upton and John Michael Vlach. Athens: University of Georgia Press.

Lasserre, Guy. 1961. *La Guadeloupe.* 2 vols. Bordeaux: Union Français d'Impression.

Lathrop, Barnes F. 1945. "The Pugh Plantations, 1860–1865." Ph.D. diss., University of Texas.

Laws, J. Bradford. 1902. *The Negroes of Cinclaire Central Factory and Calumet Plantations, Louisiana.* Department of Labor Bulletin 38. Washington, D.C.: Government Printing Office.

LeBlanc, Robert G. 1996. "The Acadian Migrations." In *A Refuge for All Ages: Immigration in Louisiana History.* The Louisiana Purchase Bicentennial Series in Louisiana History, vol. 10. Edited by Carl A. Brasseaux. Lafayette: Center for Louisiana Studies, University of Southwestern Louisiana.

Le Conte, Rene. 1996. "The Germans in Louisiana in the Eighteenth Century." In *A Refuge for All Ages: Immigration in Louisiana History.* The Louisiana Purchase Bicentennial Series in Louisiana History, vol. 10. Edited by Carl A. Brasseaux. Lafayette: Center for Louisiana Studies, University of Southwestern Louisiana.

Lenoir, Ellen, and T. Lynn Smith. 1937. *Rural Housing in Louisiana.* Louisiana Agricultural Experiment Station Bulletin 290. Baton Rouge: Louisiana State University.

Le Roy Ladurie, Emmanuel. 1971. *Times of Feast, Times of Famine: A History of Climate since the Year 1000.* Translated by Barbara Bray. Garden City, N.Y.: Doubleday.

Lewis, Kenneth E. 1985. "Plantation Layout and Function in the South Carolina Lowcountry." In *The Archaeology of Slavery and Plantation Life.* Edited by Theresa A. Singleton. Orlando, Fla.: Academic Press.

Lewis, Peirce F. 1976. *New Orleans: The Making of an Urban Landscape.* Cambridge, Mass.: Ballinger Publishing.

Linder, Suzanne Cameron. 1995. *Historical Atlas of the Rice Plantations of the ACE River Basin, 1860.* Columbia: South Carolina Department of Archives and History.

Lippmann, Edmund Oscar von. 1929. *Geschichte des Zukers, seit den altesten*

Zeitern bis zum Beginne de Rubenzuker-Fabrikation, 2d ed. Berlin: Julius Springer.

Lytle, S. A. 1968. *The Morphological Characteristics and Relief Relationships of Representative Soils in Louisiana.* Louisiana Agricultural Experiment Station Bulletin 631. Baton Rouge: Louisiana State University.

Maier, E. A. 1952. "A Story of Sugarcane Machinery." *Sugar Journal* 14:8.

Martin, Francois X. 1827. *The History of Louisiana, from the Earliest Period.* 2 vols. New Orleans: Lyman & Beardslee.

Matherne, R. J. 1969. "A History of Major Louisiana Sugarcane Varieties." *Proceedings of the International Society of Sugar Cane Technology* 13:1056–61.

McBride, George McCutchen. 1934. "Plantation." *Encyclopedia of the Social Sciences,* 12:148. New York: Macmillan.

McCall, Henry. 1899. "History of the Evan Hall Plantation, 1899." Manuscript, Tulane University Library, New Orleans.

McCallum, Henry D., and Francis T. McCallum. 1965. *The Wire That Fenced the West.* Norman: University of Oklahoma Press.

McDonald, Roderick A. 1993. "Independent Economic Production by Slaves on Antebellum Louisiana Sugar Plantations." In *Cultivation and Culture: Labor and the Shaping of Slave Life in the Americas.* Edited by Ira Berlin and Philip D. Morgan. Charlottesville: University Press of Virginia.

McIntire, William G. 1958. *Prehistoric Indian Settlements of the Changing Mississippi River Delta.* Baton Rouge: Louisiana State University Press.

McWhiney, Grady. 1988. *Cracker Culture: Celtic Ways in the Old South.* Tuscaloosa: University of Alabama Press.

McWilliams, Carey. 1939. *Factories in the Field: The Story of Migratory Farm Labor in California.* Boston: Little, Brown.

———. 1942. *Ill Fares the Land: Migrants and Migratory Labor in the United States.* Boston: Little, Brown.

Meitzen, August. 1895. *Siedlung und Agrarwesen der Westgermannen und Ostgermannen, der Kelten, Romer, Finnen, und Slawen.* 3 vols. and atlas. Berlin: Wilhelm Hertz.

Merrill, Gordon C. 1958. *The Historical Geography of Saint Kitts and Nevis, The West Indies.* Mexico City: Instituto Panamericano de Geografia e Historia.

Miles, William Porcher. 1883–95. Diary. University of North Carolina Library, Chapel Hill.

Mitcham, Samuel Wayne, Jr. 1986. "The Origin and Evolution of the Southwestern Louisiana Rice Region, 1880–1920." Ph.D. diss., University of Tennessee.

Mitchell, Margaret. 1936. *Gone with the Wind.* New York: Macmillan.

Moody's Investors Services. 1990. *Moody's Industrial Manual.* New York: Moody's Investors Services, 2:4239, 6301.

Moore, Ellen B., ed. n.d. *General Instructions to U.S. Deputy Surveyors.* Baton Rouge, La.: State Land Office.

Morrison, Hugh. 1952. *Early American Architecture.* New York: Oxford University Press.

Morton, Louis. 1941. *Robert Carter of Nomini Hall.* Williamsburg, Va.: Colonial Williamsburg.

National Geographic Society. 1988. *Historical Atlas of the United States.* Washington, D.C.: National Geographic Society.

Neiman, Fraser D. 1986. "Domestic Architecture at the Clifts Plantation: The Social Context of Early Virginia Building." In *Common Places: Readings in American Vernacular Architecture.* Edited by Dell Upton and John Michael Vlach. Athens: University of Georgia Press.

Newman, Randy. 1974. "Louisiana, 1927." Burbank, Calif.: Warner-Tamerlane Publishing, Randy Newman (BMI).

Newton, Milton B. 1967. "The Peasant Farm of St. Helena Parish, Louisiana: A Cultural Geography." Ph.D. diss., Louisiana State University.

Noble, Allen G., ed. 1992. *To Build in a New Land: Ethnic Landscapes in North America.* Baltimore: Johns Hopkins University Press.

Noble, Allen G., and Richard K. Cleek. 1995. *The Old Barn Book: A Field Guide to North American Barns and Other Farm Structures.* New Brunswick: Rutgers University Press.

Oakes, James. 1982. *The Ruling Race: A History of American Slave Holders.* New York: Alfred Knopf.

Office of the Federal Register, National Archives and Records Administration. 1990. *Code of Federal Regulations.* Environmental Protection Agency, 40, part 409, subpart D, pp. 207–9. Washington, D.C.: Government Printing Office.

Olmsted, Frederic Law. 1861. *Cotton Kingdom.* New York: Mason Brothers.

Orser, Charles E., Jr. 1988. *The Material Basis of the Postbellum Tenant Plantation.* Athens: University of Georgia Press.

Oszuscik, Philippe. 1992a. "African-Americans in the American South." In *To Build in a New Land: Ethnic Landscapes in North America.* Edited by Allen G. Noble. Baltimore: Johns Hopkins University Press.

———. 1992b. "French Creoles on the Gulf Coast." In *To Build in a New Land: Ethnic Landscapes in North America.* Edited by Allen G. Noble. Baltimore: Johns Hopkins University Press.

———. 1992c. "Passage of the Gallery and Other Caribbean Elements from the French and Spanish to the British in the United States." *Pioneer America Society Transactions,* 15:1–14.

———. 1994. "Comparisons between Rural and Urban French Creole Housing." *Material Culture* 26, no. 3: 1–36.

Overdyke, William D. 1965. *Louisiana Plantation Homes.* New York: Architectural Book Publishing.

Owens, Jeffrey Alan. 1997. "Naming the Plantation: An Analytical Survey from Tensas Parish, Louisiana." In *Agriculture and Economic Development in Louisiana.* The Louisiana Purchase Bicentennial Series in Louisiana History, vol. 16. Edited by Thomas A. Becnel. Lafayette: Center for Louisiana Studies, University of Southwestern Louisiana.

Owens, Leslie H. 1976. *This Species of Property: Slave Life and Culture in the Old South.* Oxford: Oxford University Press.

Parenton, Vernon J. 1938. "A Sociological Study of a Negro Village in the French Section of Louisiana." Master's thesis, Louisiana State University.

Pares, Richard. 1963. *War and Trade in the West Indies, 1739–1763.* London: Frank Cass.

Peevy, W. J., and R. H. Brupbacher. 1962. "Fertility Status of Louisiana Soils, No. 3: Phosphorus." *Louisiana Agriculture* 5, no. 4: 12–13.

Persac's Map, or Norman's Chart. 1931. *Plantations on the Mississippi River from Natchez to New Orleans, 1858.* New Orleans: Pelican Publishing.

Phillips, U. B. 1929. *Life and Labor in the Old South.* Boston: Little, Brown.

Pierce, G. W. 1851. "Historical and Statistical Collections of Louisiana: Terrebonne." *DeBow's Review* 11:601–11.

Pillsbury, Richard. 1989. "Landscape, Cultural." In *Encyclopedia of Southern Culture.* Edited by Charles Reagan Wilson and William Ferris. Chapel Hill: University of North Carolina Press.

Porteous, Laura L., trans. 1926. "Inventory of DeVaugine's Plantation in the Attakapas on Bayou Teche, 1773." *Louisiana Historical Quarterly* 9:570.

Post, Lauren C. 1962. *Cajun Sketches: From the Prairies of Southwestern Louisiana.* Baton Rouge: Louisiana State University Press.

Prichard, Walter, ed. 1938. "Inventory of the Duvernay Concession in Louisiana, 1726." *Louisiana Historical Quarterly* 21:979–94.

———. 1939. "The Effects of the Civil War on the Louisiana Sugar Industry." *Journal of Southern History* 5:315–32.

Prunty, Merle, Jr. 1955. "The Renaissance of the Southern Plantation." *Geographical Review* 45:459–90.

Pugh, W. W. 1888. "Bayou Lafourche from 1840–1850: Its Inhabitants, Customs, Pursuits." *Louisiana Planter and Sugar Manufacturer* 1:143–67.

———. 1889. "Width of Cane Rows." *Louisiana Planter and Sugar Manufacturer* 2:62.

Pulliam, Linda, and M. B. Newton, Jr. 1973. "Country and Small-Town Stores in Louisiana: Legacy of the Greek Revival and the Frontier." In *Mélanges,* vol. 7. Baton Rouge: Museum of Geoscience, Louisiana State University.

Pulsipher, Lydia M. 1991. "Galways Plantation, Montserrat." In *Seeds of Change.* Edited by Herman J. Viola and Carolyn Margolis. Washington, D.C.: Smithsonian Institution Press.

Pusser's Ltd. n.d. Tortola, British Virgin Islands: Pusser's Ltd.

Reed, Susan M. 1915. "British Cartography of the Mississippi Valley in the Eighteenth Century." *Mississippi Valley Historical Review* 2:213–24.

Rehder, John Burkhardt. 1971. " Sugar Plantation Settlements of Southern Louisiana: A Cultural Geography." Ph.D. diss., Louisiana State University.

Rehder, John B. 1973. "Sugar Plantations in Louisiana: Origins, Dispersals, and Responsible Location Factors." In *Geographic Perspectives on Southern Development.* West Georgia College Studies in the Social Sciences, vol. 12. Edited by John C. Upchurch and David C. Weaver. Carrollton: West Georgia College.

———. 1978. "Diagnostic Landscape Traits of Sugar Plantations in Southern Louisiana." In *Man and Environment in the Lower Mississippi Valley.* Geoscience and Man, vol. 19. Edited by Sam B. Hilliard. Baton Rouge: School of Geoscience, Louisiana State University.

———. 1989a. "Sugar Plantations." In *Encyclopedia of Southern Culture.* Edited by Charles Reagan Wilson and William Ferris. Chapel Hill: University of North Carolina Press.

———. 1989b. "Plantation Morphology." In *Encyclopedia of Southern Culture.* Edited by Charles Reagan Wilson and William Ferris. Chapel Hill: University of North Carolina Press.

———. 1992. "The Scotch-Irish and English in Appalachia." In *To Build in a New Land: Ethnic Landscapes in North America.* Edited by Allen G. Noble. Baltimore: Johns Hopkins University Press.

Revert, Eugene. 1949. *La Martinique: Étude géographique.* Paris: Nouvelles Éditions Latines.

Richardson, E. D. 1886. "The Teche Country Fifty Years Ago." *Southern Bivouac* 4:593.

Rifkin, Carole. 1980. *A Field Guide to American Architecture.* New York: Plume Books, Division of Penguin Publishing.

Rightor, Henry, ed. 1900. *Standard History of New Orleans, Louisiana.* Chicago: Lewis Publishing.

Rillieux, Norbert. 1848. "Sugarmaking in Louisiana." *DeBow's Review* 5:285–93.

Robertson, James A., trans. and ed. 1911. *Louisiana under the Rule of Spain, France, and the United States, 1785–1807.* 2 vols. Cleveland: Arthur H. Clark.

Robin, C. C. [1807] 1966. *Voyage to Louisiana.* Translated by Stuart O. Landry Jr. New Orleans: Pelican Publishing.

Roland, Charles P. 1957. *Louisiana Sugar Plantations during the American Civil War.* Leiden, The Netherlands: E. J. Brill.

Russell, Richard J. 1936. "Physiography of the Lower Mississippi River Delta." Louisiana Geological Survey Bulletin, No. 8. New Orleans: Louisiana Geological Survey.

Russell, William H. 1863. *My Diary North and South.* New York: Harper & Brothers.

Rutledge, Archibald. 1941. *Home by the River.* Indianapolis: Bobbs Merrill.

Rykels, Brenda Barger. 1991. "Our River Road Heritage: The Politics of Land Use. A Study of the History and Preservation Process at Whitney Plantation, St. John-the-Baptist Parish, Louisiana." Undergraduate thesis, Department of Landscape Architecture, Louisiana State University.

St. James Parish. 1809–1981. *Conveyance Records.* Convent, La.

St. John-the-Baptist Parish. 1820. *Original Acts.* Edgard, Louisiana.

———. 1820–1990. *Conveyance Records.* Edgard, La.

St. Mary Parish. 1812–1981. *Conveyance Records.* Franklin, La.

Sauer, Carl O. 1931. "Cultural Geography." *Encyclopedia of the Social Sciences* 6:621–23. Also in Wagner, P. L., and M. W. Mikesell, eds. 1962. *Readings in Cultural Geography.* Chicago : University of Chicago Press.

Scheuring, Ann Foley, ed. 1983. *A Guidebook to California Agriculture.* Berkeley: University of California Press.

Schlüter, Otto. 1899. "Bemerkungen zur Siedlungsgeographie." *Geographische Zeitschrift* 5:65.

Schmitz, Mark. 1979. "The Transformation of the Southern Cane Sugar Sector, 1860–1930." *Agricultural History* 53:270–85.

Scoates, Daniels. 1937. *Farm Buildings.* College Station, Tex.: D. Scoates.

Scroggs, William O. 1916. "Rural Life in the Lower Mississippi Valley about 1803." *Proceedings of the Mississippi Valley Historical Association* 8:262–77.

Sestini, Aldo. [1947] 1962. "Regressive Phases in the Development of the Cultural Landscape." In *Readings in Cultural Geography*. Philip L. Wagner and Marvin W. Mikesell, eds. Chicago: University of Chicago Press.

Sheridan, Richard B. 1974. *Sugar and Slavery: An Economic History of the British West Indies, 1623–1775*. Baltimore: Johns Hopkins University Press.

Shugg, Roger Wallace. 1937. "Survival of the Plantation System in Louisiana." *Journal of Southern History* 3:311–25.

Shuker, Iain G. W., Arthur M. Heagler, and Brian A. Chapman. 1986. *Economic Analysis of the Cost Structure of Commercial Cane Farms in Louisiana*. DAE Research Report 651. Baton Rouge: Department of Agricultural Economics and Agribusiness, Louisiana State University.

Shurtleff, Harold R. 1939. *The Log Cabin Myth*. Cambridge: Harvard University Press.

Silliman, Benjamin. 1833. *Manual on the Cultivation of the Sugarcane and the Refinement of Sugar*. Washington, D.C.: F. P. Blair.

Singleton, Theresa A. 1985. *The Archaeology of Slavery and Plantation Life*. Orlando, Fla.: Academic Press.

Sitterson, J. Carlyle. 1948. "Hired Labor on Sugar Plantations of the Antebellum South." *Journal of Southern History* 14:192–205.

———. 1953. *Sugar Country: The Cane Sugar Industry in the South, 1753–1950*. Lexington: University of Kentucky Press.

Smith, J. Frazer. 1941. *White Pillars: Early Life and Architecture of the Lower Mississippi Valley Country*. New York: William Helburn.

Stampp, Kenneth. 1956. *The Peculiar Institution: Slavery in the Ante-Bellum South*. New York: Vintage Press.

Stanley, John P. 1979. *Farming by Design*. Fresno, Calif.: Agribusiness Investment Management Services International.

Stein, Robert Lewis. 1988. *The French Sugar Business in the Eighteenth Century*. Baton Rouge: Louisiana State University Press.

Stephenson, Wendell Holmes. 1934. *Alexander Porter, Whig Planter of Old Louisiana*. Baton Rouge, Louisiana State University Press.

———. 1938. *Isaac Franklin; Slave Trader and Planter of the Old South*. Baton Rouge: Louisiana State University Press.

Stone, Kirk. 1965. "The Development of a Focus for the Geography of Settlement." *Economic Geography* 41:346–55.

Stubbs, W. C. 1901. *Sugarcane: A Treatise on Its History, Botany, and Agriculture*. Savannah, Ga.: D. G. Purse.

"Sugarcane Loading Devices." 1906. *Louisiana Planter and Sugar Manufacturer* 36:294.

Surrey, Nancy M. Miller. 1916. *The Commerce of Louisiana during the French Regime, 1699–1763*. Columbia University Studies in History, Economics, and Public Law, vol. 71. New York: Columbia University Press.

Surveyor General. n.d. *Surveyor's Township Plats*. Baton Rouge, La.: State Land Office.

Taggart, William G. 1933. "Agronomic Practices and Their Influence on the Developments of the Louisiana Sugar Industry." Master's thesis, Louisiana State University.

Taylor, James W. 1950. "Louisiana Land Survey Systems." *Southwestern Social Science Quarterly* 31:275–82.

Terrebonne Parish. 1828–1990. *Conveyance Records*. Houma, La.

Thomas, Hugh. 1997. *The Slave Trade: The Story of the Atlantic Slave Trade, 1440–1820*. New York: Simon & Schuster.

Thompson, Edgar T. 1935. *The Plantation*. Chicago: University of Chicago Press.

———. 1941. "The Climatic Theory of the Plantation." *Agricultural History* 15:49–60.

Thorpe, T. B. 1853. "Sugar and the Sugar Region of Louisiana." *Harper's New Monthly Magazine* 7:746–67.

Trudeau, Carlos. 1803. *Plan del local de las tierras que Rodean la Cuidad Nueva Orleans*. Duplicate map, Works Progress Administration; original map, Howard Library.

Tuan, Yi-Fu. 1974. *Topophilia: A Study of Environmental Perception, Attitudes, and Values*. Englewood Cliffs, N.J.: Prentice-Hall.

Upton, Dell. 1986. "Vernacular Domestic Architecture in Eighteenth-Century Virginia." In *Common Places: Readings in American Vernacular Architecture*. Edited by Dell Upton and John Michael Vlach. Athens: University of Georgia Press.

U.S. Department of Agriculture. 1959. *Soil Survey of Saint Mary Parish, Louisiana*. Series 1952, No. 3. Washington, D.C.: Government Printing Office.

U.S. Department of Agriculture, Agricultural Stabilization and Conservation Service. 1975. *Sugar Reports* 273:9.

U.S. Department of Agriculture, Economic Research Service. 1987. *Sugar and Sweetener Situation and Outlook* 12, no. 2: 25.

———. 1990. *Sugar and Sweetener Situation and Outlook* 15, no. 2:28.

———. 1993. *Sugar and Sweetener Situation and Outlook* 18, no. 2:18–44.

———. 1994. *Sugar and Sweetener Situation and Outlook* 19, no. 3:13.

———. 1997. *Sugar and Sweetener Situation and Outlook* 22, no. 6:8–9, 35.

U.S. Department of Commerce. 1913. *The Sugar Industry.* Department of Commerce Misc. Series 9. Washington, D.C.: Government Printing Office.

U.S. Department of Commerce, Bureau of the Census. 1964, 1969, 1974, 1978, 1982, 1987, 1992. *Census of Agriculture, Louisiana.* Washington, D.C.: Government Printing Office.

———. 1994. *1990 Census of Population and Housing: Ancestry of the Population of the United States.* Washington, D.C.: Government Printing Office.

Vernon, Arelia W. 1993. *African Americans at Mars Bluff, South Carolina.* Baton Rouge: Louisiana State University Press.

Vlach, John Michael. 1976. "The Shotgun House: An African Architectural Legacy." *Pioneer America: Journal of Historic American Material Culture* 8:47–70.

———. 1986. "The Shotgun House: An African American Legacy." In *Common Places: Readings in American Vernacular Architecture,* 58–78. Edited by Dell Upton and John Michael Vlach. Athens: University of Georgia Press.

———. 1993. *Back of the Big House: The Architecture of Plantation Slavery.* Chapel Hill: University of North Carolina Press.

Voorhies, Marcel, and W. M. Grayson. 1939. "An Outline of Recent Sugar Control Programs and Their Effect on the Louisiana Sugar Industry." *International Society of Sugarcane Technologists: Proceedings of the Sixth Congress.* Baton Rouge: Franklin Press.

Wade, Michael G. 1995. *Sugar Dynasty: M. A. Patout & Son Ltd., 1791–1993.* Lafayette: Center for Louisiana Studies, University of Southwestern Louisiana.

Wagner, Philip L., and Marvin W. Mikesell, eds. 1962. *Readings in Cultural Geography.* Chicago: University of Chicago Press.

Waibel, Leo. 1941. "The Tropical Plantation System." *Scientific Monthly* 52:156–60.

———. 1942. "The Climatic Theory of the Plantation." *Geographical Review* 32:307–10.

Wall, R. T. 1981. "The Vanishing Tenant Houses of Rural Georgia." *Georgia Historical Quarterly* 65:251–62.

Warf, Barney. 1988. "Regional Transformation, Everyday Life, and Pacific Northwest Lumber Production." *Annals of the Association of American Geographers* 78:326–46.

Watkins, Maguerite E. 1939. "History of Terrebonne Parish to 1861." Master's thesis, Louisiana State University.

Watson, Andrew M. 1983. *Agricultural Innovation in the Early Islamic World.* Cambridge: Cambridge University Press.

Watson, Henry. 1848. *Narrative of Henry Watson, a Fugitive Slave.* Boston: B. Marsh.

Whelan, James Patrick, Jr. 1983. "Plantation Slave Subsistence in the Old South and Louisiana." In *Plantation Traits in the New World.* Edited by Roland E. Chardon. Baton Rouge: Louisiana State University Press.

Whittlesey, Derwent S. 1929. "Sequent Occupance." *Annals of the Association of American Geographers* 19:162–65.

Wilkinson, Alec. 1989. *Big Sugar: Seasons in the Cane Fields of Florida.* New York: Knopf, Random House.

Wilkinson, J. B. 1889. *The Diffusion Process in Louisiana and Texas.* New Orleans: L. Graham & Son.

Wilkinson, R. A. 1847. "Cultivation of Sugarcane." *DeBow's Review* 4:229.

Wilson, Samuel, Jr. 1980. "Architecture of Early Sugar Plantations." In *Green Fields: Two Hundred Years of Louisiana Sugar.* Lafayette: Center for Louisiana Studies, University of Southwestern Louisiana.

———. 1987. *The Architecture of Colonial Louisiana: Collected Essays.* Edited by J. M. Farnsworth and A. M. Masson. Lafayette: Center for Louisiana Studies, University of Southwestern Louisiana.

Wolpert, Julian. 1964. "The Decision Process in Spatial Context." *Annals of the Association of American Geographers* 54:537–58.

Woodman, Harold D., ed. 1966. *Slavery and the Southern Economy: Sources and Readings.* New York: Harcourt Brace & World.

Woofter, T. J., Jr., Gordon Blackwell, Harold Hoffsommer, J. G. Maddox, J. M. Massell, B. O. Williams, and Waller Wynne Jr. 1936. *Landlord and Tenant on the Cotton Plantation.* Research Monograph 5, Works Progress Administration. Washington, D.C.: Government Printing Office.

Writer's Program of the Works Progress Administration (WPA). 1941. *Louisiana: A Guide to the State.* New York: Hastings House.

Yoder, P. A. 1919. *Sugarcane Culture for Syrup Production in the United States.* Department of Agriculture Bulletin 486. Washington, D.C.: Government Printing Office.

Zeichner, Oscar. 1939. "The Transition from Slave to Free Agricultural Labor in the Southern United States." *Agricultural History* 13:22–32.

Index

Related Titles in the Series

CHARLES S. AIKEN
The Cotton Plantation South since the Civil War

ALVAR W. CARLSON
The Spanish-American Homeland: Four Centuries in New Mexico's Rio Arriba

ROBERT F. ENSMINGER
The Pennsylvania Barn: Its Origin, Evolution, and Distribution in North America

FRANK GOHLKE
Measure of Emptiness: Grain Elevators in the American Landscape

TERRY G. JORDAN AND MATTI KAUPS

The American Backwoods Frontier: An Ethnic and Ecological Interpretation

TERRY G. JORDAN, JON T. KILPINEN, AND CHARLES F. GRITZNER

The Mountain West: Interpreting the Folk Landscape

GABRIELLE M. LANIER AND BERNARD L. HERMAN

Everyday Architecture of the Mid-Atlantic: Looking at Buildings and Landscapes

ALLEN G. NOBLE, EDITOR

To Build in a New Land: Ethnic Landscapes in North America

PAUL F. STARRS

Let the Cowboy Ride: Cattle Ranching in the American West

MARTHA A. STRAWN

Alligators, Prehistoric Presence in the American Landscape

About the Author

John Burkhardt Rehder was born in Wilmington, North Carolina, in 1942 and completed his high school years in Winston-Salem, North Carolina. He received a B.A. degree in geography from East Carolina College and an M.A. and Ph.D. in geography from Louisiana State University. In 1967, he was hired as the first cultural geographer at the University of Tennessee, where he is now a professor. John Rehder has published more than sixty journal articles, book chapters, and other publications. He has also served as president of the Mid South Region and as a two-term director on the Board of Directors for the Amcrican Society for Photogrammetry and Remote Sensing (ASPRS); as chairman of the Remote Sensing Committee for the Association of American Geographers (AAG); and two terms on the Remote Sensing Committee of the National Council for Geographic Education (NCGE).

Library of Congress Cataloging-in-Publication Data

Rehder, John B.
Delta sugar : Louisiana's vanishing plantation landscape / John B. Rehder.
p. cm. – (Creating the North American landscape)
Includes bibliographical references and index.
ISBN 0-8018-6131-4 (alk. paper)
1. Plantations–Louisiana–History. 2. Plantation life–Louisiana–History.
3. Sugar growing–Social aspects–Louisiana–History. 4. Landscape changes–Louisiana–History. 5. Human geography–Louisiana. 6. Material culture–Louisiana. 7. Landscape design–Louisiana. I. Title. II. Series.
F370.R45 1999
306.3'49'09763–dc21 99-20199